Next Generation Mobile Networks and Ubiquitous Computing

Samuel Pierre
École Polytechnique de Montréal, Canada

Information Science REFERENCE

INFORMATION SCIENCE REFERENCE
Hershey · New York

Director of Editorial Content: Kristin Klinger
Director of Book Publications: Julia Mosemann
Acquisitions Editor: Lindsay Johnston
Development Editor: Christine Bufton, Juli Mosemann
Publishing Assistant: Milan Vracarich
Typesetter: Travis Gundrum, Deanna Jo Zombro
Production Editor: Jamie Snavely
Cover Design: Lisa Tosheff

Published in the United States of America by
Information Science Reference (an imprint of IGI Global)
701 E. Chocolate Avenue
Hershey PA 17033
Tel: 717-533-8845
Fax: 717-533-8661
E-mail: cust@igi-global.com
Web site: http://www.igi-global.com

Library of Congress Cataloging-in-Publication Data

Next generation mobile networks and ubiquitous computing / Samuel Pierre, editor.
p. cm.
Includes bibliographical references and index. Summary: "This book provides a comprehensive and unified view of the latest and most innovative research findings on the many existing interactions between mobile networking, wireless communications, and ubiquitous computing"--Provided by publisher. ISBN 978-1-60566-250-3 (hardcover) -- ISBN 978-1-60566-251-0 (ebook) 1. Ad hoc networks (Computer networks) 2. Ubiquitous computing. I. Pierre, Samuel.

TK5105.77.N49 2010
621.382--dc22

2010010158

British Cataloguing in Publication Data
A Cataloguing in Publication record for this book is available from the British Library.

Editorial Advisory Board

Table of Contents

Preface xvii

Acknowledgment xxii

Chapter 1
Cooperation-Based Routing Protocol for Mobile Ad Hoc Network 1
Abdelhalim Baaziz, École Polytechnique de Montréal, Canada
Samuel Pierre, École Polytechnique de Montréal, Canada

Chapter 2
Ambient Networks 17
Alex Galis, University College London, UK
Bertrand Mathieu, Orange Labs, France

Chapter 3
Overview and Challenges of Multi-Interface and Multi-Channel Multi-Hop Wireless Networks 28
Carine Toham, Orange Labs, France
François Jan, Orange Labs, France

Chapter 4
Cross Layer Design for Multimedia Transmission over Wireless Networks 35
Christos Bouras, University of Patras and Research Academic Computer Technology Institute, Greece
Apostolos Gkamas, Research Academic Computer Technology Institute and University of Patras, Greece
Georgios Kioumourtzis, University of Patras, Greece

Chapter 5
Charging in Mobile Telecommunication Networks 43
Bálint Ary, Budapest University of Technology and Economics (BUTE), Hungary
Sándor Imre, Budapest University of Technology and Economics (BUTE), Hungary

Chapter 6
Local Algorithms for Topology Control in Ad-Hoc Networks 51
Evangelos Kranakis, Carleton University, Canada
Jorge Urrutia, Universidad Nacional Autonoma de Mexico, Mexico

Chapter 7
The Four Facets of Multimedia Streaming 59
Florence Agboma, University of Essex, UK
Antonio Liotta, Eindhoven University of Technology, The Netherlands

Chapter 8
Real-Time Communications in Wireless Sensor Networks 69
Isabelle Augé-Blum, Université de Lyon, France
Fei Yang, Université de Lyon, France
Thomas Watteyne, Université de Lyon & Orange Labs, France

Chapter 9
A Survey of IP Layer Mobility Management Protocols in Next Generation Wireless Networks 79
Li Jun Zhang, Geninov Inc., Canada
Liyan Zhang, Dalian Jiaotong University, China
Laurent Marchand, Ericsson Canada Inc., Canada
Samuel Pierre, Ecole Polytechnique de Montreal, Canada

Chapter 10
Auditory Feedback for Network Intrusion Detection 94
M.A. Garcia-Ruiz, University of Colima, Mexico
M. Vargas Martin, University of Ontario Institute of Technology, Canada
R. Aquino-Santos, University of Colima, Mexico
O. Alvarez-Cardenas, University of Colima, Mexico
A. Edwards, University of Colima, Mexico
M.G. Mayoral-Baldivia, University of Colima, Mexico

Chapter 11
Peer-to-Peer over Mobile Ad-Hoc Networks 105
Nadia N. Qadri, University of Essex, UK
Antonio Liotta, Eindhoven University of Technology, The Netherlands

Chapter 12
Middleware Technologies for Ubiquitous Computing 122
Noha Ibrahim, Université de Lyon, France
Frédéric Le Mouël, Université de Lyon, France
Stéphane Frénot, Université de Lyon, France

Chapter 13
Routing in Asymmetric Wireless Ad-Hoc Networks 132
Pramita Mitra, University of Notre Dame, USA
Christian Poellabauer, University of Notre Dame, USA

Chapter 14
Smartphones to Access Hybrid Information Spaces 146
Spyros Veronikis, Ionian University, Greece
Kyriaki Zoutsou, Ionian University, Greece
Giannis Tsakonas, Ionian University, Greece
Christos Papatheodorou, Ionian University, Greece

Chapter 15
Location-Dependent and Context-Aware Computing 156
Stan Kurkovsky, Central Connecticut State University, USA

Chapter 16
MAC Protocols for Wireless Sensor Networks 165
Torsten Braun, University of Bern, Switzerland
Markus Anwander, University of Bern, Switzerland
Philipp Hurni, University of Bern, Switzerland
Markus Wälchli, University of Bern, Switzerland

Chapter 17
Mechanisms for Automatic Web Service Composition 175
Quan Yuan, Florida Atlantic University, USA
Mihai Fonoage, Florida Atlantic University, USA
Ionut Cardei, Florida Atlantic University, USA

Chapter 18
A Robust Clustering Algorithm for Mobile Ad-Hoc Networks 187
Zhaowen Xing, University of Oklahoma, USA
Le Gruenwald, University of Oklahoma, USA
K.K. Phang, University of Malaya, Malaysia

Chapter 19
Emerging Mobility Applications of Host Identity Protocol 201
Szabolcs Nováczki,Budapest University of Technology and Economics (BUTE), Hungary
László Bokor, Budapest University of Technology and Economics (BUTE), Hungary
Gábor Jeney, Budapest University of Technology and Economics (BUTE), Hungary
Sándor Imre, Budapest University of Technology and Economics (BUTE), Hungary

Chapter 20
Service Discovery in Mobile Ad-Hoc Networks 217
Haidar Safa, American University of Beirut, Lebanon
Zeina Koteiche, American University of Beirut, Lebanon

Chapter 21
Mobility Prediction in Mobile Ad-Hoc Networks 226
Damianos Gavalas, University of the Aegean, Greece
Charalampos Konstantopoulos, University of Piraeus, Greece
Basilis Mamalis, Technological Educational Institution of Athens, Greece
Grammati Pantziou, Technological Educational Institution of Athens, Greece

Chapter 22
Calling Police Using SMS 241
Mohammad Shirali-Shahreza, Sharif University of Technology, Iran
M. Hassan Shirali-Shahreza, Amirkabir University of Technology, Iran

Chapter 23
Vehicular Communications Networks: Current Trends and Challenges 251
Moez Jerbi, France Telecom R&D, Core Network Laboratories, France
Sidi-Mohammed Senouci, France Telecom R&D, Core Network Laboratories, France
Yacine Ghamri-Doudane, Networks and Multimedia Systems Research Group, France
Mohamed Cherif, France Telecom R&D, Core Network Laboratories, France

Compilation of References 263

About the Contributors 293

Index 305

Detailed Table of Contents

Preface xvii

Acknowledgment xxii

Chapter 1
Cooperation-Based Routing Protocol for Mobile Ad Hoc Network 1
Abdelhalim Baaziz, École Polytechnique de Montréal, Canada
Samuel Pierre, École Polytechnique de Montréal, Canada

This chapter describes a hybrid routing protocol for mobile ad hoc network based on the packet relaying cooperation and zone-based network architecture. In each zone, a specific node called Principal manages the routing processing. The Principal receives, periodically from its zone's nodes, two values that express their rate of success and their rate of cooperation, and then calculates the average rate of success, and the average rate of cooperation of the zone. After what, it retransmits these averages to all nodes. The nodes use the two averages values to define the action to be undertaken, cooperate or not. The Principal, knowing the topology of its zone, receives route requests and calculates the path between the source and the destination according to the best way based on cooperation. The Principal's features are so that it is simple to realize. It does not impose constraints and it allows the information gathering that is necessary for the model solution. Such the routing protocol and the nodes (to undertake the actions that make them possible to cooperate) use this information. Simulations show the efficiency of the protocol opposite to the pro-active and reactive protocols in the context of cooperation in mobile ad hoc networks.

Chapter 2
Ambient Networks 17
Alex Galis, University College London, UK
Bertrand Mathieu, Orange Labs, France

This chapter introduces the Ambient Network (AN) which is a network integration solution that aims at fostering dynamic co-operation between the next generation, heterogeneous wireless and wired networks, in order to gather resources within and across networks to provide new services to end users. First, it depicts the main characteristics of AN before describing the Ambient Control Space, which is the space where functional entities managing the networks and the services are active, and the interfaces. Then it focuses on the SATO (Service-aware Adaptive Transport Overlays), which is a service overlay networks deployed on top of the underlying heterogeneous physical networks and tailored according to the services' requirements and users' context. This chapter aims at giving the readers a good overview of what Ambient networks are since they will be present in the next years everywhere.

Chapter 3
Overview and Challenges of Multi-Interface and Multi-Channel Multi-Hop Wireless Networks 28
Carine Toham, Orange Labs, France
François Jan, Orange Labs, France

The throughput improvement problem in wireless networks is considered. The main causes of that issue is the shared wireless medium and node spatial problems which either prevent some nodes from transmitting or jam their transmissions. The authors consider the use of both nodes equipped with multiple network cards and multiple channels as means to improve the throughput in wireless networks. That approach implies to handle some challenges namely channel distribution between the node interfaces and routing in such channel-diverse networks. This chapter analyzes some solutions proposed to deal with these challenges so that the throughput is improved. Some possible orientations are also discussed based on the advantages and drawbacks observed in existing works.

Chapter 4
Cross Layer Design for Multimedia Transmission over Wireless Networks 35
Christos Bouras, University of Patras and Research Academic Computer Technology Institute, Greece
Apostolos Gkamas, Research Academic Computer Technology Institute and University of Patras, Greece
Georgios Kioumourtzis, University of Patras, Greece

Over the last years a number of new protocols have been developed for multimedia applications in the whole OSI layer's scale. In addition wireless communications and networking fast occupy centre stage in research and development activity in the area of communication networks. In order to support better wireless user the cross layer design paradigm has been proposed. This chapter presents the challenges in design and implementation of cross layer adaptation schemes for multimedia transmission over wireless networks. In addition this chapter presents the most important parameters and constrains that should be taken into consideration when attempting cross layer adaptation in wireless networks that involves different protocols in the overall protocol stack.

Chapter 5
Charging in Mobile Telecommunication Networks 43
Bálint Ary, Budapest University of Technology and Economics (BUTE), Hungary
Sándor Imre, Budapest University of Technology and Economics (BUTE), Hungary

The tariff packages in the mobile telecommunication industry got more and more complex in the last few years. The telecommunication companies have introduced several different services along with different discounts and allowances. The price calculation of the services and the understanding of prices became harder for the subscribers. This chapter describes the basic architecture and the major flows of a billing system in the telecommunication area. It introduces a novel concept for price calculation, which could aid the Advance of Charge and Income Prediction functionalities.

Chapter 6
Local Algorithms for Topology Control in Ad-Hoc Networks 51
Evangelos Kranakis, Carleton University, Canada
Jorge Urrutia, Universidad Nacional Autonoma de Mexico, Mexico

This chapter presents a survey of recent techniques for local topology control in location aware Unit Disk Graphs including local algorithms for Routing, Traversal, Planar Spanners, Dominating and Connected Dominating Sets, and Vertex and Edge Coloring. In addition to investigating trade-offs for these problems, the authors discuss open problems that will play an important role in the future development of the subject.

Chapter 7
The Four Facets of Multimedia Streaming 59
Florence Agboma, University of Essex, UK
Antonio Liotta, Eindhoven University of Technology, The Netherlands

Recent advances in media coding techniques and network access technologies have made multimedia streaming practicable and affordable in both fixed and mobile environments. Multimedia streaming services from anywhere and at anytime is fast becoming a reality. This chapter provides a snapshot of the facets involved in delivering multimedia streaming contents to the end user's terminal. The facets discussed in the chapter extend from the streaming paradigms, media coding techniques, the network support for multimedia streaming services to the techniques used to evaluate the user's quality of experience for multimedia streaming applications. Each facet is also discussed for both fixed and mobile environments.

Chapter 8
Real-Time Communications in Wireless Sensor Networks 69
Isabelle Augé-Blum, Université de Lyon, France
Fei Yang, Université de Lyon, France
Thomas Watteyne, Université de Lyon & Orange Labs, France

This chapter presents the state-of-the-art of real-time communication in the challenging topic of Wireless Sensor Networks (WSNs). In real-time communication, the duration between the event which initiates the sending of a message, and the instant this message is received must be smaller than a known delay. Because topologies are extremely dynamic and not known priori, this type of constraint is very hard to meet in WSNs.

Chapter 9
A Survey of IP Layer Mobility Management Protocols in Next Generation Wireless Networks 79
Li Jun Zhang, Geninov Inc., Canada
Liyan Zhang, Dalian Jiaotong University, China
Laurent Marchand, Ericsson Canada Inc., Canada
Samuel Pierre, Ecole Polytechnique de Montreal, Canada

This chapter provides a survey of IP layer mobility management protocols in next generation wireless networks. In all-IP-based next generation wireless networks, mobile nodes freely change their points of attachment to the network while communicating with correspondent nodes. Hence, mobility management consists of a critical issue, which is to track mobile users' current location and to efficiently route packets to them. This chapter elaborates the mobility management procedure for the protocols such as mobility support in IPv6 (MIPv6), hierarchical mobile IPv6 mobility management (HMIPv6), fast handovers for mobile IPv6 (FMIPv6), fast handover for hierarchical mobile IPv6 (F-HMIPv6) and proxy mobile IPv6 (PMIPv6). Furthermore, future trends about mobility management are described as well. He authors hope that understanding these mobility management protocols can not only help the researchers to find more advanced solutions in this field, but also provide a training toolkit within the mobile operators.

Chapter 10
Auditory Feedback for Network Intrusion Detection 94
M.A. Garcia-Ruiz, University of Colima, Mexico
M. Vargas Martin, University of Ontario Institute of Technology, Canada
R. Aquino-Santos, University of Colima, Mexico
O. Alvarez-Cardenas, University of Colima, Mexico
A. Edwards, University of Colima, Mexico
M.G. Mayoral-Baldivia, University of Colima, Mexico

Network intrusion (the unauthorized access to a computer system perpetrated by persons or a piece of software) is a major concern of network administrators. Carrying out network intrusion detection (NID) manually is a tiring, complex and time-consuming task, often overloading the visual sensory channel. To overcome this, real-time auditory alarms and auditory information provide immediate feedback, helping to identify trends or patterns of the attacks in network logs and in real-time network accesses, allowing professionals make rapid decisions, thus representing an interesting alternative to conventional NID. The purpose of this chapter is to describe literature research on the benefits and challenges of Auditory Display (the use of non-speech sound to represent meaningful information at a computer interface) applied to wireless, wired, and mobile network intrusion detection. This chapter also highlights further work on Auditory Display and multimodal interfaces to support intrusion detection in wireless mesh networks.

Chapter 11
Peer-to-Peer over Mobile Ad-Hoc Networks 105
Nadia N. Qadri, University of Essex, UK
Antonio Liotta, Eindhoven University of Technology, The Netherlands

This chapter reviews various approaches for the convergence of Peer-to-Peer (P2P) and Mobile Ad hoc Networks (MANETs), identifying strengths and weaknesses, and putting things in perspective. P2P and MANETs are among the most active research topics in pervasive computing. The convergence of P2P networks and MANETs would allow existing P2P applications such as P2P file sharing and P2P streaming to benefit from the ubiquitous connectivity of ad-hoc. A P2P network over an ad-hoc infrastructure is a powerful combination that provides users with means to access different kinds of information anytime and anywhere. Realizing such a system is, however, not straightforward.

Chapter 12
Middleware Technologies for Ubiquitous Computing 122
Noha Ibrahim, Université de Lyon, France
Frédéric Le Mouël, Université de Lyon, France
Stéphane Frénot, Université de Lyon, France

Middleware handles many important functionalities for ubiquitous computing. The authors distinguish various middleware technologies providing key elements for all applications' requirements such as discovery, adaptation/composition, context management, and management of ubiquitous applications. In this chapter, the authors propose a classification for some of the most employed ubiquitous middleware. The classification was established upon the challenges raised by ubiquitous computing – effective use of smart spaces, invisibility, and localized scalability - and upon how the various ubiquitous middleware respond to them in terms of interoperability, discoverability, location transparency, adaptability, context awareness, scalability, security, and autonomous management. This classification shows that if many middleware are mature enough and offer specific functionalities respecting the properties of ubiquity, a real lack is noticed in having an interoperable, autonomous and scalable middleware for the execution of ubiquitous applications. The development of the service-oriented paradigm, the semantics, the Web middleware, and the ambient intelligence shows the new trend the middleware research field is engaged in.

Chapter 13
Routing in Asymmetric Wireless Ad-Hoc Networks 132
Pramita Mitra, University of Notre Dame, USA
Christian Poellabauer, University of Notre Dame, USA

The presence of asymmetric links is a common and non-negligible phenomenon in many ad-hoc networks, including MANETs and sensor networks. Asymmetry is caused by node mobility, heterogeneous radio technologies, and irregularities in radio ranges and packet loss patterns. Most existing ad-hoc routing protocols either assume fully symmetric networks or simply ignore any asymmetric links. In the first case, route discovery can fail when the symmetry assumption does not hold true, e.g., many reactive routing protocols rely on a two-phase communication process, where the same path is used to communicate between a sender and a receiver. If a single link on this path is asymmetric, the route establishment may fail. In the second case, asymmetric links are identified and explicitly ignored in the route establishment phase. This can lead to route discovery failure if there is no symmetric path between a sender and a receiver or it can lead to less than optimal routes. This document provides an overview of routing protocols that explicitly consider asymmetric links in the route discovery phase and introduces robust mechanisms that bypass asymmetric links to ensure successful route establishment.

Chapter 14
Smartphones to Access Hybrid Information Spaces 146
Spyros Veronikis, Ionian University, Greece
Kyriaki Zoutsou, Ionian University, Greece
Giannis Tsakonas, Ionian University, Greece
Christos Papatheodorou, Ionian University, Greece

This chapter presents how mobile phones nowadays can be used as a handy and always around medium to interact with information collections of both physical and electronic form. It presents their evolution into powerful computing devices capable of communicating with other computing infrastructures, like the World Wide Web, and assisting them in finding the information of interest in the most convenient form, right when the need arises. It also provides a description of the current interaction modes between users, devices and information objects as well as some examples of first and second-generation mobile services. Furthermore, the authors hope that understanding the potential introduced by mobile phones in the modern information landscape can bring some insight to new information seeking strategies, that enhance exploration and not just querying. This understanding can be used to create new, innovative and novel services to enhance the seeking experience while interacting with hybrid information spaces.

Chapter 15
Location-Dependent and Context-Aware Computing 156
Stan Kurkovsky, Central Connecticut State University, USA

Systems with context-aware features are typically able to adapt their behavior to the changes of the environment surrounding the system and/or its users without any explicit intervention of the user. Therefore, context-aware systems tend to increase the usability and effectiveness by taking into account the context of their environment. Location-based and context-aware systems are typically associated with mobile devices; they react to reflect their current physical location, time, other surrounding systems, their user(s), as well as other features of their environment, and adapt their operation to the changes of the environment. This information about the many aspects of the system's environment is referred to as context, which can be obtained from a variety of sources, such as RFID and other sensors, wireless network information, physical device status, maintaining information about the user(s) and many other sources.

Chapter 16
MAC Protocols for Wireless Sensor Networks 165
Torsten Braun, University of Bern, Switzerland
Markus Anwander, University of Bern, Switzerland
Philipp Hurni, University of Bern, Switzerland
Markus Wälchli, University of Bern, Switzerland

The chapter describes related work on medium access control protocols for wireless sensor nodes. The authors focus on scheduled and contention-based protocols that have been proposed by the research community during the last few years. In particular, the authors evaluate the potential to save energy of several representative protocols, namely LMAC, TEEM, and WiseMAC. This has been done by measurements of implementations in real sensor networks. The measurement results show that sophisticated MAC protocol design can significantly improve the energy-efficiency and increase the lifetime of a sensor node. Real-world measurements are important to determine power consumption parameters of sensor nodes.

Chapter 17
Mechanisms for Automatic Web Service Composition 175
Quan Yuan, Florida Atlantic University, USA
Mihai Fonoage, Florida Atlantic University, USA
Ionut Cardei, Florida Atlantic University, USA

This chapter introduces the web services composition as a means of studying efficient integration of the existing web services to satisfy users' requirements. It discusses the web services composition definition, combined with the current web services composition methods, and divides those methods into two categories: AI-based methods and Non-AI methods. Also, the authors present the features and the comparison of these two categories, to assist researchers in the understanding of web service composition in a variety of contexts.

Chapter 18
A Robust Clustering Algorithm for Mobile Ad-Hoc Networks 187
Zhaowen Xing, University of Oklahoma, USA
Le Gruenwald, University of Oklahoma, USA
K.K. Phang, University of Malaya, Malaysia

To mimic the operations in fixed infrastructures and to solve the routing scalability problem in large Mobile Ad Hoc Networks (MANET), forming clusters of nodes has been proven to be a promising approach. However, when existing weighted clustering algorithms calculate each node's weight, they either consider only one metric or rely on some metrics collected from extra devices. This often leads to a higher rate of re-clustering. This chapter presents a robust weighted clustering algorithm, called PMW (Power, Mobility and Workload), to form and maintain more stable clusters. In PMW, the weight of each node is calculated by its power, mobility and workload, which can be easily collected and computed locally and cover the major factors that cause re-clustering. Clustering overhead of PMW is analyzed. The simulation results confirm that PMW prolongs lifetime of MANETs and has a lower cluster head change rate and re-affiliation rate than other existing algorithms.

Chapter 19
Emerging Mobility Applications of Host Identity Protocol 201
Szabolcs Nováczki, Budapest University of Technology and Economics (BUTE), Hungary
László Bokor, Budapest University of Technology and Economics (BUTE), Hungary
Gábor Jeney, Budapest University of Technology and Economics (BUTE), Hungary
Sándor Imre, Budapest University of Technology and Economics (BUTE), Hungary

This chapter is devoted to give an overview of the Host Identity Protocol (HIP) and to discuss two different mobility management extensions built on it. The authors introduce the basic ideas and the main paradigms behind the HIP-scheme, and show its wild range usability in next generation mobile and ubiquitous networks. The broad scale of possible advanced mobility applications of HIP will be introduced based on two extensions of the protocol, both developed by the authors. The first one is a micromobility solution based on HIP (μHIP), while the second one provides a scalable and secure network mobility solution (HIP-NEMO).

Chapter 20
Service Discovery in Mobile Ad-Hoc Networks 217
Haidar Safa, American University of Beirut, Lebanon
Zeina Koteiche, American University of Beirut, Lebanon

Service discovery plays a relevant role in mobile ad hoc environments. Indeed, upon joining a self-organizing network, mobile nodes should be able to explore the environment to learn about, locate, and share the available services. As a result, many researches were performed in this area. Recently, research is being directed towards integrating the service discovery into the routing protocols. This chapter presents the basic concepts of service discovery and their related issues. It also describes the service discovery challenges that arise due to the properties of mobile ad hoc networks. The chapter is concluded by presenting some observations and discussing the current challenges to provide guidelines for possible improvements.

Chapter 21
Mobility Prediction in Mobile Ad-Hoc Networks 226
Damianos Gavalas, University of the Aegean, Greece
Charalampos Konstantopoulos, University of Piraeus, Greece
Basilis Mamalis, Technological Educational Institution of Athens, Greece
Grammati Pantziou, Technological Educational Institution of Athens, Greece

A Mobile Ad hoc NETwork (MANET) is a collection of wireless mobile nodes forming a network without using any existing infrastructure. All mobile nodes function as mobile routers that discover and maintain routes to other mobile nodes of the network and therefore, can be connected dynamically in an arbitrary manner. The mobility attribute of MANETs is a very significant one. The mobile nodes may follow different mobility patterns that may affect connectivity, and in turn protocol mechanisms and performance. Mobility prediction may positively affect the service-oriented aspects as well as the application-oriented aspects of ad hoc networking. At the network level, accurate node mobility prediction may be critical to tasks such as call admission control, reservation of network resources, pre-configuration of services and QoS provisioning. At the application level, user mobility prediction in combination with user's profile may provide the user with enhanced location-based wireless services, such as route guidance, local traffic information and on-line advertising. The authors of this chapter present the most important mobility prediction schemes for MANETs in the literature, focusing on their main design principles and characteristics.

Chapter 22
Calling Police Using SMS 241
Mohammad Shirali-Shahreza, Sharif University of Technology, Iran
M. Hassan Shirali-Shahreza, Amirkabir University of Technology, Iran

In this chapter, a method is proposed to contact the Police with mobile phones and via SMS (Short Message Service). In this method, when a person wants to contact the Police, he must only press a special key on his mobile phone for a short time to launch a special program. This program sends current location of the person using GPS system and sends it periodically with other useful information such as name, home and work address of that person automatically to the Police Station using SMS.

Chapter 25
Vehicular Communications Networks: Current Trends and Challenges 251
Moez Jerbi, France Telecom R&D, Core Network Laboratories, France
Sidi-Mohammed Senouci, France Telecom R&D, Core Network Laboratories, France
Yacine Ghamri-Doudane, Networks and Multimedia Systems Research Group, France
Mohamed Cherif, France Telecom R&D, Core Network Laboratories, France

Vehicular communications networks (VCNs) are created by vehicles equipped with short and medium range wireless communication technology. They include vehicular ad-hoc networks (VANETs), vehicle-to-vehicle and vehicle-to-infrastructure communications. VCNs enable a plethora of important applications and services, ranging from active safety or safety of life applications to traffic information, music/maps download and multi-hop internet connection.

Compilation of References 263

About the Contributors 293

Index 305

Preface

Mobile communication networks are made possible by the convergence of several different technologies, specifically computer networking protocols, wireless/mobile communication systems, distributed computing and Internet. The need for mobility is a fundamental market factor that is driving the evolution of telecommunications networks. In telecommunications terms, mobility can be defined as the ability to access all of the services that one would normally have in a fixed wireline environment such as a home or office, from anywhere. The goal is to make services such as access to web, corporate Intranet or databases, electronic mails and teleconferencing universal, accessible from anywhere, at anytime. To support this universal service availability effectively, the networks and end-systems require flexible and efficient service-enabling platform. These networks should include the functions of accessing and sharing various types of resources, and provide generic support for service creation, service delivery and service management.

The opportunities emerging from these technologies have given rise to new paradigms such as mobile computing and ubiquitous networking. However, promising these technologies seem to be, there are still numerous challenges to solve such as reliability and quality of service, infrastructure costs, and energy efficiency of mobile devices, among others. The most exciting promises of mobile computing and ubiquitous networking stay an entire new class of applications and potential new massive markets combining personal computing and consumer electronics.

Mobile computing is distinguished from classical, fixed-connection computing by: (1) the mobility of nomadic users and their terminals; and (2) mobile resource constraints such as limited wireless bandwidth and limited battery life. The mobility of nomadic users implies that the users might connect from different access points through wireless links and might want to stay connected while roaming, despite possible intermittent disconnections. Wireless links are relatively unreliable and currently are two to three orders of magnitude slower than wireline networks. Moreover, mobile hosts powered by batteries suffer from limited battery life constraints. These limitations and constraints leave much work to be done before mobile computing is fully operational. This remains true despite recent advances in wireless data communication networks and hand-held device technologies.

The world is becoming more dependent on wireless and mobile services, but the ability of wireless network infrastructures and service-enabling technologies to handle the growing demand is questionable. As wireless and mobile services grow, weaknesses become clearer. Failure to bring forth new solutions will not only affect current voice and data use but could also limit emerging wireless applications such as e-commerce and high-bandwidth Internet access.

The mission of this book is to present the latest and most innovative research findings on mobile networking to researchers from academia and industry. The overall objectives are:

- To present novel and innovative solutions in the mobile networking field;
- To serve as a basis for future research in the field.

Chapter 1, titled "*Cooperation-Based Routing Protocol for Mobile Ad-Hoc Networks*", describes a hybrid routing protocol for mobile ad hoc networks based on the packet relaying cooperation and zone-based network architectures. Several solutions have been proposed to encourage the nodes to cooperate and to contribute to the common services of such networks. These solutions are grouped into two sets; detection and reputation techniques. This chapter discusses the limits of such solutions and presents a new routing protocol. Simulations results show the efficiency of the protocol opposite to the pro-active and reactive protocols in the context of cooperation in mobile ad hoc networks.

Chapter 2 introduces the Ambient Network (AN) which is a network integration solution that aims at fostering dynamic co-operation between the next generation, heterogeneous wireless and wired networks, in order to gather resources within and across networks to provide new services to end users. Applicable for fixed, wireless and mobile networks, the AN concept is based on a uniform and flexible control overlay, the Ambient Control Space (ACS), which capitalises on current Internet technology. The ACS will unify access to network services, regardless of the underlying network type. With this novel approach to control, routing and transport, all networking stakeholders can autonomously add such functionality within their networks.

Chapter 3, titled "*Overview and Challenges of Multi-Interface and Multi-Channel Multi-Hop Wireless Networks*", presents an overview of multi-interface multi-channel wireless networks and some related challenges to deal with so as to fully take advantage of their potential. It analyzes some solutions proposed to deal with these challenges discusses some possible orientations based on the advantages and drawbacks observed in existing works. With multiple available channels, the network is split into smaller collision domains thereby reducing the probability of interference occurrence. A node equipped with multiple interfaces can transmit and receive data packets at the same time which avoids the throughput decrease when data have to be forwarded over multiple hops. So, the number of simultaneous transmissions is higher in this kind of networks than in classical ones, which favors the throughput improvement.

Chapter 4, titled "*Cross Layer Design for Multimedia Transmission over Wireless Networks*", presents the challenges in design and implementation of cross layer adaptation schemes for multimedia transmission over wireless networks. It also presents the most important parameters and constrains that should be taken into consideration when attempting cross layer adaptation in wireless networks that involves different protocols in the overall protocol stack.

Chapter 5, titled "*Charging in Mobile Telecommunication Networks*", describes the basic architecture and the major flows of a billing system in the telecommunication area, and discusses charging and billing systems in this area. It also presents the detailed algorithm and flow of offline (post-paid) and online (pre-paid) charging, shows the problems and issues generated by the new players in the service offering chain (3rd party providers and virtual network operators). Finally, it summarizes the problems of the legacy charging systems, and introduces a novel concept for price calculation, which could aid the Advance of Charge and Income Prediction functionalities.

Chapter 6, titled "*Local Algorithms for Topology Control in Ad-Hoc Networks*", presents a survey of recent techniques for local topology control in location aware Unit Disk Graphs. This survey includes recent local algorithms for Routing, Traversal, Planar Spanners, Dominating and Connected Dominating Sets, and Vertex and Edge Coloring. It also investigates trade-offs for the previously mentioned issues and several remaining open problems which will play important role in the future development of the subject, like solutions for three dimensional ad hoc networks, study of power assignments in the physical interference model and the inclusion of realistic models of mobility.

Chapter 7, titled "*The Four Facets of Multimedia Streaming*", provides a snapshot of the facets involved in delivering multimedia streaming contents to the end user's terminal. The facets discussed in the chapter extend from the streaming paradigms, media coding techniques, the network support for multimedia streaming services to the techniques used to evaluate the user's quality of experience for multimedia streaming applications. Each facet is also discussed for both fixed and mobile environments. The chapter concludes with some considerations on the evolution of streaming techniques by identifying potential areas of research that focus on three themes: ubiquitous streaming, peer-to-peer streaming, and streaming based on quality of experience management.

Chapter 8, titled "*Real-Time Communications in Wireless Sensor Networks*", presents the state-of-the-art of real-time communication in Wireless Sensor Networks (WSNs). It also discusses the different communication protocols proposed in the literatures, together with their respective advantages and drawbacks, by focusing on MAC and routing key layers in real-time communication. As most existing protocols are not suitable under realistic constraints where sensor nodes and wireless links are unreliable, the chapter ends with some insights about future trends in designing real-time protocols.

Chapter 9, titled "*A Survey of IP Layer Mobility Management Protocols in Next Generation Wireless Networks*", provides a survey of IP layer mobility management protocols in next generation wireless networks. More precisely, it elaborates on the mobility support in IPv6 (MIPv6), hierarchical mobile IPv6 mobility management (HMIPv6), fast handovers for mobile IPv6 (FMIPv6), fast handover for hierarchical mobile IPv6 (F-HMIPv6) and proxy mobile IPv6 (PMIPv6). Finally, it describes future trends about mobility management.

Chapter 10, "*Auditory Feedback for Network Intrusion Detection*", describes literature research on the benefits and challenges of Auditory Display applied to wireless, wired, and mobile network intrusion detection. It also highlights further work on Auditory Display and multimodal interfaces to support intrusion detection in wireless mesh networks.

Chapter 11, titled "*Peer-to-Peer over Mobile Ad-Hoc Networks*", reviews various approaches for the convergence of Peer-to-Peer (P2P) and Mobile Ad hoc Networks (MANETs), identifying strengths and weaknesses, and putting things in perspective. After looking at each technology independently, considering their commonalities and differences, it surveys existing literature on the topic of P2P over MANETs, then discusses the state-of-the-art that leads to the identification of the key areas requiring further research.

Chapter 12, "*Middleware Technologies for Ubiquitous Computing*", proposes a classification for some of the most employed ubiquitous middleware. This classification is established upon the challenges raised by ubiquitous computing – effective use of smart spaces, invisibility, and localized scalability. It raises the following question: how the various ubiquitous middleware respond to these challenges in terms of interoperability, discoverability, location transparency, adaptability, context awareness, scalability, security, and autonomous management? This classification shows that if many middleware are mature enough and offer specific functionalities respecting the properties of ubiquity, a real lack is noticed in having an interoperable, autonomous and scalable middleware for the execution of ubiquitous applications.

Chapter 13, titled "*Routing in Asymmetric Wireless Ad-Hoc Networks*", presents an overview of routing protocols that explicitly consider asymmetric links in route construction and route discovery phases in wireless ad-hoc networks. Such networks are found in numerous situations including sensor networks, mobile ad-hoc networks, or networked embedded control systems. The chapter introduces robust mechanisms that bypass asymmetric links to ensure successful route establishment. Routing protocols that assume that all links are symmetric can fail in their route discovery when this assumption does not hold true. As a result, it is necessary that asymmetric links are considered in route establishment.

Chapter 14, titled "*Smartphones to Access Hybrid Information Spaces*", focuses on utilizing mobile computing in hybrid information spaces for describing the usage of portable computing devices as a means of interaction with physical and digital spaces. It also provides background information on smartphones, on the challenges of mobile computing and on web-based architectures. Then, it discuss interaction issues with the device and describes their usage in hybrid spaces. Finally, it outlines the future trends and conclusions.

Chapter 15, "*Location-Dependent and Context-Aware Computing*", presents an overview of systems with context-aware features. These systems are typically able to adapt their behavior to the changes of environment surrounding themselves and/or its users without any explicit intervention of the user. The chapter introduces the concept of context and context-awareness, discusses possible architectures of context-aware and location-based systems, and provides examples of existing systems documented in the literature.

Chapter 16, titled "*MAC Protocols for Wireless Sensor Networks*", gives an overview of medium access control (MAC) protocols for Wireless Sensor Networks (WSN) including both scheduled and contention-based schemes. It focuses on scheduled and contention-based protocols that have been proposed by the research community during the last few years. Based on the classification into scheduled and contention-based protocols, the implementation of some classes showed that sophisticated protocols significantly increase the lifetime of a sensor node.

Chapter 17, "*Mechanisms for Automatic Web Service Composition*", introduces the Web services composition as a means of studying efficient integration of the existing Web services to satisfy users' requirements. It also discusses the Web services composition definition, combined with current Web services composition methods like Artificial Intelligence (AI)-based methods and Non-AI methods. The chapter ends with an analysis of the advantages and disadvantages of such methods.

Chapter 18, titled "*A Robust Clustering Algorithm for Mobile Ad-Hoc Networks*", proposes a robust weighted clustering algorithm, called PMW (Power, Mobility, Workload), to form and maintain more stable clusters in MANETs. In PMW, the weight of a node is calculated by three parameters: remaining power, mobility prediction and workload. These three metrics are computed locally, independent of extra devices, and cover the major causes of re-clustering. Simulation results confirm that PMW prolongs lifetime of MANETs and has a lower cluster head change rate and re-affiliation rate than other existing algorithms.

Chapter 19, "*Emerging Mobility Applications of Host Identity Protocol*", gives an overview of the Host Identity Protocol (HIP) and discusses two different mobility management extensions built on it. It introduces the basic ideas and the main paradigms behind the HIP-scheme, and shows its wild range usability in next generation mobile and ubiquitous networks.

Chapter 20, titled "*Service Discovery in Mobile Ad-Hoc Networks*", presents the basic concepts of service discovery and their related issues. It also describes the service discovery challenges that arise due to the properties of mobile ad hoc networks. The chapter concludes by presenting some observations and discussing the current challenges to provide guidelines for possible improvements.

Chapter 21, "*Mobility Prediction in Mobile Ad-Hoc Networks*", presents the most important mobility prediction schemes for MANETs in the literature. It classifies them according to the specific prediction methods they employ and/or according to their applications well-known mobility prediction schemes.

Chapter 22, titled "*Calling Police Using SMS*", proposes a method to contact the Police with mobile phones and via SMS (Short Message Service). In this method, when a person wants to contact the Police, he must only press a special key on his mobile phone for a short time to launch a special program. This program sends current location of the person using GPS system and sends it periodically with other useful information such as name, home and work address of that person automatically to the Police Station using SMS. This method is implemented using Java Platform Micro Edition programming language and tested on a Nokia N71 mobile phone by using an 'Evermore BT-R700' GPS receiver.

Chapter 23, titled "*Vehicular Communications Networks: Current Trends and Challenges*", presents an overview of the Vehicular Communications Networks and explores the unique features and challenges that characterise these highly dynamic networks as well as their requirements with respect to applications, types of communication, self-organization and security. It discusses various forwarding and routing strategies focussing on position-based techniques including "anchor-based routing".

Samuel Pierre, P.Eng., Ph.D.

Acknowledgment

This project saw the day thanks to the involvement of several persons for help, assistance, review, collaboration, for their comments and suggestions. I wish to acknowledge the help of all involved in the collation and review process of the handbook, without whose support the project could not have been satisfactorily completed.

Deep appreciation and gratitude is due to all reviewers who spent their time to provide their all suitable comments and suggestions: Georges Abou-Khalil, Ronald Beaubrun, Abderrahim Benslimane, Ali Chamam, Mieso Denko, Abdellatif Ezzouhairi, Maher Gaha , Hichem Harhira, Maode Ma, Hakim Naifer, Stéphane Ouellette, Alejandro Quintero, Angelo Rossi, Haidar Safa, Méral Shirazipour, Fabrice Valois, Paul Vital.

Special thanks also go to the publishing team at IGI Global, whose contributions throughout the whole process from inception of the initial idea to final publication have been invaluable. In particular to Ms. Rebeccca Beitsline and Ms. Julia Mosemann, who continuously prodded via e-mail for keeping the project on schedule and to Mr. Ali Chamam from École Polytechnique in the first part of that project- and at the end has been pursued by Mrs. Sabine Kébreau also from École Polytechnique.

In closing, I wish to thank all of the authors for their insights and excellent contributions to this handbook.

Samuel Pierre, P.Eng., Ph.D.
August 2009

Chapter 1
Cooperation-Based Routing Protocol for Mobile Ad-Hoc Network

Abdelhalim Baaziz
École Polytechnique de Montréal, Canada

Samuel Pierre
École Polytechnique de Montréal, Canada

ABSTRACT

This chapter describes a hybrid routing protocol for mobile ad hoc network based on the packet relaying cooperation and zone-based network architecture. In each zone, a specific node called Principal manages the routing processing. The Principal receives, periodically from its zone's nodes, two values that express their rate of success and their rate of cooperation, and then calculates the average rate of success, and the average rate of cooperation of the zone. After what, it retransmits these averages to all nodes. The nodes use the two averages values to define the action to be undertaken, cooperate or not. The Principal, knowing the topology of its zone, receives route requests and calculates the path between the source and the destination according to the best way based on cooperation. The Principal's features are so that it is simple to realize. It does not impose constraints and it allows the information gathering that is necessary for the model solution. Such the routing protocol and the nodes (to undertake the actions that make them possible to cooperate) use this information. Simulations show the efficiency of the protocol opposite to the pro-active and reactive protocols in the context of cooperation in mobile ad hoc networks.

INTRODUCTION

The stimulation of the cooperation in mobile ad hoc networks is one of the problems that were raised by the researchers these last years (Chlamtac, Conti & Liu, 2003). They proposed several solutions to encourage the nodes to cooperate and to contribute to the common services of the network, especially to the packet relaying. These solutions are grouped into two sets; detection and reputation techniques (Marti, Giuli, Lai & Baker, 2000) (Buchegger & LeBoudec, 2002a) (Buchegger & LeBoudec, 2002b) (Michiardi & Molva, 2002) and economic system (Buttyàn & Hubaux,

DOI: 10.4018/978-1-60566-250-3.ch001

2003) (Buttyàn & Hubaux, 2001) (Zhong, Chen & Yang, 2003). All these solutions operate in a heuristic way to provide a reliable diagram for the reinforcement of the cooperation (Félegyházi, Hubaux & Buttyán, 2006). Some researchers claimed that under certain specific conditions, the cooperation could emerge without using incentives techniques. These studies are based on the modeling of the cooperation by analytic models using game theory (Michiardi & Molva, 2003) (Fèlegyhàzi, Buttyàn & Hubaux, 2003a, 2003b) (Srinivasan, Nuggehalli, Chiasserini & Rao, 2005) (Urpi, Bonuccelli & Giordano, 2003) (Félegyházi, Hubaux & Buttyán, 2006). The studies on cooperation and game theory presented until now depict the problem of the information collection by the various nodes. They assume that information is available but they do not specify how it is obtained. These studies address distributed methods but they do not take into account the problem of flooding and/or synchronization they engender. All these studies are based on assumptions that the authors suppose acquired and which are not realistic. However, the routing protocols used in these solutions are far from satisfying their assumptions. In this chapter, we propose a new routing protocol adapted to an analytic model of cooperation and which offers to the model the tools to be realistic. The preceding solutions always raised the problem of the distributed collect of information, however in our solution; we propose a centralized collect that makes it possible to have a total and coherent vision of the network.

The remainder of this chapter is organized as follows. First, a focus on related works is presented, followed by a discussion of their limits. Then, we describe our objectives and assumptions. After that, we present in details our routing protocol, followed by a study of its performance, evaluation and some results of simulations. Finally, a conclusion is presented.

RELATED WORK

By definition, in mobile ad hoc networks, every node is supposed to relay packets for the benefit of other nodes. This may be the most fundamental characteristic of this type of network compared to conventional networks. Indeed, in military or emergency networks where the network is under the control of a single responsible, cooperation problems do not exist as the collective interest of the mission surpasses the individual interest of the nodes. However, if the nodes are their own chiefs, every node tries to obtain the most possible benefits from the network, while saving its own resources. Thus, nodes with selfish behaviors appear, resulting in non-involvement behaviors when common interest operations (routing, packet transmission, mobility management, etc) must be executed (Fèlegyhàzi, Buttyàn & Hubaux, 2003).

To maintain an efficient mobile ad hoc network and to guarantee a certain threshold of operationality, we must insist that all nodes be involved with the basic functions of the network.

Several solutions have been proposed:

- **Detection and reputation approach**: This solution is based upon two systems, detection and reputation systems. These two systems provide to a node the possibility to detect in its neighborhood the node that do not cooperate, estimate its reputation and inform the neighbors about it, and thus isolate it if necessary (to be ignored or refused its requests). The major problem of these methods is that they are based on the listening technique of the neighborhood and the information diffusion. However, listening is not sure in a wireless environment and the diffusion floods the network.
- **Economic approach:** In this approach, an economic model is used to encourage the cooperation. There is not neighborhood listening and each node is managed independently. The solution consists of

introduction of a virtual currency. Each cooperating node is paid for the packets that it relayed and which are not its own. The source node pays for the relaying of its packets to reach the destination. The node that does not have enough money to transmit its packets, is refused the service. The question is who does control the quantity of money available and who does give the permission to the node to transmit or not. The economic model approach does not offer the possibility to get around the non co-operating nodes during route establishment. Nevertheless, it encourages the nodes to cooperate more than the other approaches, since the nodes pay for the transmission of their packets. However, it is difficult to realize this solution because of the use of additional devices (Buttyàn & Hubaux, 2003) or to have third entities that are located separately from the ad hoc network (Zhong, Chen & Yang, 2003).

- **Game theory based model approach:** Several studies tried to model the phenomenon of cooperation to find analytical solutions. The analytical models try to find, in the absence of incentive and stimulation tools, the strategies that make nodes cooperating. These solutions provide the conditions under which an equilibrium based on the cooperation exists. Almost of these studies are based on game theory. Game theory is a powerful means for modeling the interactions between the players who are concerned only by their profit, and for the prediction of their strategy choice. The application of game theory in the mobile ad hoc networks is justified because we can compare the nodes to players.

We note that several essential points were never discussed in the models that adopted game theory to analyze the cooperation mathematically:

- *The relation between the protocol and the cooperation*: The researchers never established a relation between their solution and the used routing protocol. However, there is a close relationship between the used protocol and the solution to the cooperation problem. Indeed, if the routing protocol deals with the constraints of the cooperation, it results from it a more global solution, which deals with the interest of the nodes themselves and the interest of the network.
- *The effect of the node action on the rest of the nodes*: The majority of these solutions propose that the node acts according to its direct neighbors' actions, whereas the latter probably acted according to their direct neighbors' actions also, which shows the transitive character of the phenomenon. However, in these solutions, only the direct effect of the neighbors is taken into account. These solutions see the problem from a narrow angle, whereas it must be seen in a larger view, by taking into account the effect of a node on the remainder of the nodes (all the network).
- *Information collection needed by the node to take an action*: The approaches of resolutions suggested until now specify neither how the solution must be implemented nor how the nodes communicate between them the necessary information. Moreover, in the majority of the solutions, the decisions are calculated and taken by the nodes themselves in a distributed method without giving a proof of their efficiency.

However, all the suggested solutions until now apply traditional routing protocols for mobile ad hoc networks (Abolhasan, Wysocki & Dutkiewicz, 2004). These protocols do not offer the means that enable us to carry out the points discussed above. A routing protocol that resolves these issues becomes inevitable so that a based game theory

Figure 1. Mobile ad hoc network with two zones

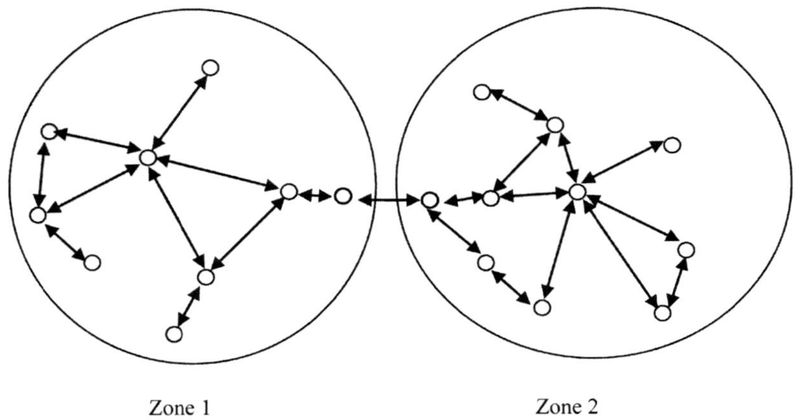

cooperation model can function correctly. In this model, the information gathering is important and the choice of the action depends on the state of the network. The model offers parameters to the protocol so it calculates the best ways according to the behavior of the nodes.

OBJECTIVES AND ASSUMPTIONS

Having a based game theory cooperation model, our objective is to offer an effective routing protocol that permit to the model to function correctly. We have established that the existing routing protocols do not offer the necessary means to exploit our model. Our analytic model consists of modeling the behavior of cooperating nodes as a play, where each player (node) takes the decision to send its packet or to relay the packets of other nodes while increasing its profit in a rational way, to arrive at the end at equilibrium between the various nodes of the network.

Let *L(s,d)* the path established between the source node *s* and the destination node *d*, and which passes through the intermediate nodes I_i for $i = 1..N$. *N* indicates the number of intermediate nodes between the source *s* and destination *d*. We define the cooperation factor *F* of the path *L(s,d)* by:

$$F\left(l_{(s,d)}\right) = \prod_{N} \psi_{I_i} \tag{1}$$

where ψ_i indicates the cooperation degree of a node. For our model, we define the zone, which gathers a certain concentration of nodes which interact between them and forming an ad hoc network, by $Zone_j$. We suppose that the nodes interacting between them tend to gather in the same zone (Figure 1).

Every node belonging to $Zone_j$ maintains three variables:

$P_{receive}$: the number of packets emitted by a source node, which actually arrive at their destination;
$P_{relayed}$: the number of packets relayed by a node;
$P_{transmit}$: the total number of packets transmitted by a node. Including the relayed packets and packets generated by the node.

The success rate of node *i* is defined by the value:

$$\delta_i = \frac{P_{receive}}{P_{transmit} - P_{relayed}} \tag{2}$$

if $P_{transmit} - P_{relayed} = 0$ then $\delta_i = -1$, otherwise $\delta_i > 0$ and the contribution rate corresponds to the value:

$$\sigma_i = \frac{P_{relayed}}{P_{transmit}} \quad (3)$$

and if $P_{transmit} = 0$ then $\sigma_i = -1$, otherwise $\sigma_i > 0$.

The *Principal* receives δ_i and σ_i values periodically from the different nodes and calculates the $\bar{\delta}$ and $\bar{\sigma}$ values which are respectively the average success rate and the average contribution rate. They are defined as follows:

$$\bar{\delta} = \frac{\sum_{\forall i \in Zone_j,\ \delta_i \neq -1} \delta_i}{\sum_{\forall i \in Zone_j,\ \delta_i \neq -1} i^0};$$

$$\bar{\sigma} = \frac{\sum_{\forall i \in Zone_j,\ \sigma_i \neq -1} \sigma_i}{\sum_{\forall i \in Zone_j,\ \sigma_i \neq -1} i^0}$$

For each node *i*, the *Principal* calculates its degree of cooperation ψ_{I_i} using the following formula:

$$\psi_{I_i} = \begin{cases} \dfrac{\delta_i}{\sum_{k \in Zone_j,\ \delta_k \neq -1} \delta_k} \times \left(1 - \dfrac{\sigma_i}{\sum_{k \in Zone_j,\ \sigma_k \neq -1} \sigma_k}\right) & \text{if } \delta_i \neq -1 \text{ and } \sigma_i \neq -1 \\ 1 - \dfrac{\sigma_i}{\sum_{k \in Zone_j,\ \sigma_k \neq -1} \sigma_k} & \text{if } \delta_i = -1 \text{ and } \sigma_i \neq -1 \\ 1 & \text{if } \sigma_i = -1 \end{cases}$$

where $\dfrac{\delta_i}{\sum_{k \in Zone_j,\ \delta_k \neq -1} \delta_k}$ is the engaging factor of the node compared to its success rate and where $1 - \dfrac{\sigma_i}{\sum_{k \in Zone_j,\ \sigma_k \neq -1} \sigma_k}$ is the contribution factor of the node compared to its contribution to the network.

The engaging factor means that if a node has a higher success rate than the network's average success rate, it is more judicious for it to cooperate than that has a lower success rate. Similarly, if a node has a smaller contribution rate than the network's average contribution rate, it is more judicious for it to cooperate than that has a higher contribution rate. Indeed, a node has a utility function that allows it to calculate acquired gain. It is considered that a node wins when its packets reach their destination and loses when it relays others' packets. It is assumed that two factors α and β are common for all nodes and they represent respectively node profit factor and loss factor. The utility function is defined as follows:

$$U_i = \alpha\delta_i - \beta\sigma_i$$

Thus, the node chooses actions that will allow maximizing its utility function. The node transmits its own packets with rate λ and relays other nodes' packets with rate μ. Actions that the node undertakes are either increase or decrease these transmission rates. Obviously, a node can maximize its utility while decreasing its contribution, thus it would have an impact upon gains achieved by other nodes, which would oblige them to decrease their cooperate contribution and consequently affect our node's gain. For this reason, it is considered that a reasonable node would attempt to maximize its profit while taking into account that it will not affect the profits gained by other nodes.

Each node chooses a strategy that allows it to take actions to deal with a situation among the following:

Situation I:

$$\left.\begin{matrix} \delta_i < \bar{\delta} \\ \sigma_i < \bar{\sigma} \end{matrix}\right\}$$

In this case, both the node's success rate and contribution rate are below average.

Action I:

The node must increase its contribution (relay more packets). Increasing the node's contribution can influence the other nodes' contribution to be larger, which would increase the node's success rate.

Situation II:

$$\left.\begin{array}{l}\delta_i < \overline{\delta} \\ \sigma_i > \overline{\sigma}\end{array}\right\}$$

In this case, the node's success rate is below average and the contribution rate is above average.

Action II:

The node must decrease its contribution (relay fewer packets). In spite of the generous node contribution, its success rate remains above average. This is due to a lack of cooperation from other nodes. Decreasing its contribution would oblige other nodes to increase their contribution rate, consequently increasing the node's success rate.

Situation III:

$$\left.\begin{array}{l}\delta_i > \overline{\delta} \\ \sigma_i < \overline{\sigma}\end{array}\right\}$$

In this case, the node's success rate is above average, while its contribution rate remains below average.

Action III:

The node must increase its contribution (relay more packets). Providing an inadequate contribution rate could result in an unsatisfactory success rate for other nodes. It would oblige the latter to decrease their cooperation, causing a decrease of the node's success rate.

Situation IV:

$$\left.\begin{array}{l}\delta_i > \overline{\delta} \\ \sigma_i > \overline{\sigma}\end{array}\right\}$$

In this case, both the node's success rate and its contribution rate are above average.

Action IV:

The node must increase the transmission rates of its own packets and its contribution rate. It is estimated that if it increase its transmission rateλ, that permit an increase of the value of$\overline{\delta}$. Consequently, it will incite other nodes to behave similarly, thus increasing the network debit.

Situation V:

The node remains in the same situation I, II or III during a large period, even though it performs the action corresponding to this situation.

Action V:

The node has to move. After having undertaken a certain number of actions (quantity yet to be determined) to change its situation, the node always remains in the same situation. It must move to eliminate the possibility that the problem is due to network topology. The movement affects network configuration and it is assumed that the configuration has a direct effect on network output.

Increasing or reducing the transmission rate λ and μ is done so that $\frac{\mu}{\lambda + \mu}$ can be as close as possible to $\overline{\sigma}$ and the value of the utility function U_i remains satisfied (> 0). Of course, all these operations (the calculation of $\overline{\delta}$ and $\overline{\sigma}$, gathering δ_i and σ_i, the calculation of the $F\left(l_{(s,d)}\right)$ cannot be made in a distributed way because we will fall into the problem of data coherence. For this reason, we need a central entity, which does this work. It is obvious that these calculations also allow choosing the best paths that offer a satisfying rate of cooperation; this also means that these operations are part of the tasks of the routing protocol. Thus, we will integrate all these operations in the routing protocol. It is also clear that the existing routing protocols do not offer such

Figure 2. Mobile ad hoc network with zones

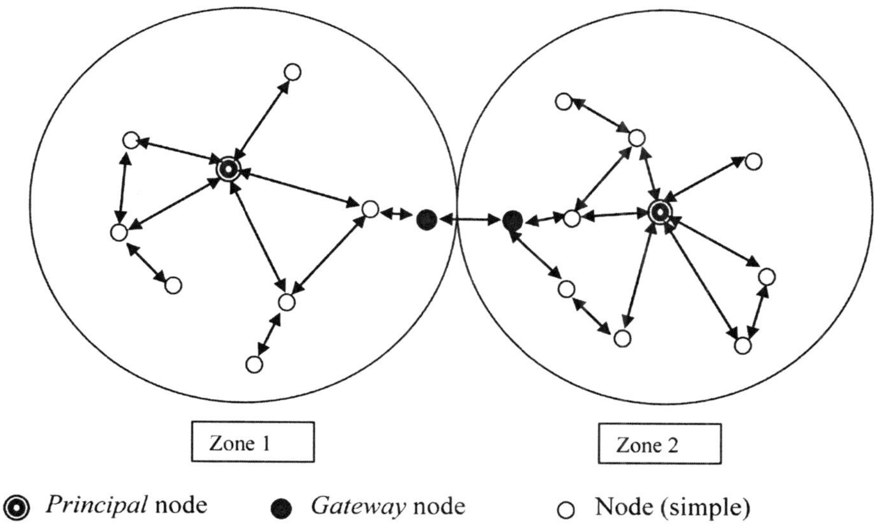

means. Consequently, defining a new protocol that satisfies our needs become necessary.

THE ROUTING PROTOCOL

Based on the assumptions described previously, it is clear that the existing routing protocols (reactive, proactive or hybrid) do not provide the tools necessary to implement our model. In this direction, we present a new hybrid routing protocol equipped with mechanisms that allow the use of the cooperation model. The metric used for the computation of paths is the nodes' aptitude for cooperation. Thus, we named our protocol as "Cooperation-Based Hierarchical Routing Protocol" (CBHRP).

PRESENTATION

CBHRP is a hybrid routing protocol, hierarchical, cooperation-based. The nodes of the network are organized in several zones. We suppose that the nodes interacting between them tend to gather in the same geographical area. Each zone is equipped with a central node that will manage the information gathering and the diffusion of the data necessary to the nodes for the cooperation, and at the same time, necessary for the routing. Thus, we can consider the general structure of the network in several zones with an administrator node for each zone. That node is called *Zone Principal* (*Principal*). The zones will be able to communicate between them via nodes bordering which are called gateway nodes (gateway) (Figure 2).

The routes are built by requesting the *Principal*. This last is equipped with a route table containing the nodes belonging to its zone. In fact, the *Principal* knows at any time the topology of its zone (the nodes members of the zone and their respective neighbors). The computation of the route is done in two stages. The first stage, which is reactive, consists of making a route request to the *Principal* by the source node with a RREQ message (Route_REQuest). The second stage, which is proactive, consists of calculating the route by the *Principal* according to the route table using the cooperation criteria. The *Principal* answers the source node with a RRES message (Route_RESponse).

Table 1. Zone members

Node_Id	*Cooperation Degree*	*Type*	*Neighbors*

ELECTION OF *PRINCIPAL*

We took as a reference the CBRP protocol (Jiang, Ji & Tay, 1999) to define the algorithm, which allows the formation of the zones. When a new node arrives, it is put in the "undecided" state. The first action that this node must make is to temporize and start to diffuse a *Hello_message* message. When a *Principal* node receives this message, it answers immediately with a *Hello_response* message indicating that it is the *Principal* of the zone. Or when a member node receives the *hello_message*, it answers with a *Hello_response* message indicating that there is a close *Principal*. After that, when the new node receives this answer, it will change its state into a "member" state, and makes a request for association (to be developed later). But when the node does not receive any message from any *Principal* or from any other member node indicating that there is a close *Principal*, it puts itself *Principal*, but only, if it has bidirectional link with one or more close nodes. Otherwise, if it does not have any link with any other node, it remains in the "undecided" state and repeats the process with the sending of the *Hello_message* message.

The *Principal* has a table of members in which is stored information about the nodes members of its zone. (see Table 1)

- The field *Node_Id* defines the member node identity of the zone.
- The field *Cooperation Degree* defines the aptitude of the node to cooperate; it is used by the routing protocol like a measure for the computation of the best route.
- The field *Type* determines whether the member node is a gateway node or a simple node.
- The field *Neighbors* defines the list of the neighbors of the node that are member of the same zone.

The *Principal* is also equipped with a second table (Table 2) defining only gateway members of its zone. This table contains the following information:

- The field *Node_Id* defines the identity of the gateway member node of the zone.
- The field *Neighbors (other zones)* defines the identity of the neighbors which are member of another zone.
- The field *Id_Principal (other zone)* defines the *Principal* of the zone to which the neighbors belong.

NODE ASSOCIATION

In the beginning, a zone is formed only by the *Principal*. Thereafter, nodes arrive and become members of the zone. To become member of a zone, the node executes the association procedure.

Table 2. Zone bordering members

Node_Id	*Neighbors (other zones)*	*Id_Principal(other zone)*

Figure 3. Direct association procedure

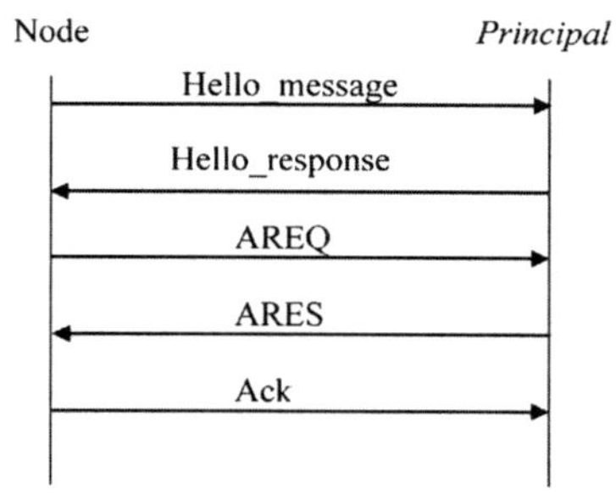

Figure 4. Indirect association procedure

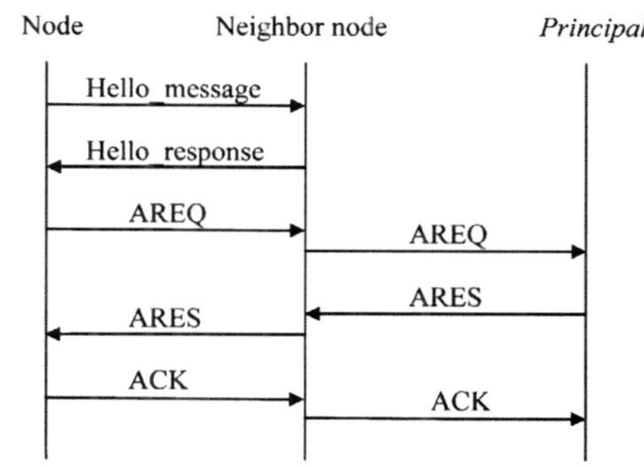

Association allows the *Principal* to determine the nodes that belong to its zone. We define the procedure as follows: For the close nodes, the *Principal* can recognize them directly (the *Principal* can receive directly the signal of close neighbors), these nodes will join the *Principal* by an association request. When the node receives the *Hello_response* message from the *Principal*, it receives at the same time the *Principal* identity and the zone where it is situated. In this case, the node formulate an association request *Association_REQuest* (AREQ) to the *Principal* to become a member of the zone, the *Principal* answers with an *Association_RESponse* (ARES) message to confirm the association (Figure 3).

The node which does not receive a signal from the *Principal*, can recognize the *Principal* of the zone via the neighbors, and it can find more than one *Principal* (the case where the node is between two zones bordering). Indeed, when a new node *i* sends *Hello_message*, which is intercepted by another node *j* that is not a *Principal*, but is already a member of a zone. The node *j* answers node *i* by transmitting its identity as well as the identity of the *Principal* with which it is associated. The choice of the *Principal* to which the node *i* will join, can be determined by several criteria, among them the number of hop between the node and the *Principal*, or numbers of neighbors who have the same *Principal* node, etc. The node recovers the identity of the *Principal* from one of these neighbors and sends an association request via these nodes to the *Principal* (Figure 4). The *Principal* node after acceptance of association, sends to the node *i* an association acknowledge and the path use for communicating with it (the *Principal* can send to the node *i* two or three paths in case of rupture).

The Association procedure allows the node to be recorded with the *Principal*; this last keeps a table containing the identity of the nodes member of its zone and their neighbors. Indeed, the member node sends also the identities of its close nodes (neighbors). Any change in the topology of the zone is reported by the concerned nodes to the *Principal*, thus the *Principal* has a global view of its zone topology at any moment.

TOPOLOGY MAINTAIN

With each change of topology, the *Principal* receives from the concerned nodes useful information; updates its tables and transmits to the nodes concerned with the changes the paths enabling them to communicate with it. Indeed, each node can detect its direct neighbors because they periodically send beacons to inform their existence and to which zone they belong. Thus, periodically or when a change occurs, each node recovers this information and inform its *Principal* (the information contains the list of the neighbors and the zone to which they belong). Thus, the *Principal* can identify which are the gateway nodes, which will enable it to communicate with the other zones.

ROUTE COMPUTATION

The requests for routes are done by the nodes to the *Principal*. This last uses its tables to compute the best route based on the cooperation, i.e., calculate the route which gathers intermediate nodes between the source node and the destination node and are most likely to cooperate in the retransmission of the packets from the source node towards the destination node.

ROUTE REQUEST

When a node wants to communicate with another node it sends a route request to the *Principal* (the node knows the path to the *Principal*). At this stage, it is the reactive part of the routing. The *Principal* checks if the destination node belonged to its zone, if it is the case, it establishes a route (the best one) from the tables and sends it to the source node, it is the proactive part of the routing (Figure 5).

If the node source is not a direct neighbor of the *Principal*, it makes its request via a nearby node by using the path, which was established to it by the *Principal* (Figure 6).

If the destination node is not in the same zone, the *Principal* sends a request to the various gateway nodes; these later transmit the same request to their respective neighbor gateway nodes of the other zones. The neighbor gateway nodes send the request to their respective *Principal*s. The *Principal* of the target zone calculates the route between the gateway and the destination and sends the result to the gateway, this latter sends the result to the gateway of the zone to which the node source belongs and the gateway in its turn sends the result to its *Principal*. After having received several results and after certain time (deadline), the *Principal* calculates the best route (according to certain criteria) and sends the result to the source node.

After the route establishment, the communication is direct between the source node and the destination node. The route does not pass by the *Principal*.

BEST ROUTE COMPUTATION

The choice of the best route is calculated by using the cooperation degree metric of the nodes. Indeed, like the *Principal* has the values δ and σ, it can by the same way calculate the degree of cooperation δ of the nodes like the nodes do for themselves, so that it can evaluate the factor of cooperation F of all possible routes and select the best route whose factor F is largest (Figure 7).

Figure 5. Direct route request

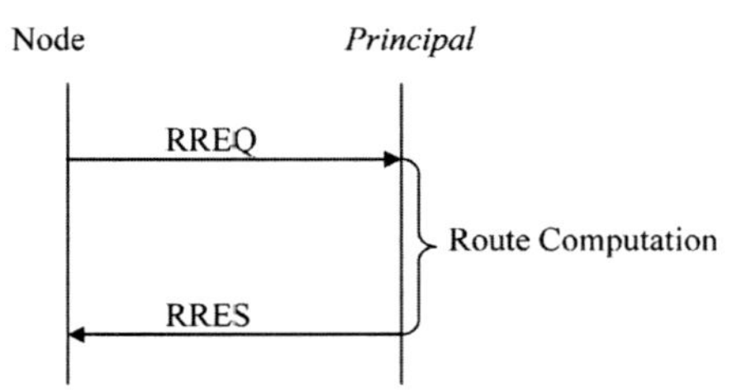

Figure 6. Indirect route request

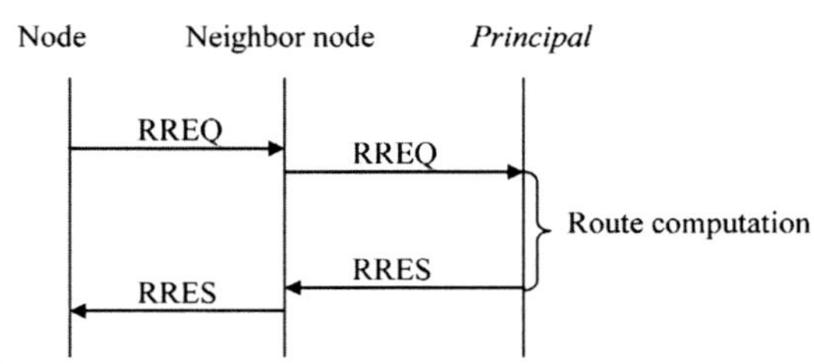

RUPTURE NOTIFICATION

During a route rupture, the node that identifies the rupture announces it directly to the *Principal* which will make an update of its table and recomposes a new route and then informs the source node so that it can proceed with the new route (Figure 8).

PERFORMANCE EVALUATION

Our performance evaluation is based on a comparison between our routing protocol, a proactive routing protocol and a reactive routing protocol. The application of our theoretical cooperation model requires the sending of the values σ and ψ_i from each node of the network for the calculation of F and δ. In a network which applies a proactive or reactive routing protocol and where there is not a central entity which calculates the values σ and $\overline{\delta}$ on behalf of the nodes, it is the nodes themselves which will have to make this calculation and consequently, they must recover the values $\overline{\sigma}$ and $\overline{\delta}$ of all the other nodes of the network.. If we apply our model to an ad hoc network which uses a proactive routing protocol (each node has a table of route towards all the destinations), we note that each node must send the values $\overline{\sigma}$ and δ to all other nodes. Thus if we have N nodes, each node must transmit these values to *N-1* remaining nodes, which gives *N (N-1)* transmission in all. In the case of a reactive routing protocol, the things change a little, because for transmitting the values σ and δ, it is necessary initially to calculate the routes towards all the other destinations, which means that the transmission of these values generates the transmission of route requests and route response. In other words, if we have N nodes and if a node wants to send the values σ and δ to the *N-1* other nodes, it must, in the worst case, initially send a *N-1* route request to the *N-1* destinations and receive *N-1* route response. In the best case, if we suppose that it already has the routes, it directly transmits the values σ and δ to the *N-1* nodes. What results, in the worst case, we have *N-1* requests and *N-1* answers and *N-1* transmissions of σ and δ, which gives *3 (N-1)* transmissions, in the best case, we have *N-1* transmissions. Thus we can say that the average of transmissions is *2 N (N-1)* for the entire network. For our routing protocol, each

Figure 8. Rupture notification

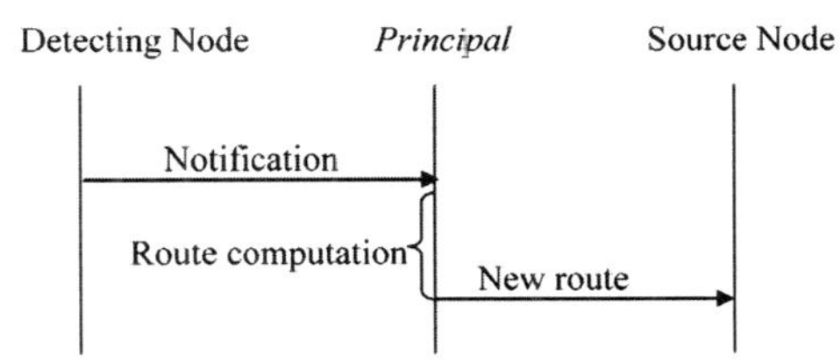

Figure 7. Interzone route request

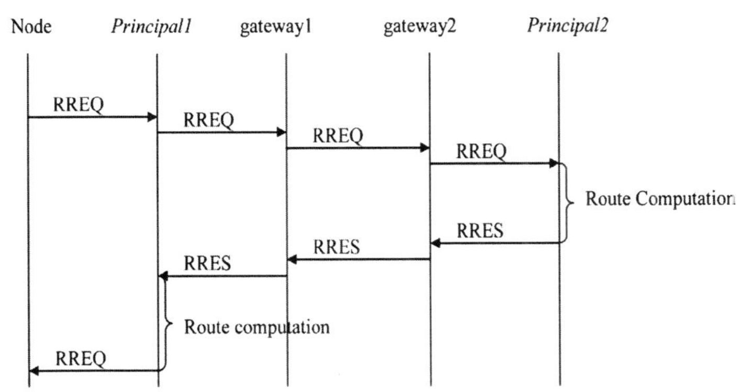

node has the path which links it to the *Principal*, thus each node sends the values σ and δ only one time to the *Principal*, this last recovers all the data, calculate the averages and retransmits them to all nodes, what means that we have *N* transmissions by nodes and *N* transmission by the *Principal*. We have in all *2N* transmission (Figure 9).

Figure 9. Number of transmission of δ and σ per protocol

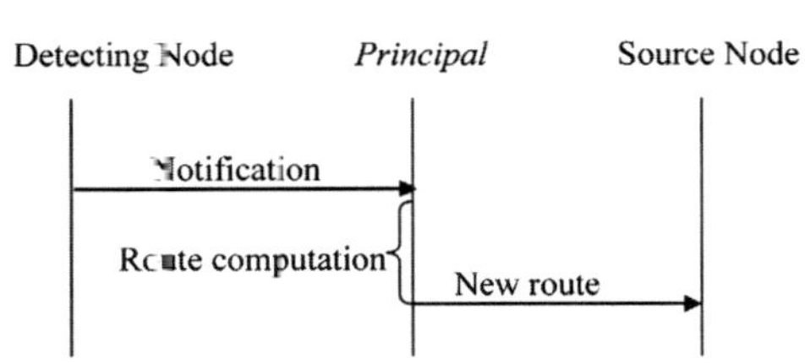

SIMULATION RESULTS

In this section, we present the performance of the proposed model to illustrate the potential benefits. Let us first describe our simulation setup. We simulate a mobile ad hoc network of 20 nodes. The nodes have been placed uniformly distributed, onto an area of 1000*m X* 1000*m* and moved according to the random waypoint model with a velocity of 1*m/s* with a transmission radius of 250 *m* for each node. The simulation time is the time needed to permit to all nodes to transmit 100000 packets. We divide this simulation time into 50 periods. During a period, each node can transmit on average of 1000 packets and relay 2000 packets for a total of 100000 packets during all simulation. The total energy of a node does not enable it to transmit more than 100000 packets, in other words, if the node arrives at the threshold of 100000 packets, it is out of use. After each period, relevant data are transmitted by the nodes to the *Principal*, which will use these data to calculate the routes, the *Principal* calculates also some data and returns them to the nodes which will use them to undertake their actions according to definite strategies.

Simulation Elaboration

1. A uniform randomly distributed nodes onto the network area;
2. For each node we affect a random destination node;
3. During a period, the nodes transmit their packets in alternate way, one packet a time, until all nodes finish transmitting the number of packets indicated for this period. The order of transmission is recomputed after each turn;
4. For reason of simplicity, when a node transmits a packet *P*, no other node can transmits its own packet before the packet *P* is relayed by the intermediate nodes towards the destination or it is rejected;
5. To illustrate the potential performance of our model, we created four different scenarios:
   a. The first uses our model of cooperation and the routing protocol that we defined. That scenario is called scenario **S+P** (strategy with protocol);
   b. The second uses our model of cooperation but without the routing protocol we defined, but rather a DSR likely routing protocol, we call this scenario **S-P** (strategy without protocol);
   c. The third scenario, we define it as the naive scenario where all nodes cooperate with hundred percent (they relay all the packets that they receive) and using a DSR likely routing protocol;
   d. The fourth scenario is the selfish scenario, where we define a partial set of nodes (in this case, we set the quart of the set of nodes like selfish nodes) that they never cooperate (they never relay the packets which they receive). On the other hand, the remainders of the nodes operate in a naive way, this

scenario uses also a DSR likely routing protocol.

6. After each period, we make data statistics for evaluating the various scenarios.
7. Simulation is carried out several times, after which, we calculate the average of all them;
8. The objectives of the evaluation are to show and prove the performances that are bring by the scenario of our model associated with our routing protocol compared to the other cases;

Results

a. Node Cooperation tendency: The simulations performed for our protocol show that the node cooperation rates tend to be stabilized on a certain level respectively to each node (Figure 10).

b. Routing protocol effect on the success rate: we compare the evolution of the average success rate of nodes for the two scenarios **S+P** and **S-P**, and we see that the **S+P** scenario performs better than the **S-P** one. This is due to the use of the protocol that we define (Figure 11).

c. Emission graph: the graph (Figure 12) allows us to see the evolution of node emissions during the simulation. We constate, that in scenarios using our analytic model, the emissions tend to augment with time until a certain period. Then, it starts to decrease. This diminution is justified because, with time, the energy of nodes starts to decrease also. So consequently, the nodes decrease their emission to be alive for a long time. In the naive and selfish scenarios, the emissions tend to decrease from the firsts periods because the nodes spend an important part of their energy to relay others' packets.

d. Relaying graph: the graph (Figure 13) allows us to see the evolution of node packet relaying during the simulation. We constate, that in scenarios using our analytic model, the packet relaying average per period is steady for a long time (until period 31). Then, it starts to decrease. This diminution is justified because, with time, the energy of nodes starts to decrease also. However, in the naive and selfish scenarios, the packet relaying average per period tends to decrease from the earliest periods (from period 9) and goes under the average of the two scenarios **S+P** and **S-P** (from the period 21). This is because the nodes that relay packets spend an important part of their energy to relay others' packets and go deceased very early so the packet relaying average go down rapidly.

Figure 10. Node Cooperation tendency

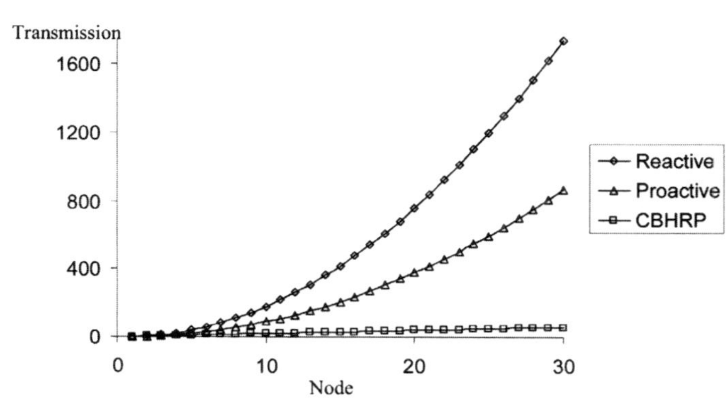

Figure 11. Effect of the routing protocol on the success

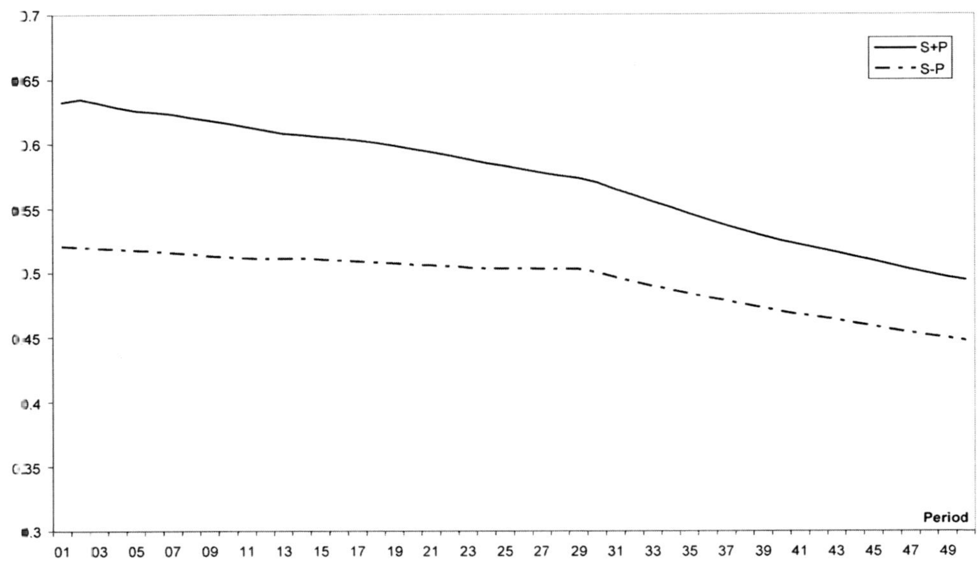

Figure 12. Emissions graph

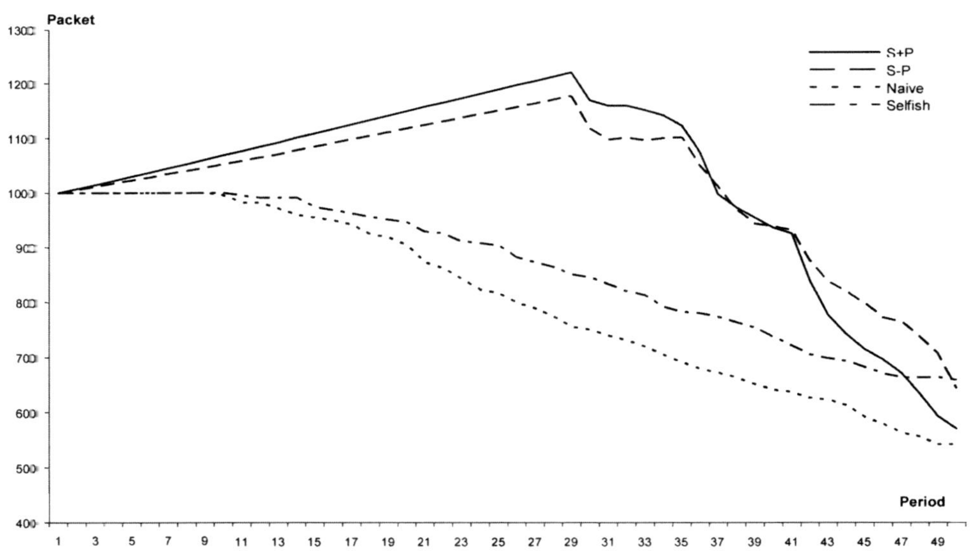

CONCLUSION

We presented a solution to the ad hoc mobile network routing protocols, which do not offer the necessary tools to cure the problem of the cooperation between the nodes. The alternative solution that we have proposed consists of a hybrid routing protocol based on packets relaying cooperation between nodes. The principal features of our protocol are that it is simple to achieve; it

Figure 13. Relaying graph

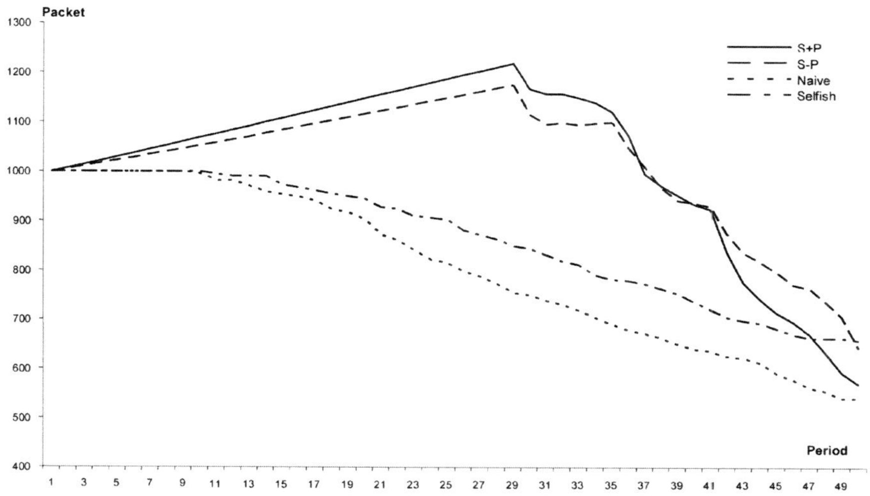

does not impose constraints; and it allows a central information collection to calculate, in a coherent way, the necessary data used by the routing protocol and necessary to the nodes to choose their strategy to allows them to have a certain threshold of cooperation. Thus, the survival of the network is guaranteed. With some performance evaluations and simulation results, we showed the effectiveness of our protocol despite the proactive and reactive protocols.

REFERENCES

Abolhasan, M., Wysocki, T., & Dutkiewicz, E. (2004). A review of routing protocols for mobile ad hoc networks. *Journal of Ad Hoc Networks, Elsevier*, *2*(1), 1–22. doi:10.1016/S1570-8705(03)00043-X

Buchegger, S., & LeBoudec, J. Y. (2002a). Performance analysis of the CONFIDANT protocol. In *Proceedings of the 3rd ACMS International Symposium on Mobile Ad Hoc Networking Computing and Computing (MobiHoc), Lausanne, Switzerland, June 09–11* (pp. 226–236).

Buchegger, S., & LeBoudec, J. Y. (2002b). Nodes Bearing Grudges: Towards Routing Security, Fairness, and Robustness in Mobile Ad Hoc Networks. In *Proceedings of the Tenth Euromicro Workshop on Parallel, Distributed and Network-based Processing* (pp. 403-410). IEEE Computer Society.

Buttyàn, L., & Hubaux, J. P. (2001). *Nuglets: a virtual currency to stimulate cooperation in self-organized ad hoc networks* (Technical report DSC/2001/001). Swiss Federal Institute of Technology, Lausanne.

Buttyàn, L., & Hubaux, J. P. (2003). Stimulating cooperation in self-organizing mobile ad hoc networks. *Mobile Networks and Applications*, *8*(5), 579–592. doi:10.1023/A:1025146013151

Chlamtac, I., Conti, M., & Liu, J. (2003). Mobile ad hoc networking: Imperatives and challenges. *Ad Hoc Network Journal*, *1*(1), 13–64. doi:10.1016/S1570-8705(03)00013-1

Fèlegyhàzi, M., Buttyàn, L., & Hubaux, J. P. (2003a). *Equilibrium Analysis of Packet Forwarding Strategies in Wireless Ad Hoc Networks – The Dynamic Case (extended version)* (Technical report Nr. IC/2003/68). Laboratory of to Computer Communications and Applications, Swiss Federal Institute of Technology, Lausanne.

Fèlegyhàzi, M., Buttyàn, L., & Hubaux, J. P. (2003b). *Equilibrium Analysis of Packet Forwarding Strategies in Wireless Ad Hoc Networks – The Static Case (extended version)* (Technical report Nr. IC/2003/33). Laboratory of to Computer Communications and Applications, Swiss Federal Institute of Technology, Lausanne.

Félegyházi, M., Hubaux, J.-P., & Buttyán, L. (2006). Nash Equilibria of Packet Forwarding Strategies in Wireless Ad Hoc Networks. *IEEE Transactions on Mobile Computing*, *5*(5), 463–476. doi:10.1109/TMC.2006.68

Jiang, M., Ji, J. & Tay, Y.C. (1999). *Cluster based routing protocol*. Internet Draft, draft-ietf-manet-cbrp-spec-01.txt, work in progress, 1999.

Marti, S., Giuli, T. J., Lai, K., & Baker, M. (2000). Mitigating routing misbehaviour in mobile ad hoc networks. In *Proceedings of the Sixth annual ACM/IEEE International Conference on Mobile Computing and Networking* (pp. 255–265).

Michiardi, P., & Molva, R. (2002). Core: A Collaborative Reputation Mechanism To Enforce Node Cooperation In Mobile AD HOC Networks. In *Proceedings of the 6th IFIP Communications and Multimedia Security Conference, Portorosz, Slovenia, September.*

Michiardi, P., & Molva, R. (2003). A game theoretical approach to evaluate cooperation enforcement mechanisms in mobile ad hoc networks. In *Proceedings of WiOpt 2003: Modeling Optimization and Mobile in, Ad Hoc and Wireless Networks, INRIA Sophia-Antipolis, France, March 3-5.*

Srinivasan, V., Nuggehalli, P., Chiasserini, C. F., & Rao, R. R. (2005). An analytical approach to the study of cooperation in wireless ad hoc networks. *IEEE Transactions on Wireless Communications*, *4*(2), 722–733. doi:10.1109/TWC.2004.842950

Urpi, A., Bonuccelli, M., & Giordano, S. (2003). Modelling cooperation in mobile ad hoc networks: A formal description of selfishness. In *Proceeding of WiOpt 2003: Modeling Optimization and Mobile in, Ad Hoc and Wireless Networks, INRIA Sophia-Antipolis, France, March 3-5.*

Zhong, S., Chen, J., & Yang, Y. R. (2003). Sprite: A Simple Cheat-Proof Credit-Based System for Mobile Ad-Hoc Networks. In *Proc. of the 22nd Annual Joint Conference of the IEEE Computer and Communications Societies (INFOCOM'03)*, *3*(30), 1987-1997.

KEY TERMS AND DEFINITIONS

Ad Hoc Mobile Network: A self-configuring network of mobile nodes connected by wireless links.

Game Theory: The study of the ways in which strategic interactions among rational players produce outcomes with respect to the preferences (or utilities).

Hybrid Routing Protocol: Combines the proactive and reactive routing approaches.

Proactive Routing Protocol: A category of mobile network routing protocols. Proactive routing protocols maintain routes to all destinations, regardless of whether or not these routes are needed.

Reactive Routing Protocol: A category of mobile network routing protocols where the route between the source and the destination is no previously established. The route is established on demand only when it is explicitly needed to forward packets.

Chapter 2
Ambient Networks

Alex Galis
University College London, UK

Bertrand Mathieu
Orange Labs, France

ABSTRACT

This chapter introduces the Ambient Network (AN) which is a network integration solution that aims at fostering dynamic co-operation between the next generation, heterogeneous wireless and wired networks, in order to gather resources within and across networks to provide new services to end users. First, it depicts the main characteristics of AN before describing the Ambient Control Space, which is the space where functional entities managing the networks and the services are active, and the interfaces. Then it focuses on the SATO (Service-aware Adaptive Transport Overlays), which is a service overlay networks deployed on top of the underlying heterogeneous physical networks and tailored according to the services' requirements and users' context. This chapter aims at giving the readers a good overview of what Ambient networks are since they will be present in the next years everywhere.

INTRODUCTION

With the new network technologies and the evolution of devices, end-users can now access any service anywhere, anytime from any device, in any situation. This came possible via the evolution of the wireless networks (3G, WiFi, WiMax, …), the Internet world and the ever increasingly offer of services. We have been entering in the "Always Connected" world. However, if successful and very helpful, there are still major technological bottleneck areas, which justify research and development for new network environments.

3GPP network related bottlenecks: The very success of 3GPP (n.d.) is now creating obstacles to innovation in the networking technology deployed. The size and scope of the public 3GPP networks now make the introduction and deployment of new network technologies and advanced services difficult and very costly. It is thus important to develop solutions that cater for the easy inclusion of the new technologies, while still keeping the

DOI: 10.4018/978-1-60566-250-3.ch002

impact on the overall network to a minimum level. This holds both for the introduction of specialized access networks under a common core and the inclusion of new functionality in a network's core domain.

Internet related technological bottlenecks: The Internet has been founded on a basic architectural premise: a simple network service is used as a universal means to interconnect intelligent end-systems. The end-to-end argument has served to maintain this simplicity by pushing complexity into the endpoints, allowing the Internet to reach an impressive scale in terms of inter-connected devices. The current structure of the Internet, though, clearly deviates in many respects from this underlying networking principle, as the Internet is broken into a larger number of interacting network domains leading to placement of functionality inside the network, especially at network domain boundaries. The missing step (Bassi, Denazis, Galis, Fahy, Serrano, & Serrat, 2007) is to finally acknowledge the need for increased complexity in the network and to provide mechanisms harmonizing the interaction across network domain boundaries.

Service related bottlenecks: Future use of Information and Communications Technologies will constitute a new service landscape, in terms of new areas of usage, new ways of service provisioning, new user behaviour patterns as well as new dependencies between services and the everyday activities of humans and devices. The expected growth in services is not well supported by the current generation of networks, which are mainly application or service agnostic. New network support for autonomic and automatic service provisioning is needed to efficiently support this development.

Ambient Network (AN) is a network integration solution (WWI-AN, n.d.; Niebert, Flinck, Hancock, Karl, & Prehofer, 2004; Johnsson, Rembarz, Schieder, Surtees, Ahmed, & Schoo, 2008; Perera, Boreli, Herborn, Georgiades, Eisl, & Hepworth, 2008; Kappler, Poyhonen, Johnsson, & Schmid, 2007) addressing the above bottlenecks and in particular a solution to the current problems of switching from one network to the other in order to keep in contact with the outside world. The solution envisaged is the development of a network software-driven control infrastructure that will run on top of all current or future network physical infrastructures to provide a way for devices to connect to each other, and through each other to the outside world and to provide seamless service provisioning and roaming. As such AN offer a solution to the key networking problem of fostering dynamic co-operation between the next generation, heterogeneous wireless and wired networks, in order to gather resources within and across networks to provide new services to end users.

AMBIENT NETWORK CHARACTERISTICS

Ambient Networks have been designed having several principles in mind, addressing the needs of future Internet and mobile and wireless systems as well as providing innovative solutions for "fixed-mobile convergence". The major characteristics of Ambient Networks are as follows:

All-IP Network-Oriented

Networking in Ambient Networks has been defined in order to remove architectural restrictions on who or what can connect to what. Compared to existing inter-networking, ANs assume that all is network and the distinction between end-nodes and networks is no longer visible. Ambient Networks are IP-based mobile networks that can be regarded as a continuation of current Internet design practices.

Heterogeneity

Ambient Networks are networks managed and administered by different operators, and the underlying network technologies may be different. It may vary between different network technologies (e.g. UMTS, WiMax, Bluetooth, xDSL,...) or differ in sizes (e.g. from small Personal Area Networks to large-scale Peer-to-Peer networks.

Composability

An Ambient Network is capable of dynamically composing (and de-composing) with one or several other network(s). The participating Ambient Networks may belong to different administrative/economic entities. Thus co-operation between composing Ambient Networks is essential to provide network services in a co-operative as well as competitive fashion. The defined ANI (Ambient Network Interface) provides the necessary facilities for communication/cooperation across different composing/decomposing ANs.

Control Interface Provisioning

Well-defined control interfaces to other Ambient Networks (through the ANI) and to other service platforms or applications (through the Ambient Service Interface (ASI)) are provisioned in AN. The Ambient Resource Interface (ARI) abstracts the specific control techniques of the underlying connectivity network and allows control functions to operate seamlessly over a heterogeneous network infrastructure.

Explicit Control Space

In today's networks, there is a large degree of homogeneity on the very basic connectivity/packet forwarding functions, but the network control is distributed over multiple layers and is specific to network technology, operator and even implementation. In AN, an explicit control space is defined, which is a provisioning of the main Ambient Control Space (ACS) functions, enhancing existing control functions. Control functions of individual ANs must also be composed during AN composition. However, individual functions must control the same resources as before composition. Resources should not become common assets after composition, but should be treated as an asset that can be traded or offered, e.g., via Service Level Agreement (SLA).

Autonomic and Self-Management

AN includes self-* (self-configuration, -optimisation, -organisation, -healing, -protecting, -adaptation, -awareness, -programmability (Galis, Denazis, Brou, & Klein, 2004)) characteristics in its internal design for the network management or service management. Autonomic system management in the ambient network environments improves and eases the maintenance instead of manual pre-configuration or off-line negotiations.

Context Provisioning Control

The Context Provisioning control function assures consistent and autonomic network context centred co-operation between all control functions within the ACS. The goal is to make Ambient Networks control functions context-aware.

Mobility

Mobility refers to the network architectures' ability to dynamically compose, as well as the mobility of user groups within ANs. An Ambient Network mobility solution will have to work well across business and administrative boundaries, which requires solutions that ensure the security of inter-domain communications.

Service Accessibility

Services becoming more and more user-centric and since context could change frequently in ANs, a way to deliver content to end-users' context is achieved using overlay techniques, where service components are dynamically "chained" to provide the global adapted service. New services that are originally not available to end users in a different domain are now accessible to end users at any time/location via Service-Aware Transport Overlays (SATOs).

THE AMBIENT CONTROL SPACE CONCEPT

The concept of the **Ambient Control Space (ACS)** is introduced to encompass all control functions in a certain Ambient Network (AN) domain., which can be used as a control plane overlay to integrate and interoperate seamlessly any existing networks (WWI-AN, n.d.). The ACS consists (see Figure 1) of a number of Functional Entities (FEs) that cooperate to implement the overall control functionality.

The ACS is supposed to be agnostic to the underlying connectivity technologies. Thus, it operates on abstract, technology independent connectivity entities provided by a *flow abstraction*. Similarly, the ACS offers *bearer abstractions* through the ASI (Ambient Service Interface) to allow the design of applications that are independent of any underlying mobility, service quality or media delivery mechanisms.

The Ambient Control Space (see Figure 1) hosts several functional entities which enable the support for mobility, self-management, Network Composition, QoS control or multi-access, and also more abstract functions like the provisioning of context information or support for media delivery overlays.

Figure 1. Illustration of the ACS with its three Reference Points

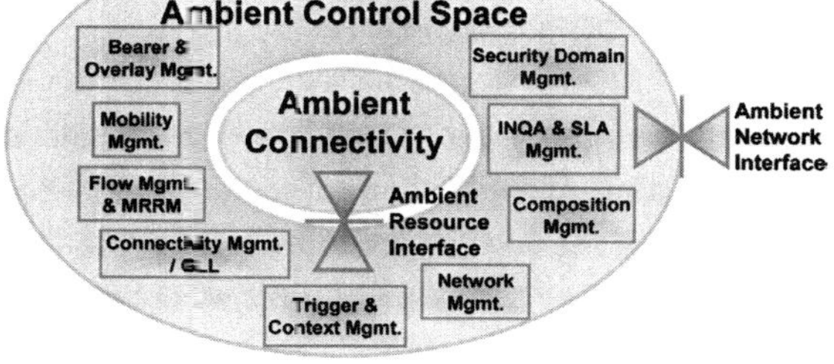

- The **Bearer and Overlay Management** functions offer end-to-end bearer services to applications through the ASI and deals with establishing and tearing down of Service-aware Adaptive Transport Overlays (SATOs). SATOs are aware of the semantics of media streams and can manipulate them along the delivery path.
- The **Multi Radio Resource Management (MRRM)** manages the resources in the access and configures and establishes flows in the access ("access flows").
- The **Connectivity Management (and Generic Link Layer)** functions detect new links and initiate the establishment of connectivity between ANs across them.
- The **Flow and Mobility Management** functions are involved in the establishment of Flows for Bearers and in ensuring that bearers remain independent and unaware of connectivity changes or movement events that occur in the underlying connectivity.
- The **Trigger and Context Management** functions coordinate the storage, access control, dissemination and aggregation of context information. Triggers, that is, notifications about mobility-related events and state changes, are collected and disseminated by the Triggering FE. Context information, such as user preferences or network status, is available in the (highly distributed) context information base. Context information can only be accessed according to explicit policies, to ensure privacy and protect business information.
- The **Composition Management** controls the composition of ANs. Co-operating ANs could potentially belong to separate

administrative or economic authorities. Composition agreements should then been defined before ACSes can compose.

THE AMBIENT NETWORK INTERFACES

In AN, three interfaces are defined:

- The **Ambient Network Interface** (ANI) connects the ACSs of different ANs (inter-ACS communication) as well as it might be used to facilitate interaction between functions residing in the same ACS (intra-ACS interaction). When inter-working between ANs that belong to different authorities is to be ensured, conformance testing on well-defined Interfaces is eventually required.
- The **Ambient Service Interface** (ASI) offers the connectivity and control functions for use by upper layer applications within a node operating in an Ambient Network. It allows applications and services to issue requests to the ACS concerning the establishment, maintenance and termination of end-to-end connectivity between functional instances connecting to the ASI.
- The **Ambient Resource Interface** (ARI) is the reference point through which functions of the ACS control and use services of the underlying connectivity. The ARI is located inside an Ambient Network node and also provides an abstraction of the resources in the underlying connectivity infrastructure.

The ARI encapsulates the capabilities of the underlying infrastructure into abstracted information. Because these abstractions are not tied to any specific network technology, the ACS functions are portable between different network types; this decoupling enables a simple migration path from current systems to the Ambient Networks approach.

SERVICE-AWARE ADAPTIVE TRANSPORT OVERLAYS (SATOS)

In today's world, users request personalised service. An overall architecture enabling the user-centred delivery of service, any time, everywhere, whatever the device and the network are is then required. The SATO concept (Mathieu, Song, Galis, Cheng, Jean, Ocampo, Lai, Brunner, Stiemerling, & Cassini, 2007; Mathieu, Galis, Jean, Ocampo, Lai, Stiemerling, Cano, Kampmann, Tariq, Balos, Ahmed, Busropan, & Prins, 2007; Mathieu, Song, Galis, Cheng, Jean, Ocampo, Lai, Brunner, M., Stiemerling, M., Cassini, M., & Kampmann, 2007) enables the provisioning of value-added services, where the value is not only added in the endpoints, but also in the network. SATO is useful for customization or adaptation of content to end-users' context but also for providing new network value-added services like: virus scan, mitigation of SPAM or SPAM over IP Telephony (SPIT), peer-to-peer services such as Voice over IP (VoIP) communication using the Session Initiation Protocol (SIP), for instance.

The Ambient Service Interface (ASI) has been defined as an "upper layer" interface of the ACS that is accessible to applications to define their requirements and specify how the service should be delivered (in terms of QoS, security, and connectivity). As such, Service Providers can make service requests to the ACS(s) through the ASI, and the ACS will process the request, by locating, reserving, and managing the end user services in a transparent fashion from the end users' point of view. This is done by the creation and maintenance of a service-specific overlay network that fulfils the service provider requirements to manage the service delivery to end-users while adapting to user and network context. As such, a means to

Figure 2. SATO Network Architecture with AN Interfaces

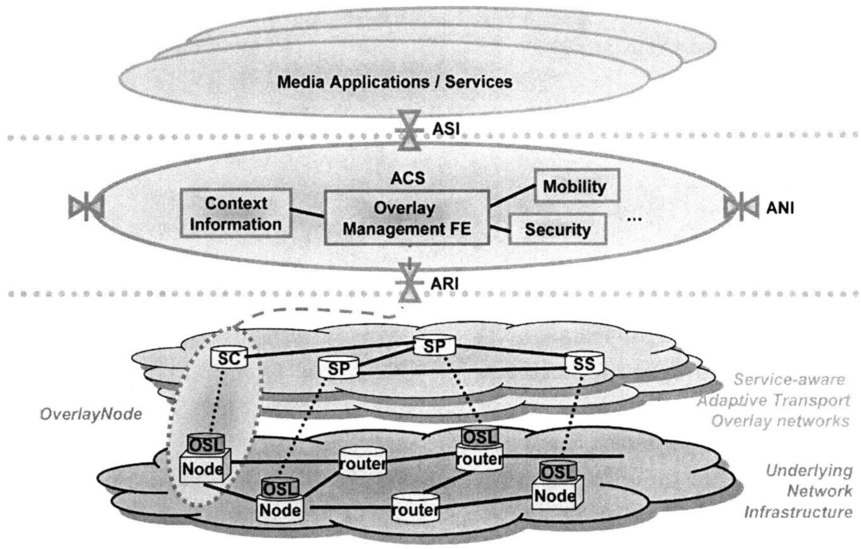

manage context is required and communication between all entities needed.

In a SATO, one or multiple end devices will fulfil the role of clients and one or multiple of a server (see Figure 2). Whereas in some service scenarios the roles are not that clearly distinguishable, i.e., in pure peer-to-peer services where any party is client and server at the same time. The clients are called SATOClient (SC) and the servers are called SATOServer (SS). Intermediate nodes will host the so-called SATOPorts (SP), which are the components that process the data (e.g., running a video transcoder…). All these elements form a SATO on top of the underlying network infrastructure.

The SATOPort is the fundamental building block of a SATO, providing atomic functions that are aggregated to form a complete end-to-end transport overlay. For providing the global end-to-end service, a QoS-aware overlay routing protocol is defined in order to chain the SPs in the right order in order to provide the global service, ensuring the QoS required by the Service Provider, when requesting the setup of a SATO.

Such intermediate nodes are able to host not only a single SATOPort at a given time, but also multiple SATOPorts; either SATOPorts of a different type or even from the same type. Therefore, isolation between processing modules is ensured, via the use of secured isolated execution environments. A deployment module is associated with the execution environment, since, sometimes, when establishing the overlay, a SP offering the required service may not be found.

The SATO itself is self-adaptive. After changes in network context have happened, the SATO manager is capable of carrying out dynamic reconfiguration, which enables the SATO to adapt to the new network context. The change of context can be detected by the own components of the SATO, like the SATO measurements or triggered by external entities of the ACS, like the context or the mobility Functional Entities.

SATO SYSTEM DEVELOPMENT

Traditional IP networks use IP addresses to identify a node within a network. This IP address is used for routing purposes and depends on the localization of the node. Because of the node mobility and subsequent IP address changes, which could be frequent in Ambient Networks, a new naming has been specified within the project (Johnsson, et al. n.d.). For strong relationship between identities and security purposes, it is strongly recommended that those identifiers carry some cryptographic properties. This is mainly why we are considering identifiers as the hash of a public key in a similar way as considered in HIP (Host Identity Protocol) (Moskowitz, & Nikander, 2006). HIP is using cryptographic Identifiers between the application and the IP layer. Applications are talking to cryptographic Identifiers instead of IP addresses, and a HIP module is providing a binding between Cryptographic identifiers and IP Addresses. In the deployment phase, this is this identifier, based on HIP that is used instead of the IP address in order not to be concerned about retrieving the current IP address, whatever the attached network is.

Of course, the deployed module should be verified and should be trusted before its potential deployment in the overlay network. For this, it is assumed that only some parties (such as the network operator or trusted third-parties like software components providers) can deploy their modules in this architecture. For this, we assume that the deployment process starts by an authentication mechanism to authenticate both the software provider and the node. Indeed, the software provider should be authenticated to be sure it has the permission to deploy its own code in such nodes. But the node should also be authenticated to be sure that the code is not deployed in another node, which could be a malicious node, aiming at avoiding the good delivery of service or aiming at catching the software in order to analyze it. By using cryptographic Identifiers, and HIP procedure of attachment, we are providing proof-of-ownership of exchanged data, which means that one node can check that data has been sent by the nodes identified by a specific hash of its public key or Host Identity Tag (HIT).

Those HIT and the binding between them and the IP address should be stored in a trusted database; HIP suggests the use of a DNS server, together with a Rendezvous Server and possibly a DNSSEC server instead of a simple DNS server, for securing the binding. In real deployment, this DNSSEC server could be hosted by a network operator.

When designing this solution, a challenge/response mechanism for exchanging a "secret" during the authentication phase, such as a Diffie-Helmann algorithm (Diffie, & Helman, 1976) has been considered. This secret is used for messages between the service provider and the node deployment module.

Furthermore, the deployed software should not be corrupted or modified during the deployment phase. The code is then transmitted over a secure connection such as IPsec. Indeed, The HIT attachment procedure involves IPsec (Moskowitz, & Nikander, 2006) ESP Security Association (SA) negotiation. This can at least provide an integrity check, and eventually confidentiality of the transaction.

Finally, the execution environment (EE) of the node should be secure in a way that the SPs should not interfere with other deployed SPs (if no "normal" interaction is required). The EE should also enable the dynamic update (or deployment) of software without the need of a stop/restart of the EE or the node itself. Since SPs are service components, the EE should map this concept to real software modules.

The OSGI (Open Services Gateway Initiative) framework (2007), specified in the OSGi alliance, has been chosen as the technology to implement this EE. Indeed, this solution is well suited to our requirements and needs for several networks:

- It defines an architecture allowing the deployment of the services on a wide range of networks, from WAN to PANs via LANs, which resembles to what ANs are.

- The OSGI framework hosts components of services (which are called bundles in the OSGi terminology) which are requested by the users, and which are similar to what SPs are. These bundles are "automatically installable", and can be remotely downloaded and removed on request.
- The OSGI framework is usable in equipment with limited memory: it has a small footprint.
- The services can be remotely managed
- The dynamic update of a service should not generate period of unavailability and should be without consequences for the other already deployed services.

Figure 3. Dynamic Deployment in SATO

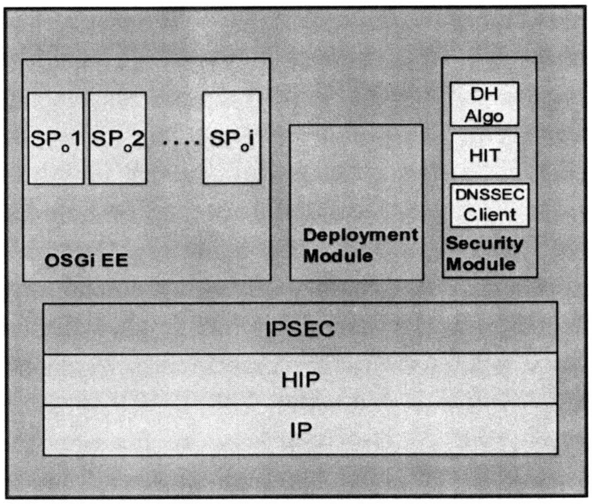

The deployment module and the security issues are then integrated in the OSGi framework to provide this complete secure dynamic deployment function. Figure 3 depicts this architecture.

STANDARDISATION PROCESS

Ambient networks encompasses a lot of research areas, including network composition, context-awareness, multi-radio technologies, resources identifiers, delivery of composed services.... Then, different standardisation bodies are involved in the standardisation process of ambient networks.

The 3GPP (3rd Generation Partnership Project) proposes, for instance, the technical report named "TR 22.980", which includes integration of networks with different administrative domains, and the dynamic and flexible integration of ad-hoc networks, PANs, WLANs etc. A uniform network Composition procedure is particularly explored, independent of what kind of network is "composed" with the 3GPP system.

The IETF (Internet Engineering Task Force) is largely involved in this process with the HIP group which aims at standardising the HIP protocol, and its interaction with others technologies such as the traversal issues related to NAT (Network Address Translation) and Firewall (IETF Internet draft, draft-ietf-hip-nat-traversal) or its extension with DNS (Domain Name Server) (RFC 5205) and the IPv6 prefix identifiers for overlay routable identifiers (IETF RFC 4843). The IETF is also concerned by aspects related to the mobility. For example, the Mipshop Working group aims at optimizing the handoff of devices and proposes several RFCs for use in Ambient Networks.

The IEEE (Institute of Electrical and Electronics Engineers) is also involved, mainly in the access control issues in 802.21 group (the media independent handoff working group). For instance, the multi-radio power management group investigates issues very useful for ambient networks in this sub-group.

CONCLUSION

The AN concept is based on a uniform and flexible control overlay, the Ambient Control Space (ACS), which capitalises on current Internet technology. The ACS will turn today's network of networks into service-networks which benefit from self-organisation, self-optimisation and self-management, and will make network management reactive and capable of swift response to changing contexts with minimal human intervention. The ACS will unify access to network services, regardless of the underlying network type. With this novel approach to control, routing and transport, all networking stakeholders can autonomously add such functionality within their networks.

This concept is applicable for fixed, wireless and mobile networks. Through its new features and cost-efficient solutions, AN technologies would empower businesses and end users.

The Ambient Network principles can be seen as the foundations or the building blocks for a Network of the Future, e.g. the Future Internet, where the network paradigms evolve to a dynamic composition, cooperation between heterogeneous networks to provide abstraction to end-users enjoying services, and to an autonomic management of both the networks and the services. Furthermore, it introduces what the network stack could be, giving up the OSI 7 layers concept to adopt a limited vertical network structure, enabling cross-layer mechanisms, composed of 3 strata, where the AN interfaces could be the reference points between strata.

AN will then significantly enhance the competitiveness of the telecommunication industry, through the development of a next-generation infrastructure to support complex services across heterogeneous domains.

REFERENCES

Bassi, A., Denazis, S., Galis, A., Fahy, C., Serrano, M., & Serrat, J. (2007). *Autonomic Internet: A Perspective for Future Internet Services Based on Autonomic Principles*. Paper presented at IEEE 3rd International Week on Management of Networks and Services End-to-End Virtualization of Networks and Services (Manweek 2007) / MACE 2007 2nd IEEE International Workshop on Modelling Autonomic Communications Environments, 29 October – 2 November 2007, San José, USA. Retrieved from http://magellan.tssg.org/2007/mace/mace.php

Diffie, W., & Helman, M. (1976, November). New directions in cryptography. *IEEE Transactions on Information Society, 22*(6), 644–654. doi:10.1109/TIT.1976.1055638

Galis, A., Denazis, S., Brou, C., & Klein, C. (Eds.). (2004). *Programmable Networks for IP Service Deployment*. Artech House Books.

3GPP. (n.d.). 3rd Generation Partnership Project. Retrieved from http://www.3gpp.org/

Johnsson, M., et al. (n.d.). *Draft System Description*. IST Project Ambient Networks D7. Retrieved from http://www.ambient-networks.org/Files/deliverables/D7-A.2_PU.pdf

Johnsson, M., Rembarz, R., Schieder, A., Surtees, A., Ahmed, K., & Schoo, P. (2008). *Ambient Networks – An Extensible Architecture for Self-Managing and Composable Systems in Heterogeneous and Dynamic Environments*. IEEE Communication Magazine.

Kappler, C., Poyhonen, P., Johnsson, M., & Schmid,S. (2007). Dynamic network composition for beyond 3G networks: a 3GPP viewpoint. *Network IEEE, 21*(1).

Mathieu, B., Galis, A., Jean, K., Ocampo, R., Lai, Z., Stiemerling, M., et al. (2007). *Dynamic Adaptable Overlay Networks for Personalised Service Delivery*. Paper presented at IEEE 7th International Symposium on Communications and Information Technologies - http://www.elec.uow.edu.au/ISCIT2007/ M2NM 2007- First Ambient Networks Workshop on Mobility, Multiaccess, and Network Management; 16-19 Oct 2007, Sydney, Australia; and http://nicta.com.au/research/projects/ambient_networks/m2nm-2007

Mathieu, B., Song, M., Galis, A., Cheng, L., Jean, K., Ocampo, R., et al. (2007). *Autonomic Management of Context-Aware Ambient Overlay Networks*. Paper presented at IEEE International Conference on Communications and Networking in China (IEEE CHINACOM 2007), 15 - 17 August 2007, Shanghai. Retrieved from www.chinacom.org/2007

Mathieu, B., Song, M., Galis, A., Cheng, L., Jean, K., Ocampo, R., et al. (2007). *Self-Management of Context-Aware Overlay Ambient Networks*. Paper presented at 10th IFIP/IEEE International Symposium on Integrated Management (IM 2007); 21-25 May 2007, Munich, Germany. Retrieved from http://www.im2007.org

Moskowitz, R., & Nikander, P. (2006, May). *Host Identity Protocol (HIP) Architecture, RFC 4423.*

Niebert, N., & Flinck, H. Hancock, R. Karl, H. & Prehofer, C. (2004). *Ambient Networks – Research for Communication Networks Beyond 3G.* Paper presented at 13th IST Mobile and Wireless Communications-Summit 2004, 27-30 June 2004, Lyon. Retrieved from www.mobilesummit2004.org

OGSI. (2007). *About the OSGi Service Platform* (Technical White paper). Retrieved from www.osgi.org

Perera, E., Boreli, R., Herborn, S., Georgiades, M., Eisl, J., & Hepworth, E. (2008). A mobility toolbox architecture for all-IP networks: an ambient networks approach. *Wireless Communications IEEE, 15*(2).

WWI-AN. (n.d.). Ambient Networks Project / Reports. Retrieved from www.ambient-networks.org

KEY TERMS AND DEFINITIONS

AN: Ambient Network (AN) offer a solution to the key networking problem of fostering dynamic co-operation between the next *generation, heterogeneous* wireless and wired networks, in order to gather resources within and across networks to provide new services to end users.

ACS: The Ambient Control Space (ACS) is introduced to encompass all control functions in a certain Ambient Network domain, which can be used as a control plane overlay to integrate and interoperate seamlessly any existing networks. It hosts several functional entities which enable the support for mobility, self-management, Network Composition, QoS control or multi-access, and also more abstract functions like the provisioning of context information or support for media delivery overlays.

ANI: The Ambient Network Interface (ANI) connects different ANs (inter-ACS communication) as well as it might be used to facilitate interaction between functions residing in the same ACS (intra-ACS interaction). When inter-working between ANs that belong to different authorities is to be ensured, conformance testing on well-defined Interfaces is eventually required.

ASI: The Ambient Service Interface (ASI) offers the connectivity and control functions for use by upper layer applications within a node operating in an Ambient Network. It allows applications and services to issue requests to the ACS concerning the establishment, maintenance and termination of end-to-end connectivity between functional instances connecting to the ASI.

ARI: The Ambient Resource Interface (ARI) is the reference point through which functions of the ACS control and use services of the underlying connectivity. The ARI is located inside an Ambient Network node and also provides an abstraction of the resources in the underlying connectivity infrastructure.

HIP: Because of the node mobility and subsequent IP address changes, which could be frequent in Ambient Networks, the Host Identity Protocol (HIP) is used to enable applications to talk to cryptographic Identifiers instead of IP addresses. A HIP module is providing a binding between Cryptographic identifiers and IP Addresses.

SATO: The Service-aware Adaptive Transport Overlays (SATO) is the overall architecture enabling the user-centred delivery of service, any time, everywhere, whatever the device and the network are. It enables the provisioning of value-added services, the so-called SatoPorts (SP), such as customization or adaptation of content to end-users' context but also new network value-added services, where the value is not only added in the endpoints, but also in the network.

SP: The SATOPort (SP) is the fundamental building block of a SATO, providing atomic functions (e.g., content adaptation, transcoder, ads insertion…) that are aggregated to form a complete end-to-end transport overlay.

Chapter 3
Overview and Challenges of Multi-Interface and Multi-Channel Multi-Hop Wireless Networks

Carine Toham
Orange Labs, France

François Jan
Orange Labs, France

ABSTRACT

The throughput improvement problem in wireless networks is considered. The main causes of that issue is the shared wireless medium and node spatial problems which either prevent some nodes from transmitting or jam their transmissions. We consider the use of both nodes equipped with multiple network cards and multiple channels as means to improve the throughput in wireless networks. That approach implies to handle some challenges namely channel distribution between the node interfaces and routing in such channel-diverse networks. This chapter analyzes some solutions proposed to deal with these challenges so that the throughput is improved. Some possible orientations are also discussed based on the advantages and drawbacks observed in existing works.

INTRODUCTION

As far as network transmissions are concerned, people used to communicate through wired networks. Satisfying requirements such as sporadic cheap on-demand connections anywhere were hardly achieved with these networks. So, the wireless networks came as means to solve that issue with a wired equivalent performance. Users can build wireless local area networks without a great amount of deployment charges, just for a meeting or a congress.

However, considering the throughput performance, wireless networks are not as efficient as the wired ones. They generate some issues related to their wireless nature. First of all, the data are transmitted on a wireless medium called a channel. So, the transmissions are sensitive to any natural disturbance such as human movements or other

DOI: 10.4018/978-1-60566-250-3.ch003

device processing. These uncontrolled interferences limit the throughput. In addition, the channel is shared by all nodes in the network. If two nodes use it at the same time, it ends with packet collision. The latter also favors the throughput decrease. This issue gets worse when the node number increases. Another reason for the limited throughput in wireless networks is the problem of hidden terminals (Tobagi, 1975). Assume that there are two nodes A and C which are outside the carrier sensing range of each other but have a common neighbor, say B. B can hear signals from A and C but A and C cannot hear each other. Packets from A and C collide at B's if they are transmitted simultaneously. Node C is a hidden terminal for node A and vice versa. The transmissions of both the two nodes are thus delayed. The throughput decrease is a consequence of that situation. The same issue occurs in another use case wherein there are exposed stations (Iyer, 2006). Let us consider again nodes A, B, C and another node D. Assume that they build a chain topology so that A and C are B neighbors but are not in the carrier sensing range of each other. B and D are C's neighbors but cannot directly communicate with each other. If node A is transmitting data to node B, the exposed station C is prevented from using the channel until this transmission is complete because it hears signals from B. However, if nodes B and C are enough far away from each other, they can use the channel simultaneously with few packet losses. Multi-hop communications also reduce the throughput in wireless networks. The data forwarding is slow down because the packets are relayed by intermediate nodes over multiple hops; and these nodes have to wait until the packet reception is complete before starting to forward them towards the destination.

So, the throughput improvement in multi-hop wireless networks is a great challenge. Using multiple network cards per node and dedicating a channel to each of them is a method to handle that challenge. But that approach mandates to manage some aspects like channel distribution between node network cards and data forwarding in such channel-diverse networks.

BACKGROUND

Lets us firstly define some key terms which are used later.

A collision domain is a set of nodes that are in the interference range of each other. These neighbor nodes cannot use the channel simultaneously. So they compete for accessing it.

The channel assignment is the process of distributing channels between node network cards. It is run in a network in which multiple channels are available. Later, the term *interface* refers to a network card.

The deafness problem occurs when a node neighbor cannot join it on their expected communication channel.

Routing is the process of forwarding data packets from a source node to a destination one. The entity that handles this process is called a routing protocol. To convey the data, these routing protocols have to select a sequence of intermediate nodes (a path) based on different criteria. These paths are called routes and the criteria are the metrics. A metric example is the shortest-path which means that the used criterion is the minimum hop number.

Existing solutions about improving the throughput of multi-hop wireless networks can be classified into three categories. The first one consists of those which assume that all nodes are equipped with one interface and share a unique channel. The main followed approach is to bound the interferences in order to reduce the collisions on the channel. A way to achieve this goal is to implement a power control mechanism (ElBatt, 2000, Muqattash, 2005). This latter allows nodes to adjust their transmission power so as to only reach a specific set of nodes in their vicinity and not interfere with the others. This set of nodes is called the node connectivity range. In another implementation, nodes negotiate the transmission power of an intended transmission through RTS and CTS packets so as not to disturb ongoing neighbor communications.

The second class is composed of solutions designed for nodes equipped with one interface that hops from one channel to another. So (2004) suggests using a synchronization window at the beginning of each slot of the IEEE 802.11 Distributed Coordination Function in the MAC layer. This window allows nodes to listen to a default channel so as to negotiate and reserve a data channel for establishing a communication with another node. Bahl (2004) replaces the synchronization window by a rendezvous channel for easier node synchronization. Thus, there are multiple available channels that split the network into smaller collision domains thereby reducing the interferences in node vicinity. In the same spirit, Jain (2001) negotiates the data channels through control packets such as RTS and CTS on a dedicated control channel.

These first two presented categories have some significant drawbacks. Concerning the solutions based on power control, they sometimes need to have a global topological view in order to adjust the transmission power in an optimal way. Then to achieve this goal, a lot of control messages are exchanged. That generated overhead harms the throughput. Moreover, since each node is equipped with one interface, the maximum achievable throughput of multi-hop communications is limited as mentioned before.

The third category of existing works deal with the limits of the first two ones. The solutions belonging to this class are designed for nodes equipped with multiple interfaces in a network wherein multiple channels are available (Kyasanur, 2005, Kumar, 2005, Alicherry, 2005). These nodes are called Multi-interface and Multi-Channel (MIMC) nodes. The interferences are limited because the network is naturally split into multiple smaller collision domains by the multiple channels. Besides, a node can use different channels with its multiple interfaces for simultaneously emitting and receiving packets. So, multi-hop parallel transmissions can be established with an efficient throughput. In the next sections, we focus on this third class.

MULTI-INTERFACE MULTI-CHANNEL WIRELESS NETWORKS

We previously mentioned that equipping nodes with multiple interfaces and multiple channels is a worth studying approach for throughput improvement purpose. However, the Multi-Interface and Multi-Channel (MIMC) networks have to be managed well so that more issues are not generated. Many challenges should be handled in order to benefit from their potential.

First of all, the channel assignment must fulfil many requirements. It should firstly mitigate the deafness problem. Solutions that suggest dedicating a channel to control packets and using it for negotiating the data channels (Wu, 2000) satisfy that condition. But that control channel may become a bottleneck if many transmission nodes need to use it for negotiating their data channel. Another approach is to permanently tune the node interfaces to a set of channels (Kyasanur, 2005, Kumar, 2005). These latter interfaces can be combined with some other ones, aptly called switchable interfaces, which switch between different channels (Wu, 2002). But if a node is only equipped with a unique switchable interface, it can be seen as unreachable by its neighbour because its interface is tuned to a channel different from the neighbour's. Secondly, the channel assignment should minimize the broadcast cost. The aim of a broadcast application is to send the packets to all the source neighbours whatever the channel they use. In a MIMC network it is expensive to achieve this goal if a copy of the packet is thus sent on each channel. Kyasanur (2005) suggests dedicating a third interface, tuned to a fixed channel, to broadcasting. Although that approach reduces the broadcast cost, there is a risk to have the interface or the channel underused. Finally, taking the network state into account during the channel assignment process is efficient. Since there are multiple available channels, a node can select a channel based on the load (Adya, 2004) or on the Signal to Interference plus Noise Ratio (SINR) of the different channels (Jain, 2001).

A channel assignment that uses fixed interfaces and dedicates a channel to broadcasting if such applications are likely to frequently occur, is the best trade-off. Switchable interfaces are generally used to avoid having one of the available channels being underused. But in MIMC networks, it is useless because the node number is higher than the channel number. So, the different channels are distributed between nodes for establishing simultaneous transmissions. To take into account the network state, the best trade-off can be to select the channels based on a combination of different characteristics such as load, SINR, Packet Error Rate in order to integrate the environment impact when choosing a channel. This can be done by building up a condition that sums up different weighted values which quantify the different retained characteristics. This approach is a flexible one because it allows the channel assignment to adapt itself to the considered use case by giving more or less importance to certain aspects.

Another challenge that should be handled as far as MIMC wireless networks are concerned, is the design of a suitable routing technique. In existing routing protocols the trend is to use the shortest path metric. This metric is not as efficient in MIMC wireless networks as in classical ones because it does not take the channel diversity into account. A R1 route that uses 5 hops and 3 channels could provide a better throughput than a R2 route that uses 3 hops and one channel; but the shortest path will always select the R2 route. Draves (2004) proposes the WCETT (Weighted Cumulative Expected Transmission Time) metric. This solution assumes that the channel assignment uses fixed interfaces. The WCETT metric takes into account the transmission delay over the route links and the channel diversity on the considered route. Kyasanur (2005) extends the WCETT metric to the case wherein there are both fixed and switchable interfaces. An additional aspect is then integrated into that metric: the switching cost. This latter represents the latency due to the fact that a packet has to wait until the switchable interface is tuned to the suitable channel before being transmitted. The main limit of these metrics is the fact that they ensure the use of multiple channels in a route, but not their alternation on that route.

The channel alternation and the channel diversity are complementary; the first allows nodes to simultaneously transmit and receive data on the different channels brought by the second. So, an efficient approach is to ensure both the channel diversity and the channel alternation in the selected paths. Some works merge channel assignment and routing. Alicherry (2005) makes a mathematical formulation of the joint channel assignment and routing problem for the throughput optimization in wireless mesh networks. The routing is presented as a flow graph problem for a given number of channels and nodes that have different traffic demands to satisfy. Raniwala (2004, 2005) assumes that the load traffic profile is known. So, the channels are assigned to node interfaces and the routes are selected with the aim of satisfying these traffic requirements. They build up a circular dependence between the channel assignment and the routing algorithm. The route selection takes into account the link capacity which depends on the channel assignment and this latter cares about the considered route. The traffic load is computed either globally or locally. That approach is interesting because it firstly takes the network requirements into account. Then it operates with any routing protocol. But one can wonder about the convergence of the algorithm and the reliability of the prediction/computation of the traffic profile. Another different approach might be to adapt existing routing protocols so that they take into account the channel diversity, instead of designing new ones.

The last but not the least challenge in MIMC wireless networks is to pay attention to the management of the multiple interfaces and channels. The application layer generally uses network card IP addresses of the source and destination nodes to identify them. A node equipped with multiple interfaces has multiple IP addresses a priori since each interface has its own IP address. Thus when the packets are forwarded through a route wherein multiple channels and multiple interfaces are involved, multiple IP addresses can refer to a unique node thereby confusing the application layer. A management process has to be designed in order to make that channel and interface switching transparent to the application layer. Thus existing applications can be run in MIMC networks. Chereddi (2005) designed an architecture that introduced an interface and channel abstraction layer between the network layer and the MAC layer. That module exposes a single virtual interface to the higher layers. Adya (2004) proposes a solution in the same spirit. This idea of virtualizing the interfaces is a worth studying approach to cope with this considered challenge. Another interesting approach is to identify the nodes with a global ID used by the application layer. That node ID will federate all the IP addresses corresponding to the multiple node interfaces.

FUTURE TRENDS

In order to fully exploit the potential of MIMC networks, different approaches are considered.

A first trend is to design a channel assignment for a specific routing protocol (Qu, 2006). The Optimized Link State Routing protocol is extended for handling channel assignment and channel switching. It is achieved through hello messages which are used for piggybacking information about the channels. The main drawback of this work orientation is that it is not a modular one since the solution is designed for a specific configuration even if it performs well in the considered context.

A more progressive orientation is to design an independent channel assignment (Subramanian, 2007). In this latter, the channel assignment aims at minimizing the overall network interferences. Channels are assigned to each communication link in order to get the network interference closer to a specified lower bound. This approach requires to well model the overall network interference or at least to get a reliable approximation.

Another relevant approach is to design multi-path routing protocols for MIMC wireless networks (Tam, 2007). The proposed solution suggests assigning channels to transmission links so that transmission packets are forwarded through multiple parallel paths which use different channels.

A worth studying work could be to try to build up a generic solution that adapts itself to existing routing protocols. It could be in the shape of a module integrated into any existing routing protocol in order to allow this latter to perform well in the context of MIMC wireless networks. This module will firstly provide the routing protocol with cross layer information about the channels. Then, it will manage some tasks such as channel assignment in coordination with the routing protocol so as to enhance the throughput.

CONCLUSION

This chapter presents an overview of multi-interface multi-channel wireless networks and some related challenges to deal with so as to fully take advantage of their potential.

Concerning the throughput provided for multi-hop communications, multi-interface multi-channel wireless networks came as the best solution to get close to a wired equivalent performance. With multiple available channels, the network is split into smaller collision domains thereby reducing the probability of interference occurrence. A node equipped with multiple interfaces can transmit and receive data packets at the same time which avoids the throughput decrease when data have to be forwarded over multiple hops. So, the number of simultaneous transmissions is higher in this kind of networks than in classical ones, which favors the throughput improvement.

However, there are some aspects that should be managed well during the design process in order to get the previously mentioned improvement. A main but not exhaustive list of these aspects contains elements such as channel assignment, routing technique and the compliance of a multi-interface-enabled MAC layer with the existing applications.

REFERENCES

Adya, A., Bahl, P., Padhye, P., Wolman, A., & Zhou, L. (2004). A multi-radio unification protocol for IEEE 802.11 wireless networks. *IEEE International Conference on Broadband Communications, Networks and Systems, Broadnets '2004.*

Alicherry, M., Bhatia, R., & Li, L. (2005). Joint channel assignment and routing for throughput optimization in multi-radio wireless mesh networks. *ACM International Conference on Mobile Computing and Networking, Mobicom '2005.*

Bahl, P., Chandra, R., & Dunagan, J. (2004). SSCH: Slotted Seeded Channel Hopping for capacity improvement in IEEE 802.11 ad-hoc wireless networks. *ACM International Conference on Mobile Computing and Networking, Mobicom '2004* (pp. 216–230). Philadelphia, PA, USA.

Chereddi, C., Kyasanur, P., So, J., & Vaidya, N. H. (2005). Demo: Multi-Channel Multi-Interface Wireless Mesh Testbed. *ACM International Conference on Mobile Computing and Networking, Mobicom '2005*, Cologne, Germany.

Draves, R., Padhye, J., & Zill, B. (2004). Routing in Multi-Radio, Multi-Hop Wireless Mesh Networks. *ACM International Conference on Mobile Computing and Networking, Mobicom '2004.*

ElBatt, T. A., Krishnamurthy, S. V., Connors, D., & Dao, S. (2000). Power management for throughput enhancement in wireless ad-hoc networks. *IEEE International Conference on Communications* (pp. 1506–1513).

Iyer, A., & Rosenberg, C. (2006). Understanding the Key Performance Issues with MAC Protocols for Multi-hop Wireless Networks. *Wireless Communications and Mobile Computing, 6*(6).

Jain, N., Das, S., & Nasipuri, A. (2001). A Multichannel CSMA MAC Protocol with Receiver-Based Channel Selection for Multihop Wireless Networks. *IEEE International Conference on Computer Communications and Networks, IC3N '2001.*

Kumar, D. A., Alazemi, H., Vijaykumar, R., & Roy, S. (2005). Optimization Models for Fixed Channel Assignment in Wireless Mesh Networks with Multiple Radios. *IEEE Communications Society Conference on Sensor, Mesh and Ad Hoc Communications and Networks, SECON '2005*, Santa Clara, USA.

Kyasanur, P., & Vaidya, N. H. (2005). Routing and Interface Assignment in Multi-Channel Multi-Interface Wireless Networks. *IEEE Wireless Communications and Networking Conference, WCNC '2005*, New Orleans, LA.

Muqattash, A., & Krunz, M. (2005). POWMAC: A single-channel power-control protocol for throughput enhancement in wireless ad hoc networks. *IEEE Journal on Selected Areas in Communications, 23*(5), 1067. doi:10.1109/JSAC.2005.845422

Qu, Y., Lung, C.-H., & Srinivasan, A. (2006). Multi-channel OLSR with Dedicated Control Interface. *International Symposium on Performance Evaluation of Computer and Telecommunication Systems, SPECTS '2006,* Calgary, Alberta (pp. 155–162).

Raniwala, A., & Chiueh, T.-C. (2005). Architecture and Algorithms for an IEEE 802.11-Based Multi-Channel Wireless Mesh Network. *IEEE Conference on Computer Communications INFOCOM '2005.*

Raniwala, A., Gopalan, K., & Chiueh, T.-C. (2004). Centralized Channel Assignment and Routing Algorithms for Multi-Channel Wireless Mesh Networks. *Mobile Computing and Communications Review, 8*(2), 50–65. doi:10.1145/997122.997130

So, J., & Vaidya, N. H. (2004). Multi-channel MAC for Ad Hoc Networks: Handling Multi-Channel Hidden Terminals using a Single Transceiver. *ACM International Symposium on Mobile Ad Hoc Networking and Computing, Mobihoc '2004.*

Subramanian, A. P., Gupta, H., & Das, S. (2007). Minimum Interference Channel Assignment in Multi-Radio Wireless Mesh Networks. *IEEE Communications Society Conference on Sensor, Mesh and Ad Hoc Communications Networks, SECON '07*, San Diego California, USA.

Tam, W.-H., & Tseng, Y.-C. (2007). Joint multi-channel link layer and multi-path routing design for wireless mesh networks. *IEEE Conference on Computer Communications INFOCOM 2007.*

Tobagi, F., & KleinRock, L. (1975). Packet Switching in Radio Channels: Part II-the Hidden Terminal Problem in Carrier Sense Multiple-Access Modes and the Busy-Tone solution. *IEEE Transactions on Communications, 23*(12). doi:10.1109/TCOM.1975.1092767

Wu, S.-L., Lin, C.-Y., Tseng, Y.-C., & Sheu, J.-P. (2000). A New Multi-Channel MAC Protocol with On-Demand Channel Assignment for Multi-Hop Mobile Ad Hoc Networks. *International Symposium on Parallel Architectures, Algorithms and Networks, ISPAN '00.*

Wu, S.-L., Tseng, Y.-C., Lin, C.-Y., & Sheu, J.-P. (2002). A Multi-Channel MAC Protocol with Power Control for Multi-Hop Mobile Ad Hoc Networks. [SCI]. *The Computer Journal, 45*(1), 101–110. doi:10.1093/comjnl/45.1.101

KEY TERMS AND DEFINITIONS

Collision Domain: a set of nodes that are in the interference range of each other.

Channel: a frequency band used as a wireless medium for exchanging data in wireless networks.

Interface: a wireless network card.

Channel Assignment: the process of setting a channel to a wireless network card. It is run in a network where multiple channels are available.

Deafness Problem: when a node neighbor cannot join it on their expected communication channel because the node does not have its corresponding interface tuned to that channel.

Hop: the link between two nodes that can directly communicate with each other. A route can be composed of multiple hops if some intermediate nodes are used to forward the data from the source to destination.

Metric: A criterion or a set of criteria used for selecting a route among multiple candidate ones.

Routing: the process of forwarding data packets from a source node to a destination node. The entity that handles this process is called a routing protocol.

Route: The path selected by the routing protocol to convey the data packets.

Shortest-Path: represents the fact that the criterion used by the routing protocol for selecting a route is the minimum hop number.

Chapter 4
Cross Layer Design for Multimedia Transmission over Wireless Networks

Christos Bouras
University of Patras and Research Academic Computer Technology Institute, Greece

Apostolos Gkamas
Research Academic Computer Technology Institute and University of Patras, Greece

Georgios Kioumourtzis
University of Patras, Greece

ABSTRACT

Over the last years a number of new protocols have been developed for multimedia applications in the whole OSI layer's scale. In addition wireless communications and networking fast occupy centre stage in research and development activity in the area of communication networks. In order to better support wireless user the cross layer design paradigm has been proposed. This chapter presents the challenges in design and implementation of cross layer adaptation schemes for multimedia transmission over wireless networks. In addition, this chapter presents the most important parameters and constrains that should be taken into consideration when attempting cross layer adaptation in wireless networks that involves different protocols in the overall protocol stack.

INTRODUCTION

This chapter examines the issue of cross layer design for multimedia transmission over wireless networks. In wireless networks multimedia data transmission inherits all the characteristics and constrains related to the propagation to the free space. One first striking difference between wired and wireless networks is the cause of packet losses. Packet losses in wired networks mainly occur due to congestion in the path between the sender and the receiver, while in wireless networks packet losses mainly occur due to corrupted packets as a result of the low Signal to Noise Ratio (SNR), the multi-path signal fading and the interference from neighboring transmissions. A second difference between wired and wireless networks is the "mobility factor". Mobility in wireless networks

DOI: 10.4018/978-1-60566-250-3.ch004

introduces a number of additional barriers in multimedia data transmission.

As wireless communications and networking fast occupy centre stage in research and development activity in the area of communication networks, the suitability of the layered protocol architecture is coming under close scrutiny from the research community. Although layered protocol architectures have served well for wired networks, they seem not to be suitable for wireless networks. To illustrate this point, researchers usually present what they call a "cross layer design" proposal. Thus, there have been a large number of cross layer design proposals in the literature recently (Raisinghani, 2004), with some of them focusing on multimedia transmission (Van der Schaar, 2005, Exposito, 2008, Kofler, 2007). Generally speaking, cross layer design refers to protocol design done by actively exploiting the dependence between protocol layers to obtain performance gains. This is unlike layering, where the protocols at the different layers are designed independently.

This chapter presents the challenges in designing and implementation of cross layer adaptation schemes for multimedia transmission over wireless networks. The remaining of this chapter is structured as follows: The next section presents the international experience in the area of cross layer design for multimedia transmission over wireless networks. The third section presents the challenges in cross layer design for multimedia transmission over wireless networks. The fourth section presents the future trends in the area. Finally, the fifth section concludes this chapter.

BACKGROUND

Over the last years a number of new protocols have been developed for multimedia applications in the whole OSI layer's scale. The RTP and RTCP protocols (Schulzrinne, 2003), which operate on the transport layer usually on top of the UDP protocol, have been especially designed for multimedia data transmission. The RTSP (Schulzrinne, 1998) protocol offers control mechanisms over real time multimedia transmission whereas SIP (Schulzrinne, 2002) and H.323 are used in multimedia conferencing.

Apart from the above developments there have been a number of proposals for improving Quality of Service (QoS) in multimedia applications through cross layer adaptation strategies. In Van der Schaar, 2005 the need of a cross layer optimization is examined and an adaptation framework is proposed amongst the APP, the MAC and the Physical (PHY) layers. In Shakkottai, 2003 the issue of cross layer design in wireless networks is addressed. The focus is on the way that higher layers share knowledge of the PHY and MAC layers conditions in order to provide efficient methods to allocate network resources over the Internet. In Van der Schaar, 2003 a joined APP and MAC adaptation is proposed with the use of MPEG-4 and the latest Fine Granularity Scalability (FGS) extension. In this work, packets containing multimedia data are classified into different classes and in the light of poor network conditions only packets with high value are transmitted. The network conditions are jointly measured by combining the information obtained by the retransmission number of a lost MAC frames (Automatic Repeat-reQuest, ARQ) and the information provided by the RTCP protocol. Signaling issues between the layers for cross layer optimization over wireless networks are examined in Wang, 2003. The authors propose a new signaling framework in which signaling can be done between two non-neighboring layers, through light-weighted messages and the use of a message control mechanism to avoid message dissemination overflow. Although this proposal avoids heavy the Internet Control Message Protocol (ICMP) messages for out-bound signaling between the layers that is proposed in Sudame, 2001, it introduces very high complexity. In

Chen, 2004 a joined adaptation scheme of the APP, MAC and PHY layers is presented. Packet transmission is made with a novel scheduling algorithm at the MAC layer whose function is based on the user and application priority levels. Priorities are assigned to users on the basis of paid services, in which users are classified into groups with different QoS levels. Finally, Carneiro, 2004 outlines the need for new cross layer architecture to address known problems of mobility, packet losses and delay that are observed in wireless networks. The main idea of a cross layer manager is discussed in which all layers send notification messages to the manager who is responsible for intra layer co-ordination.

The cross layered architecture available to the literature can be divided to the following categories (Srivastava, 2005):

- Creation of new interfaces (Dimic, 2004): Several cross layer designs require creation of new interfaces between the layers. The new interfaces are used for information sharing between the layers at runtime.
- Merging of adjacent layers (Choudhury, 2006): Another way to perform cross layer adaptation is to design two or more adjacent layers together such that the service provided by the new layer is the union of the services provided by the constituent layers.
- Design coupling without new interfaces (Tong, 2004): Another category of cross layer design involves coupling two or more layers at design time without creating any extra interfaces for information sharing at runtime.
- Vertical calibration across layers (Liu, 2004): Adjusting parameters that span across layers.

CHALLENGES IN CROSS LAYER DESIGN FOR MULTIMEDIA TRANSMISSION OVER WIRELESS NETWORKS

Cross layer adaptation is a very challenging process due to the numerous parameters involved in the whole procedure. This section outlines the most important parameters and constrains what should be taken into consideration when attempting cross layer adaptation in wireless networks that involves different protocols in the overall protocol stack.

Network Elements Involved in the Adaptation Process

In multimedia transmission three entities can be distinguished that take part in the information exchange procedure: the sender, the core network elements (links, routers) and the receiver. The term sender includes either a multimedia server or an individual host which participates in a multimedia data exchange with another remote host. There has been a detailed discussion whether or not all three elements should be involved in an adaptation scheme, targeting at improving the QoS offered to the end user. The most challenging and maybe the most beneficial approach would be the participation of all three elements in the adaptation mechanism, especially when the multimedia data are transmitted among various network domains. However, even in the same network domain the designer of the adaptation strategy has to decide whether or not both the sender and the receiver should participate in the adaptation process. The designer should also consider that the complexity increases when inter-domain adaptation and policies are to be implemented. Therefore, the cross layer adaptation scheme that is related to the participation of the entities involved in the multimedia transmission can fall into the next four categories:

- Sender based: The sender performs the cross layer adaptation. This approach has the advantage of easy deployment due to the fact that it does not require any support from the network or the receivers. On the other hand this approach has limited capabilities.
- Receiver based: The receiver performs the cross layer adaptation. This approach also has the advantage of easy deployment due to the fact that it does not require any support from the network or the sender. Again this approach has limited capabilities.
- Network supported: The network elements are involved in the cross layer adaptation. In a heterogeneous environment such as the Internet, agreements have to be set up amongst the various network domains to ensure any cross layered implementation across the path between the sender and the receiver. In the same domain the administrator can define their own policies-mechanisms.
- Hybrid: A combination of two or more of the above approaches. This approach is the most complicated but has the potential to provide better performance.

Layers Involved in the Cross Layer Adaptation (Inter-Layer Optimization)

Most of the available bibliography focuses on a jointly PHY and MAC layers adaptation. This bibliography (Choudhury, 2006, Verikouris, 2005 and Chen, 2004) has proven that PHY and MAC layers are very important especially in wireless networks and must be taken into account during cross layer adaptation and optimization. Moreover, the APP layer has been used in several cross layer adaptation schemes (Radha, 2001 and Ahmed, 2005). While the above mentioned layers (PHY, MAC and APP) have been extensively researched in cross layer adaptation schemes there has been little work done in the whole protocol stack. The transport/session layer can play important role in cross layer adaptation for wireless networks, as a number of adopting mechanisms in this layer (like TFRC (Handley, 2003) for example) have been extensively evaluated in wired networks, revealing adaptation opportunities in wireless networks. Although, the network layer cannot be used straightforwardly for cross layer adaptation it can be used for indirectly cross layer adaptation through QoS schemes implemented at the network layer (Hwang, 2008).

Parameters Involved in Cross Layer Adaptation (Intra-Layer Optimization)

Each layer offers a number of different parameters through which the desired adaptation can be achieved. The optimization of each layer parameters includes the selection of the applicable parameters which could lead to better results. At this point, we should mention that the adaptation of a parameter in one layer may and most likely, will influence the parameters in other layers. Therefore, the adaptation of the parameters in each layer should be done by taking into account the above mentioned assumption. By summarizing the above, any approach for optimal selection of the adaptation parameters should consider the following two actions:

- Optimization of the parameters that only affect the layer in which they appear
- Optimization of the parameters that affect two or more layers

Table 1 shows the various parameters that can be involved in cross layer adaptation.

Signaling Amongst the Various Layers

Signaling amongst the various layers is another important aspect. A communication mechanism amongst the various layers is required in order to

Table 1. Parameters for cross layer adaptation in wireless networks

Layer	Parameters
PHY	Signal modulation
MAC	ARQ, FEC, QoS (802.11e)
Network	QoS (Diffserv, IntServ), IPv6
Transport / Session	Adaptive Transmission Rates (TFRC, DCCP, other mechanisms (Van der Schaar, 2007))
Application	Encoding parameters (Layered Encoding, MPEG4)

implement a cross layer adaptation scheme. There are various approaches for cross layer signaling:

- Existing protocols / Packet headers: Under this approach, the existing protocol headers can be used for signaling. One appealing approach is the use of IPv6 (network layer) and the extension header mechanism by defining one new type of extension header which will be used for cross layer parameters notification. A second approach is the exploitation of APP packets of the RTCP (transport layer) protocol, which is defined in Schulzrinne, 2003. The APP packets can carry application specific parameters that can be used for signaling in a cross layer adaptation scheme. However, the APP packets are exchanged between the receiving and the sending transmission layers. Therefore, there must be a mechanism in place in order to disseminate these parameters to other layers.
- ICMP messages: ICMP (Internet Control Message Protocol) is a widely deployed signaling protocol in IP-based networks. In this approach, desired parameters, measured by corresponding layers can be disseminated via ICMP messages. A new ICMP message is generated only when a parameter has been changed. However, as the ICMP message is always encapsulated in an IP packet, it has to pass by the Network Layer even when signaling is only desired between the MAC and the APP layer.
- Network Services: In this approach a separate network service is implemented and it is responsible for collecting parameters from the various layers. These parameters that are collected by the network service can be available to APP (or other layer(s)) in order to perform cross layer adaptation.
- Local Profile: In this approach each layer stores important parameters to local profiles (file system) in the mobile device. Each layer accesses these profiles in order to obtain the information needed to perform cross layer adaptation.

By summarizing the above mentioned approaches, the first approach utilizes the already exchangeable packets as in-band signaling over the network. The second approach makes use of extra packets as out-band signaling over the network. The third approach stores parameters to a network server and the fourth approach stores parameters to mobile device.

Adaptation Strategy

Another important issue is how the adaptation strategy could be realized (Xiuyuan, 2008). There are the following approaches in this field:

- Bottom-up approach: In this approach the lower layers (PHY and MAC) provide the upper layers with optimal services by reducing the transmission errors (not efficient for multimedia transmission).

- Top-down approach: In this approach the APP layer informs the lower layers for the importance of each data packet and the lower layers treat each data set with a different way, based on QoS criteria. The higher layer protocols optimize the parameters and the strategies of the next lower layer.
- MAC-centric approach: In this approach the APP layer passes its traffic information and requirements to the MAC layer that decides which APP layer packets/flows should be transmitted and at what QoS level.
- Integrated approach: This approach is the most challenging because the adaptation strategy is decided jointly by all the layers.

The above cross layer approaches exhibit different advantages and drawbacks for wireless multimedia transmission, and the best solution depends on the application requirements, the used protocols, the algorithms at the various layers, the complexity and finally the limitations. To summarize the above mentioned approaches, we can say that the most appealing approach is the integrated one. However, this approach is difficult to implement due to the increased complexity, as a direct result of the big number of possible strategies and the associated parameters involved.

Devices Constrains

The decision on the above mention design issues must be done under the following constrains:

- Device constrains: Mobile devices have many limitations when compared to desktop systems. These include display limitations, CPU resources and power consumption.
- Network constrains: Network constrains include available bandwidth, delay, Round Trip Time (RTT) and QoS support.
- Application constrains: Application constrains include maximum and acceptable delay, maximum and acceptable delay jitter (especially for interactive applications), maximum and acceptable packet loss ratio and finally bandwidth constrains.

In conclusion, the main objective of the optimization process is the optimal selection of the above described parameters in order to provide the best multimedia experience to the end user by taking into account the above described constrains.

FUTURE TRENDS

Nowadays, we are moving from the static connectivity of the wired networks to the "anytime anywhere mobile applications". In addition, we are facing important increase in the usage of wireless access networks either in the form of PAN (Personal Area Networks e.g. Bluetooth), LAN (Local Area Networks e.g. IEEE 802.11) and MAN (Metropolitan Area Networks e.g. IEEE 802.16) or in the form of current 3G and future 4G mobile networks and important increase of mobile multimedia applications like voice over IP, Video on Demand, videoconference, Media streaming, etc. The cross layer design will facilitate the above important changes by providing a unified scheme which will allow the incessant usage of networked media by adapting the media transmission to the specifically needs of the wireless networks and the mobile terminal (e.g. laptop, PDA, mobile phone). In addition, cross layer adaptation will allow smooth operation of mobile multimedia applications during the transition form one wireless network technology to other.

CONCLUSION

This chapter presents the challenges in designing and implementing cross layer adaptation schemes for multimedia transmission over wireless networks. Cross layer adaptation for multimedia transmission will have important impact both in the research community and in the industry. More particularly, cross layer adaptation will allow better access to media content for users in a variety of locations, contexts and mobility scenarios.

REFERENCES

Ahmed, T. (2005). Adaptive *p*acket video streaming over IP networks: A cross-layer approach. *IEEE Journal on Selected Areas in Communications*, *23*(2). doi:10.1109/JSAC.2004.839425

Carneiro, G., Ruela, J., & Ricardo, M. (2004). Cross-layer design in 4G wireless terminals. *IEEE Wireless Communications*, *11*(2), 7–13. doi:10.1109/MWC.2004.1295732

Chen, J., & Zheng, T. (2004). Joint cross-layer design for wireless QoS content delivery. *IEEE International Conference on Communications* (Vol. 7, pp. 4243- 4247).

Choudhury, S., & Gibson, J. (2006). Joint PHY/MAC Based Link Adaptation for Wireless LANs with Multipath Fading. *IEEE Wireless Communications and Networking Conference, 2006* (Vol. 2, pp. 757-762).

Dimic, G., Sidiropoulos, N. D., & Zhang, R. (2004). Medium Access Control — Physical Cross-Layer Design. *IEEE Sig. Proc.*, *21*(5), 40–50. doi:10.1109/MSP.2004.1328087

Exposito, E., Van Wambeke, N., Chassot, C., & Drira, K. (2008). Introducing a cross-layer interpreter for multimedia streams, Computer Networks. *The International Journal of Computer and Telecommunications Networking*, *52*(6), 1125–1141.

Handley, M. et al, (2003). *TCP Friendly Rate Control (TFRC).* IETF RFC 3448.

Hwang, J., Huang, C., & Chang, C. (2008). Cross-Layer End-to-End QoS for Scalable Video over Mobile WiMAX. In Chen, K., Roberto, J., & Marca, B. (Eds.), *Mobile WiMAX*. John Wiley & Sons, Ltd.doi:10.1002/9780470723937.ch13

Kofler, I., Timmerer, C., Hellwagner, H., & Ahmed, T. (2007). Towards MPEG-21-Based Cross-Layer Multimedia Content Adaptation. In *Proceedings of the Second International Workshop on Semantic Media Adaptation and Personalization* (pp. 3-8).

Liu, Q., Zhou, S., & Giannakis, G. B. (2004). Cross-Layer Combining of Adaptive Modulation and Coding with Truncated ARQ Over Wireless Links. *IEEE Transactions on Wireless Communications*, *3*(5), 1746–1755. doi:10.1109/TWC.2004.833474

Radha, H., Van der Schaar, M., & Chen, Y. (2001). The MPEG-4 Fine-Grained Scalable video coding method for multimedia streaming over IP. *IEEE Transactions on Multimedia*, *3*(1). doi:10.1109/6046.909594

Raisinghani, V. T., & Iyer, S. (2004). Cross-Layer Design Optimizations in Wireless Protocol Stacks. *IEEE Comp. Commun.*, *27*, 720–724. doi:10.1016/j.comcom.2003.10.011

Schulzrinne, H. et al (2002). *SIP: Session Initiation Protocol*. IETF RFC 3261.

Schulzrinne, H. et al (2003). *RTP: A Transport Protocol for Real-Time Applications*. IETF RFC 3550.

Schulzrinne, H., Rao, A., & Lanphier, R. (1998). *Real Time Streaming Protocol (RTSP).* IETF RFC 2326.

Shakkottai, S. Rappaport & T.S. Karlsson, P.C. (2003). Cross-layer design for wireless Networks. *IEEE Communications Magazine*, *41*(10). doi:10.1109/MCOM.2003.1235598

Srivastava, V., & Motani, M. (2005). Cross-layer design: a survey and the road ahead. *IEEE Communications Magazine*, *43*(12), 112–119. doi:10.1109/MCOM.2005.1561928

Sudame, P., & Badrinath, B. (2001). On providing support for protocol adaptation in mobile wireless networks. *Mobile Networks and Applications*, *6*(1), 43–55. doi:10.1023/A:1009861720398

Tong, L., Naware, V., & Venkitasubramaniam, P. (2004). Signal Processing in Random Access. *IEEE Sig. Proc.*, *21*(5), 29–39. doi:10.1109/MSP.2004.1328086

Van der Schaar, M., Krishnamachari, S., & Choi, S. (2003). Adaptive Cross-Layer Protection Strategies for Robust Scalable Video Transmission Over 802.11 WLANs. *IEEE Journal on Selected Areas in Communications*, *21*(10). doi:10.1109/JSAC.2003.815231

Van der Schaar, M., & Sai Shankar, D. (2005). Cross-layer wireless multimedia transmission: Challenges, principles and new paradigms. *IEEE Wireless Communications*, *12*(4), 50–58. doi:10.1109/MWC.2005.1497858

Van der Schaar, M., & Turaga, D. S. (2007). Cross-Layer Packetization and Retransmission Strategies for Delay-Sensitive Wireless Multimedia Transmission. *IEEE Transactions on Multimedia*, *9*(1), 185–197. doi:10.1109/TMM.2006.886384

Verikouris, C., Alonso, L., & Giamalis, T. (2005). *Cross-layer optimization for wireless systems: A European research key challenge*. Global Communications Newsletter.

Wang Q. & Abu-Rgheff, M.A., (2003). Cross-Layer Signalling for Next-Generation. *Wireless Systems, IEEE WCNC 2003* (Vol. 2, pp. 1084–89).

Xiuyuan, L., Xiaojuan, W., Wenming, L., & Xiaogang, W. (2008). An Adaptive Cross-Layer Scheduling Algorithm for Multimedia Networks. *International Conference on Intelligent Information Hiding and Multimedia Signal Processing* (pp. 52-55).

KEY TERMS AND DEFINITIONS

Delay Jitter: Delay jitter is defined to be the mean deviation (smoothed absolute value) of the difference in packet spacing at the receiver compared to the sender for a pair of packets.

Forward Error Correction: Forward error correction is a system of error control for data transmission, whereby the sender adds redundant data to its messages, which allows the receiver to detect and correct errors without the need to ask the sender for additional data.

Multimedia Data: Multimedia data refers to data that consist of various media types like text, audio, video and animation.

Packet Loss Rate: Packet loss rate is defined as the fraction of the total transmitted packets that did not arrive at the receiver

Quality of Service (QoS): The ability to provide specific guarantees to traffic flows regarding the network characteristics such as packet loss, delay and jitter experienced by the flows.

RTP/RTCP: Protocol which is used for the transmission of multimedia data. The RTP (Real-time Transport Protocol) performs the actual transmission and the RTCP (Real-time Transport Control Protocol) is the control and monitoring transmission.

RTSP: The Real-Time Streaming Protocol (RTSP) allows control multimedia streams. Control includes absolute positioning within the media stream, recording and possibly device control.

SIP: Session Initiation Protocol (SIP) is an application-layer control (signaling) protocol for creating, modifying, and terminating sessions with one or more participants.

Signal to Noise Ratio: Signal-to-noise ratio is an electrical engineering concept, defined as the ratio of a signal power to the noise power corrupting the signal.

Chapter 5
Charging in Mobile Telecommunication Networks

Bálint Ary
Budapest University of Technology and Economics (BUTE), Hungary

Sándor Imre
Budapest University of Technology and Economics (BUTE), Hungary

ABSTRACT

The tariff packages in the mobile telecommunication industry got more and more complex in the last few years. The telecommunication companies have introduced several different services along with different discounts and allowances. The price calculation of the services and the understanding of prices became harder for the subscribers. This chapter describes the basic architecture and the major flows of a billing system in the telecommunication area. It introduces a novel concept for price calculation, which could aid the Advance of Charge and Income Prediction functionalities.

INTRODUCTION

The technological evolution and the market competition of the telecommunication industry lead to the change of approach in the price calculation of the different services in the last decade. At the very beginning, the price of the mobile telecommunication service depended mainly only on the time of the call, the called party and the duration of the call. The latter was a simple linear component in the price calculation. Nowadays, the available tariff packages are much more complex. The new, advanced technologies brought new dimensions in services and in measurable units; volume or quality for example. Currently, the price of the service depends on a lot of parameters, such as the called party, the calling party, the accumulated service consumption in the current billing period, the delivered services, the type of the customer, the duration of the call and many others. Not even mentioning the different allowances and discounts that can be purchased separately or can be bundled with other services. The tariff packages are so complex, that it is actually hard to compare them to one another.

Since the legacy charging systems of the network operators are mostly unable to handle the large variety of new services and discounts,

DOI: 10.4018/978-1-60566-250-3.ch005

these systems need to be updated, revised or exchanged. From the billing system point of view, there are several possible solutions to keep up with the market driven evolution of complexity (Schwartz, 2003). One is to keep the old, legacy billing system and try to squeeze the complex (ever changing) pricing mechanism into it. Another one is to redesign the whole system, and take the new circumstances and the new marketing approaches into consideration. Of course, there are several companies (like Amdocs, Siemens or Ericsson) that are selling their own solutions / products for this problem (Johnson, 2005). These new, off-the-shelf billing systems are quite flexible, and provide the possibility to the mobile network operators to develop almost any marketing driven price calculation with minimal manual work. Sadly, this flexibility does not necessarily come together with a suitable mathematical model, and does not provide the possibility to do extra calculations with the tariff packages.

The huge number of available functions and the standardized interfaces of UMTS give the possibilities to 3rd party providers to use the infrastructure of the telecommunication system, and offer value-added services to the subscribers (UMTS forum, 2002). Because these providers are financially isolated from the network operator, and because of the continuous growth of the number of pre-paid services and subscribers, an accurate and fast (or real-time) charging mechanism is indispensable.

The following sections describe the corresponding standards and the basic architecture and functionalities of a billing system. The main section also introduces a novel, mathematical model for price calculation.

BACKGROUND

Although UMTS introduced lately and the charging protocol hasn't changed significantly since GPRS (General Packet Radio Service) was introduced somewhere at the end of the 20th century, different phases of the standard are still under development. 3GPP (3rd Generation Partnership Project) specifies guidelines and recommendations for the network elements, and the standards are detailing two different types of charging methods: *offline* and *online charging* (3GPP, 2004).

An actual service request might be pre-paid or post-paid, depending on whether the price of the requested service is paid before, or after the consumed service. Most telecommunication company groups their subscribers into two sets. Pre-paid subscribers may only require services in pre-paid mode, so they have to top up some amount of money to their balance, and may only require services until their balance covers the price of those requests. Post-paid users may gain services anytime (at least until their un-paid balance does not exceeds a given security threshold), and may pay the prices of those services in predefined periods. However, these subscribers types are straightforward and most common, telecommunication companies may introduce novel subscriber types and may set up or dismiss restrictions regarding the type of services and payment mode. The connection between the payment mode (pre-paid and post-paid) and charging method (offline and online) are detailed later.

During offline charging, the network elements (switches) are sending charging information (Charging Data Record - CDR) to the billing system in a predefined format. The CDRs contain all the necessary information, which is needed to compute the actual price of the requested service. This includes the calling party, called party, type of service, amount of the requested service, quality of service and many others. Because these charging records carry information about the services requested, the functionality of the CDRs extends beyond charging, and it's possible to analyze service-utilization, and gain statistical information about the services, content and subscribers with them. By archiving the CDRs, the user-complaints can also be easily settled. Since this information is collected after the event / service, and sent trough a widespread network, real-time charging is not feasible.

The offline charging functionality is in most cases managed by an IT system (a UNIX based software for example), which needs more time to do the necessary computations, but gives huge freedom for the operators to develop and introduce complex allowances and discounts already mentioned. Offline charging is perfectly suitable for post-paid users, since the system does not have to restrict services, and does not have to compute the price of the service in a real-time manner.

In online mode (real-time charging), the communication between the billing system and the network elements is a socket based communication. For every service, the billing system allocates proper amount of consumable service unit (minute, kilobytes, etc.) from the subscribers' account to the serving nodes (MSC, SGSN, GGSN), and these nodes are deflating the delegated units simultaneously as they serve the requested service. If the service terminates before all the units are consumed, the network elements are retransferring the remained units to the billing system. To assure continuous service delivery, if the users do not terminate the service, a new amount of granted unit should be sent to the serving network element before the previous one runs out (3GPP, 2000). Since pre-paid subscribers have paid for the service before the actual consumption, it is necessary to limit their calls to the amount of their balance. Because of this, the real-time approach of the online charging function is more important than it is in offline mode. The necessary computation speed cannot be provided by a regular software (with a regular and sane level of hardware) and thus, online charging is most probably managed by special hardware elements (service nodes), and the business logic need to be hard-coded into these entities. Besides the computation requirements and limitations, online charging requires much more administrative traffic (network overhead) than offline charging, since the delegation messages need to be sent continuously to the network elements. Because of the huge contrast between online and offline charging, it is highly likely, that the two functions are managed by two different systems or products (Schwartz, 2004).

The architecture and main modules of the billing systems differs in offline and online case (as displayed on Figure 1), however in either case a CRM (Customer Relationship Management) module part must be presented and tied up tightly with the system in order to maintain the subscribed offers and apply the correct business logic when calculating the price of the requested service. Also, the price calculating module (called RATER in the corresponding terminology) has to be presented in both systems, however most probably (if the business logic and business flows are not too complicated) the offline system requires only a one-way rating function, while pre-paid system relies on inverse-rating as well, when calculating the consumable service units from the subscribers' accounts (Lucas, 2004). Apart from these modules, the offline billing system must have a billing module, which is responsible to calculate billing related charges and assemble the printable invoice for the subscribers / customers and a post-billing

Figure 1. Charging System Architecture

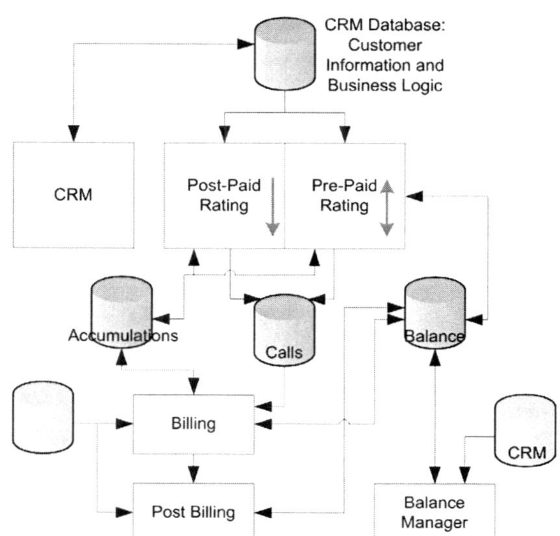

(or collection or accounts & receivable) module, which is responsible to track the payment activities of the users and trigger different collection flows if any fraud is detected. Online billing system must have a balance manager module, which handles pre-paid balance updates when the user uploads balance for his/her account (Schwartz, 2005).

The advent of the 3rd party providers and virtual network operators (VNO) requires new interfaces and new business flows. These providers shall be able to define the price of their services online, have to have some kind of wholesale charging with the network provider and have to send charging related information to the charging system (MobileIn, 2006). The network provider shall be responsible to rate, charge and bill these services for their subscribers.

From the network point of view the regular network elements (switches) are sending charging data records to the offline-, or communicating real-time with the online charging system (Figure 2). The corresponding standards define logical entities in these switches with detailed functional description. Whenever a new functionality or new architecture is introduced (such as IMS – Internet Multimedia Subsystem, MNVOs – Mobile Virtual Network Operators or 3rd party providers) the corresponding architecture elements must mimic the functionality of the legacy network elements from charging point of view and must send CDRs, or communicate online with the billing system.

Figure 2. Charging System and the Network elements

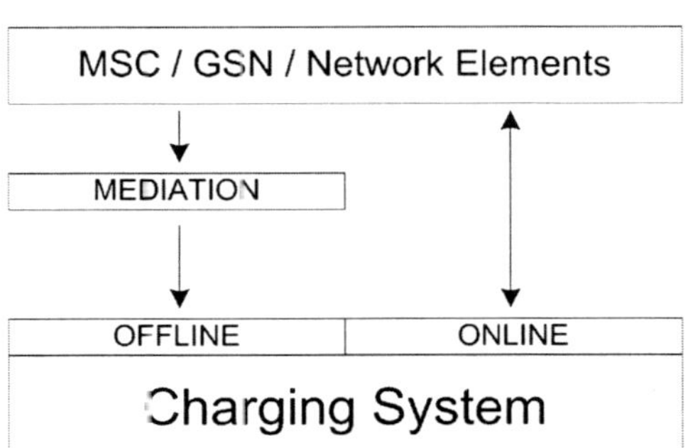

CHARGING IN MOBILE TELECOMMUNICATION NETWORKS

As mentioned in the previous section, the main module of a charging and billing system is the rater. The main task of this module is to calculate the value of a consumed service, or calculate the allowed consumable service from a given amount of money (in case of online charging). Although it seems trivial at the first sight, this is much more sophisticated on a lower level.

First of all, the amount of considerable parameters increased significantly in the last decade. Ten years ago, the price of a service depended only on the length of the service, the called number and the time of the call. Nowadays additional parameters are taken into consideration (such as the number of distinct called numbers, favorite numbers, favorite countries, quality of service, etc) together with different and sophisticated allowances and discounts, which may interfere and depend on each other. Some of these new parameters can be predicted in advance, some of them cannot be measured easily even when the user consumes the service (such as the quality of a GPRS data flow for example). In addition the service consumption alone does not determine the value of the service, but most pricing logic takes the recent behavior of the subscriber into consideration (such as amount of used allowance or total amount of spoken minutes).

Not only the number of considerable parameters has changed, but the number of the values of a requested service has expanded as well. The value of a service is no longer restricted to money, but can be loyalty points, free gifts, free allowances or anything else, which can be derived from the service consumption information.

If we would like to give a mathematical formula for the rating, we have to model two major functions: the calculation of the charges (values) and the update of the accumulated consumption information (which will be used to store information about the subscriber's recent behavior). The simple equations are as follows (Ary, 2007):

$v = r_s(c,a),$

$a' = u_s(c,a),$

Where vector a denotes the accumulated service consumption, c contains the details of the current call and v stands for the values of the call. The firs equation computes the values of the call from the current call detail and the accumulated call information. The rating function (r_s) which calculates these values depends on the rating logic (which is defined by the service providers) and depends on the offers the subscribers have bought. The latter equation computes and updates (u_s) the related accumulations and it also depends on the corresponding rating schemes.

Most charging systems use the general "write-the-code" concept when developing the r_s and u_s functions with the help of the regular flow-control methodologies (if-then-else, for, while and until). This makes rating very flexible and fast, since almost everything can be solved on a very low level, however, on the other side, the code gets very complex, as developments are done continuously on these rating functions (Garifo, 2007). Not to mention, that the price of a development (and testing) is increasing, as the codes get bigger and bigger. Apart from the self-generated complexity, this legacy concept does not allow detailed and sophisticated mathematical calculation with the rating scheme. It does not allow (easily) inverse rating, which is required for online charging, and it does not allow the pre-calculation of charges. The latter one is used to display the expected price of the call before the actual service consumption.

With the advent of 3rd party providers, the existing business models of the network / service providers had to be extended together with the existing billing and charging systems, in order to let these new players to adjust and parameterize their offers through some kind of standard interface. This functionality / possibility induct the need of the AoC (Advice of Charge) service, which allows the customers to query and inform themselves about the price of the requested services before the actual service request.

Most of the charging systems operating nowadays are more extended and upgraded versions of the original and legacy systems, rather than a mature and complete solution of the requirements of the 21st century. Our model tries to fit these new requirements with the following concept:

The rating logic (r_s) can be defined as a graph. Each state represents the service consumption with similar conditions and with the same unit price (Figure 3). The transitions between the states are triggered by the actual call detail record (denoted with c previously) and the user's accumulated usage (denoted with a in this section). The price (values) of the call can be computed with the actual unit price of the state (Ary, 2007).

The benefits of this model come from the deep mathematical background. Expected prices of the services (AoC) can be calculated, easier marketing or user analysis may be performed and the forecast of the telecommunication company's income (NIBT – Net Income Before Tax) is much simpler than with a legacy charging system.

Figure 3. Price Graph Example

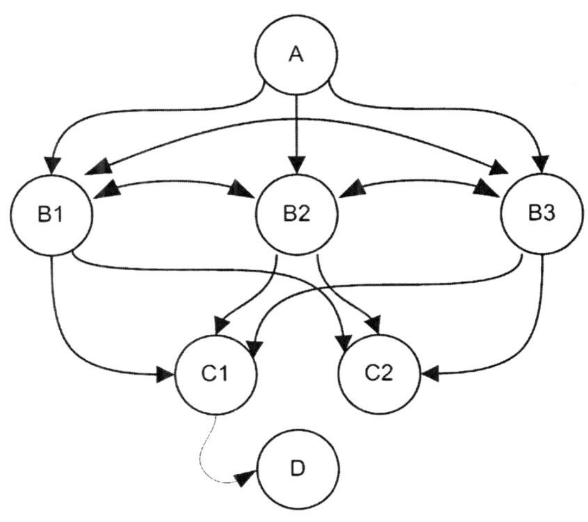

The amount of charging overhead can be reduced if we apply offline charging for prepaid subscribers when their balance is over a predefined threshold, and apply online charging otherwise. Since offline charging operates with less network overhead, additional network resources can be freed with this approach. If the threshold is computed correctly, then the subscribers can not exceed their prepaid balance. This kind of mode switching functionality can be combined with the state graph approach, since we can assign the type of charging (offline or online) to each state. If the state – in which the rating scheme is – an offline state, then offline charging is applied, if it is an online state, then the online charging mechanism is used (Ary, 2005).

FUTURE TRENDS

Billing systems are very expensive and very critical part of the telecommunication companies' IT infrastructure. Developments are continuous and short TTM (time to market) period is a must. As 3rd party providers and virtual network operators are emerging, the need of a flexible billing system is increasing, which offers open interfaces and well defined flows for the network and for the content / service provider as well. Interfaces are required to allow the 3rd party providers to define the price of their service and also for the subscribers to query the price of these services (Garifo, 2001).

Flexibility is also required to follow up the continuously increasing offer portfolio. New services with new measurable dimensions are emerging from time to time. In order to be able to charge these novel services, the billing and charging systems has to be revised. The more flexible the billing systems are the less time is required to develop, test and offer these new services to the public.

As market competition forced the service providers to develop novel and unique pricing schemes, the complexity of the billing systems grew significantly. If this trend continues, the revision and exchange of the legacy charging system is inevitable to a decent system, which should be very flexible and support the currently available (and every possible) charging dimension.

A lot has been talked and discussed about convergent charging systems, where the same system handles online and offline charging. Although the benefits of the converged and unified system is straightforward (Ericsson, 2005), the current results are not very reassuring because of the huge differences between the requirements of these two methods.

CONCLUSION

In this chapter we have discussed charging and billing systems in the telecommunication area. We have presented the detailed algorithm and flow of offline (post-paid) and online (pre-paid) charging; we have showed the problems and issues generated by the new players in the service offering chain (3rd party providers and virtual network operators). We have summarized the problems of the legacy charging systems, and we have introduced a possible new model for charging, which is capable to aid necessary functionalities, such as Advice of Charge and Income Prediction.

The current chapter was supported by the National Office for Research and Technology (NKTH) in the frame of project number 2.3 of Mobile Innovation Center (MIK).

REFERENCES

Ary, B., Debrei, G., & Imre, S. (2005). Real-time charging in third-generation mobile networks. *The 8th International Conference on Telecommunications (ConTEL 2005), 239-245.* Zagreb, Croatia: IEEE.

Ary, B., & Imre, S. (2007). *Advice of Charge in Telecommunication Services. Mobile and Wireless Communications Summit.* Budapest, Hungary: Information Society Technologies.

Ericcson (2005). Prepaid Postpaid Convergent Charging. *Whitepaper.*

UMTS Forum (2002). Charging, Billing and Payment Views on 3G Business Models. *Technical report (Report No 21).*

Garifo, C. (2001). Increasing Flexibility: Billing Systems Stretch to Handle Varieties of IP-based Services. *Xchange magazine.* Retrieved August 26, 2008, from http://www.xchangemag.com/articles/192back.html

3GPP (2004). 3rd Generation Partnership Project; Technical Specification Group Services and System Aspects; Telecommunication management; Charging management; Charging architecture and principles (Release 6). *Technical report (TS 32.240).*

3GPP (2000). Charging and Accounting Mechanisms. *Technical report (TR 122 924).*

Johnson, B. (2005). Telecom Billing Systems - Build, Buy, Outsource. Retrieved December 11, 2007, from http://job.wowak.com/introduction-telecom-billing.html.

Lucas, D. M. (2004). Where Should Rating be Implemented. *Billing World and OSS today.*

MobileIn. (2006). What is MVNO. *MobileIn.* Retrieved August 26, 2008, from http://www.mobilein.com/what_is_a_mvno.htm

Schwartz, L. (2005). *Bridging the Gap: Linking CRM and Business Support Systems.* Billing World & OSS Today.

Schwartz, S. (2003). *Next-Gen Rating: It Will Be Only As Good as the Network.* Billing World & OSS Today.

KEY TERMS AND DEFINITIONS

(Mobile) Virtual Network operator: A business player, which offers telecommunication services to its subscribers, without any telecommunication infrastructure (switched, base stations) and frequency allocation of the radio spectrum. Subscribers may be billed by the MVNO or this functionality may be outsourced to a 3rd party. The network provider and the MVNO have to be in financial relationship.

Advice of Charge: Functionality, that is responsible to predict the price of the service before the actual service consumption.

Billing or Invoicing: Refers to a logical function in the charging system, which is responsible to summarize the charges for a customer (in a given billing period), and assemble the printable invoice.

Charging or Rating: Refers to a logical function in the charging system, which is responsible to compute the price of the service for specific service consumption.

Charging System: A software and hardware system, which is responsible to process service consumption related charging information, and create the relevant bill for the end users or deduct the price of the service from the subscribers prepaid balance.

Content, Service Provider or 3rd Party Provider: A business player, which offers consumable contents (ringtones, games, themes) or consumable services (value added service, location based services) through the mobile network to the customers. The service may we billed by the network or by the content / service provider depending on their agreement.

CRM (Customer Relationship Management): A software system, which is responsible to handle customer related processes, such as activation, cancellation and service offerings.

Income Prediction: Functionality, that is responsible to predict the income of a mobile network operator.

Offline Charging: A charging method which rates CDRs in an offline manner, which assures that most of the relevant records are rated till the end of the billing period. This functionality is assured by a one way batch communication (file transfer) from the network elements to the billing system.

Online Charging: A charging method which is capable to rate CDRs in a real-time manner, and reduces the customer's balance on the fly. This functionality is assured by a bi-directional online communication between the network elements and the billing system.

Postpaid: A payment method which indicates, that the service consumption related charges are paid after the actual service, in most cases in with a monthly period.. A post-paid subscriber may request services anytime, and must settle the relevant bills accordingly.

Prepaid: A payment method which indicates, that the service consumption related charges are paid before the actual service. A pre-paid subscriber may only request services, if proper amount of balance is presented on the subscriber's balance.

Rating Scheme or Business, Pricing Logic: Defines the logic and the exact, well defined steps which are required to compute the price of the service for a specific subscriber. This logic may differ for each and every offer, which a service provider may sell to the subscribers.

Chapter 6
Local Algorithms for Topology Control in Ad-Hoc Networks

Evangelos Kranakis
Carleton University, Canada

Jorge Urrutia
Universidad Nacional Autonoma de Mexico, Mexico

ABSTRACT

In this chapter, we present a survey of recent techniques for local topology control in location aware Unit Disk Graphs including local algorithms for Routing, Traversal, Planar Spanners, Dominating and Connected Dominating Sets, and Vertex and Edge Coloring. In addition to investigating trade-offs for these problems, we discuss open problems that will play an important role in the future development of the subject.

INTRODUCTION

Ad hoc wireless networks consist of a collection of hosts of limited power communicating with each other over a wireless medium without any pre-designed or fixed infrastructure. Topology control refers to the problem of maintaining a stable and connected infrastructure among the hosts of an ad hoc network. Effective use of topology control can reduce energy consumption and increase the capacity of the network, due to reduced contention to access the wireless channels. This is accomplished by having nodes in a wireless multi-hop network define collaboratively the network topology by forming the proper neighbor relation under certain criteria.

DOI: 10.4018/978-1-60566-250-3.ch006

Despite the fact that communication tasks must be resolved only by consulting nearby hosts, algorithmic solutions must solve global computational tasks, involving, e.g., the computation of dominating and independent sets, vertex and edge colorings, and spanners. Practical considerations inspire the additional requirement for algorithms to be local in the sense that each node of the network should make decisions based only on the information obtained from nodes located a constant (independent of the size of the network) number of steps away from it. Local topology control is vital for heterogeneous network environments affected by mobile hosts, variable neighbor density, and dynamic reconfigurations. Network stability must be attained by devising local algorithms for solving traditional communication tasks like the computation of forwarding tables for routing and broadcasting, node and edge colorings for scheduling and channel assignment.

The present chapter surveys recent results on local, constant approximation, deterministic algorithms for accomplishing topology control in Unit Disk Graphs (UDGs) with location aware nodes. In more detail, we introduce important background information in Section 2 and discuss routing and traversal in Subsections 3.1 and 3.2, respectively, planar spanners in Subsection 3.3, dominating and connected dominating sets in Subsection 3.4, and vertex and edge colorings in Subsection 3.5. For all the algorithms described we also provide known trade-offs between required locality, processor memory and processing as well as optimality of the object constructed.

BACKGROUND

Algorithms devised for traditional wire-line systems are not always adequate in ad hoc networking. In dynamically changing ad hoc networks, participating hosts cannot be assumed to have knowledge of the entire system. In addition, it is often impractical or even impossible to explore the whole network prior to executing an algorithm since by the time the entire system has been examined a new change may have occurred that was not taken into account.

In this setting, locality emerges as an important concept. In local algorithms it is required that the status of a node depends only on the nodes at most a constant number (independent of the size of the network) of edges (hops) away from it. Introduced by (Linial [1992]), this model has the advantage that each node in the network need only be aware of the existence of other parts of the network that are only a constant (usually small) number of hops away from it. Algorithmic design based on locality guarantees stability (changes in the network outside a constant neighborhood do not influence the computation), consistency of solutions regardless of the order of execution, and constant termination time of the proposed algorithm. This approach was further investigated in the work of (Naor and Stockmeyer [1995]) which investigated constant-time solutions for labeling problems and the book of (Peleg [2000]) which proposes a locality-sensitive approach to distributed computing.

Nodes in wireless networks have limited transmission range and communication between two nodes depends on their Euclidean distance. A standard model of wireless network is the Unit Disk Graph (UDG) which consists of nodes with identical transmission range, say one unit. In this graph, two nodes are adjacent if and only they are within range of each other. Many graph-theoretic problems do not admit local algorithms, even when restricted to the class of UDGs. To overcome this limitation an important assumption concerns location awareness, whereby nodes are assumed to know their geographic positions obtained either from a GPS receiver or from virtual coordinates assigned by another source. Location awareness in conjunction with locality is an important paradigm for the design of efficient algorithms in ad hoc networks. Several networking problems become solvable in the local setting when the network is location aware. In this case the graph is embedded in the plane and each node knows its geographic position. Algorithms for location aware networks are sometimes easier to design and they may lead to better time complexities and/or approximation bounds.

ISSUES IN TOPOLOGY CONTROL

In this section we discuss how locality in conjunction with location awareness can be used to provide algorithms suitable for topology control in wireless ad hoc networks.

Figure 1. Discovering a route between s and t in a planar graph

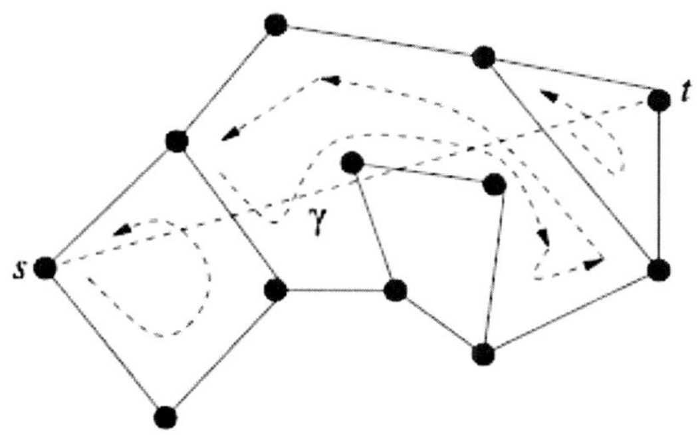

Routing

Face routing is a technique that was first proposed in (Kranakis et al. [1999]) in order to discover routes in a geometrically embedded planar subdivision. Given a source s and a destination t it discovers a route by traversing only the faces crossed by the straight line γ formed by the nodes s and t. After each face traversal it advances to a new face of the planar subdivision. It is guaranteed to succeed because each face traversal reduces the (geometric) Euclidean distance of the current position to the target. The important feature of this algorithm is locality in the route discovery process. At each step (see Figure 1) progress is made along a face of the subdivision and it is irrelevant what happens in the remaining part of the graph as long as it remains connected.

Moreover, to succeed one never has to remember anything more than the straight line from s to t and the current position, information that is easily acquired on-line by a GPS.

Because of the importance of face routing for wireless networking, efforts have been made to extend this result to richer classes of networks. It is therefore worth mentioning (Chavez et al. [2006b]) which discusses route discovery with constant memory in oriented planar geometric networks (Eulerian and Outer-planar), as well as the work of (Kranakis et al. [2006]) which discusses on-line routing in quasi-planar graphs (a class of graphs with distinct faces which allow edge crossings only within faces). There is also some recent work to specific three dimensional representations of graphs. For example

(Kranakis et al. [2006]) studies routing in polyhedral geometrically embedded graphs, (Fraser [2007]) extends face routing to geometric graphs of genus one (i.e., embedded on a torus), and (Durocher et al. [2008]) extends face routing to three dimensional graphs delimited by two parallel planes at distance $1/\sqrt{2}$.

Traversal

Network traversal is a technique widely used in networking for visiting every node of a network, using a small number of steps when required to process the nodes, edges, faces, etc, of a network in some order. For example, it may involve reporting each node, edge, and face of a planar graph exactly once, in order to apply some operation to each. As such it can be used to discover network resources, implement security policies, and report network conditions. Traversal can be used to discover routes between two hosts, but in general it will be less efficient than routing since it cannot guarantee that its discovery process will be restricted to employing only information relevant to routing.

Although DFS (Depth First Search) of the primal nodes and edges or dual faces and edges of the graph is the usual approach followed for implementing traversal, usually it cannot be implemented without using mark bits on the nodes, edges, or faces, and a stack or queue. The traversal technique from (Chavez et al. [2006b]) is applicable to the class of quasi-planar networks (this is a class of subdivisions of the plane in which we allow many edges to cross each other). The general idea of the algorithm is to define a total order on all edges thus giving rise to a unique predecessor for every quasi-face (a closed walk in the subdivision). The predecessor relationship imposes a virtual directed tree. The algorithm will search for the root of this tree and then will report quasi-faces of the graph in DFS order on the tree. For this, a well-known tree-traversal technique is used in order to traverse the tree using O(1) additional memory.

Planar Spanners

Enabling face routing provides important motivation for the design of algorithms that construct spanners (i.e., planar subgraphs) of UDGs. To be useful in an ad hoc network setting, algorithms for constructing spanners should be local. Additional important characteristics of such spanners should include connectivity (basic requirement for message delivery), low degree (eases channel allocation and frequency assignment problems and/or time multiplexing constraints), stretch factor (maximal ratio of the length–hops or Euclidean–of the shortest path in the subgraph with respect to the length of the shortest path in the original graph, and cost (total length of the edges of the subgraph–can also use squares of lengths–as compared to the length of the edge in the MST.

The Gabriel Graph (Bose et al. [2001]) was one of the first such spanners; two nodes keep their link if and only if the disk having as diameter the line with the two nodes as end-points contains no other node from the network (see Figure 2). The Local Minimum Spanning Tree (Li et al. [2004]) produces a planar spanner by having each node construct the minimum spanning tree of its distance k neighborhood; a link between two nodes remains in the spanner if and only if it belongs to the distance k spanning trees of both nodes. An extension of this result to Quasi Unit Disk Graphs (see Barriere et al. [2003]) is given in (Chavez et al. [2006a]). A similar idea also works

for Local Delaunay Triangulations, since in this case a triangle is defined by three vertices whose circle contains no other points from the point-set.

Half-Space Proximal (Chavez et al. [2006c]) is another class of subgraphs of the UDG whereby each vertex determines the closest vertex and excludes all vertices lying on the other side of the bisector of the line formed by these two nodes; it then iterates until no node is left uncovered. The resulting graph is a spanner but unlike the Yao graph, the Half-Space Proximal does not require globally consistent orientation by the nodes. Some of its nice properties include connectivity, constant stretch-factor, and maximum degree five.

Figure 2. In the Gabriel test nodes A,B forward packets to each other via node C

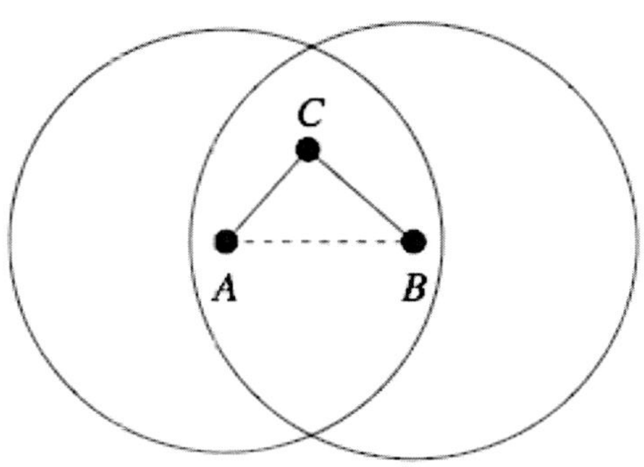

Dominating and Connected Dominating Sets

Consider a graph G. A set D of vertices dominates a vertex u if there is a vertex v in D such that {u, v} is an edge. The set D is called a dominating set for G if it dominates every vertex in G. A set D is called a minimum dominating set for G if it is a dominating set with minimum cardinality. We call a dominating set a connected dominating set if the subgraph induced by its vertices is connected. The minimum dominating set problem is concerned with finding such a dominating set.

Dominating sets are used as a backbone infrastructure and help maintain network stability, power conservation, limit interference, and reduce the number of nodes that contain routing information in ad hoc networks. This is accomplished by organizing the nodes in clusters. One vertex in each cluster takes the role of a leader (often called cluster-head) and the other vertices in the cluster are assigned to this cluster-head (see Figure 3).

Thus the cluster-heads form a dominating set and are responsible for the communication of the members of the cluster. In order to be able to send messages from one cluster to another the cluster-heads form a connected graph which results in a connected dominating set.

Despite the fact that dominating set and connected dominating set problems are NP-hard, polynomial-time approximation schemes have recently been constructed for UDGs. The resulting approximation bounds, although they appear to be better when restricted to the class of UDGs, apply mainly to the non-local setting where all the nodes have complete knowledge of the entire network. Extending the main idea of (Czyzowicz et al. [2008a]) on tiling the plane (Wiese and Kranakis [2008]) give a polynomial time approximation scheme for dominating and connected dominating sets of UDGs.

The main idea of the algorithm is to tile the plane with hexagons each assigned a class number. For some hexagons h we construct a set Th that contains all nodes in h and the nodes in a certain surrounding area. These sets Th are disjoint and have certain properties that ensure an approximation ratio which is as close to 1 as we wish. The sets Th are constructed by iterating over the class numbers of the hexagons. First we cover hexagons of class 1 by computing sets Th for all hexagons h of class 1. Assume that all hexagons of class i have already been covered. We proceed to cover all hexagons of class i + 1 whose vertices have not been completely covered so far by computing sets Th for those hexagons. We stop when all vertices in all hexagons have been covered. Moreover, the number of iterations does not exceed the total number of classes. Finally we compute for all sets Th the minimum dominating set D(Th). We output as D the union of the sets D(Th). The processing time that each vertex needs to determine whether or not it is part of the computed set is bounded by a polynomial in the number of vertices which are a constant number of hops away from it.

Vertex and Edge Coloring

Graph coloring problems have numerous applications in scheduling and channel assignment.

Figure 3. Dominating sets and cluster formation

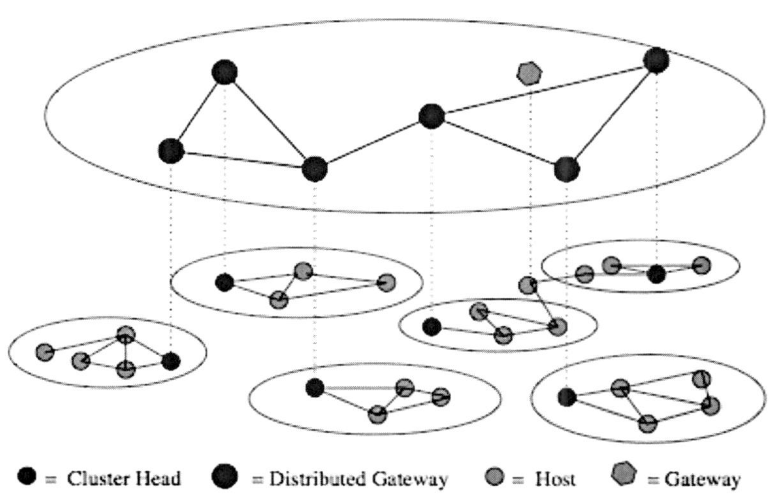

Frequency channel assignment is modeled by a graph in which two vertices are connected by an edge if the broadcasting units of their respective nodes interfere and therefore have to be assigned different channels. Since channels in the frequency band are limited and expensive resources the aim is to minimize the total number of used frequencies.

In (Czyzowicz et al. [2008b]) a local algorithm for 7-coloring planar subgraphs of UDGs is presented by using elaborate tilings of the plane. Each vertex can compute its color in a 7-coloring of the planar graph using only information on the subgraph located within at most a constant number (in our case h = 201) of hops away from it. The algorithm does not need to determine locally either what the different connected components are or even what are the local parts of a component connected somewhere far away and the complexity depends on the size of the data acquired within the specified number of hops.

In (Czyzowicz et al. [2007]) a local algorithm is presented for edge colouring (l, k)-edge/wedge subgraphs of UDGs, for all integers l, k. These are geometric graphs (see Figure 4) such that for some positive integers l, k the following property holds at each node u: if we partition the unit circle centered at u into 2k equally sized wedges then each wedge can contain at most l points different from u. An important parameter in the algorithm is the horizon distance d: a given node u never needs to be aware of the location of nodes beyond its horizon, as measured by the euclidean distance from u. The basic 2k + 1 edge coloring algorithm for (l, k)-edge/wedge subgraphs of UDGs, which is presented in this paper, uses a local horizon distance 7.81·lk.

Figure 4. Forming a wedge graph at vertex u with k = 4 and l = 3

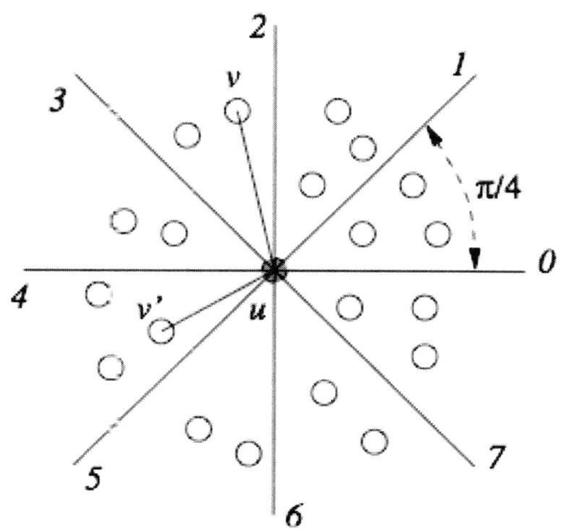

CONCLUSION

We discussed techniques for local topology control in location aware Unit Disk Graphs. Our survey included recent local algorithms for Routing, Traversal, Planar Spanners, Dominating and Connected Dominating Sets, and Vertex and Edge Coloring. In addition to investigating trade-offs for the previously mentioned issues, several remaining open problems will play important role in the future development of the subject. These include solutions for three dimensional ad hoc hoc networks, study of power assignments in the physical interference model and the inclusion of realistic models of mobility.

REFERENCES

Barriere, L., Fraigniaud, P., Narayanan, L., & Opatrny, J. (2003). Robust positionbased routing in wireless ad hoc networks with irregular transmission ranges. *Wireless Communications and Mobile Computing*, *3*(2), 141–153. doi:10.1002/wcm.108

Bose, P., Morin, P., Stojmenovic, I., & Urrutia, J. (2001). Routing with guaranteed delivery in ad hoc wireless networks. *Wireless Networks*, *7*, 609–616. doi:10.1023/A:1012319418150

Chavez, E., Dobrev, S., Kranakis, E., Opatrny, J., Stacho, L., & Urrutia, J. (2006a). Local construction of planar spanners. In *LATIN* (Valdivia, Chile, March 20-24, 2006) (LNCS 3887 pp. 286–297).

Chavez, E., Dobrev, S., Kranakis, E., Opatrny, J., Stacho, L., & Urrutia, J. (2006b). Route discovery with constant memory in oriented planar geometric networks. *Networks*, *48*, 7–15. doi:10.1002/net.20114

Chavez, E., Dobrev, S., Kranakis, E., Opatrny, J., Tejeda, H., & Urrutia, J. (2006c). Half-space proximal: A new local test for extracting a bounded dilation spanner. In *OPODIS 2005* (LNCS 3544).

Czyzowicz, J., Dobrev, S., Fevens, T., Gonzalez-Aguilar, H., Kranakis, E., Opatrny, J., & Urrutia, J. (2008a). Local algorithms for dominating and connected dominating sets of unit disk graphs. In *LATIN 2008*, April 07-11, Buzios, Brazil.

Czyzowicz, J., Dobrev, S., Gonzalez-Aguilar, H., Kranakis, E., Opatrny, J., & Urrutia, J. (2008b). Local 7-coloring for planar subgraphs of unit disk graph. In *TAMC 2008*, April 25-29, Xi'an, China (LNCS 4978, pp. 170-181).

Czyzowicz, J., Dobrev, S., Kranakis, E., Opatrny, J., & Urrutia, J. (2007). Local edge colouring of Yao-like subgraphs of unit disk graph. In *14th SIROCCO* (June 6-8, 2007, Livorno, Italy) (LNCS 4474).

Durocher, S., Kirckpatrick, D., & Narayanan, L. (2008). *On routing with guaranteed delivery in three dimensional ad hoc wireless networks* (pp. 546–557). ICDCN.

Fraser, M. J. (2007). Local routing on tori. In *Proceedings of 6th ADHONOW* (LNCS 4686, pp. 1–14).

Kranakis, E., Mott, T., & Stacho, L. (2008). Online routing in quasi-planar and quasi-polyhedral graphs. *Discrete Applied Mathematics*, *156*(18), 3430–3442. doi:10.1016/j.dam.2008.01.027

Kranakis, E., Singh, H., & Urrutia, J. (1999). Compass routing on geometric networks. In *Proceedings of the Canadian Conference on Computational Geometry (CCCG-99)* (pp. 51–54).

Li, X.-Y., Wang, Y., & Song, W.-Z. (2004). Applications of k-Local MST for topology control and broadcasting in wireless ad hoc networks. In *INFOCOM-04*, Piscataway, NJ, USA, Mar. 7–11.

Linial, N. (1992). Locality in distributed graph algorithms. *SIAM Journal on Computing*, *21*(1), 193–201. doi:10.1137/0221015

Naor, M., & Stockmeyer, L. (1995). What can be computed locally? *SIAM Journal on Computing*, 24.

Peleg, D. (2000). *Distributed Computing: A Locality-Sensitive Approach. SIAM, 2000*. Monographs in Discrete Mathematics and Applications.

Urrutia, J. (2007). Local Solutions for Global Problems in Wireless Networks. *Journal of Discrete Algorithms*, *5*(3), 395–407. doi:10.1016/j.jda.2006.05.004

Wiese, A., & Kranakis, E. (2008). Local PTAS for dominating and connected dominating set in location aware unit disk graphs. In *6th Workshop on Approximation and Online Algorithms* (LNCS).

KEY TERMS AND DEFINITIONS

Eulerian Graph: A directed graph such that for every vertex its in-degree equals its out-degree.

GPS: Geographic Positioning System.

Local Algorithm: A communication algorithm whereby messages need only propagate a constant number of hops independent of the size of the network.

Location Aware Network: A wireless network where all hosts know their geometric location.

Minimum Spanning Tree: A spanner of a graph which has no cycles and has minimum weight among all such spanners.

Outer-Planar Graph: A planar graph all of whose vertices lie on a cycle.

Planar Face: A cycle in a planar graph with no internal edges.

Planar Graph: Geometric representation of a graph so that no two edges cross.

Planar Spanner: A connected planar subgraph of a graph which spans all the vertices of the graph.

Stretch Factor of a Spanner: The worst case ratio of the length of a minimum path between two nodes in the graph divided by the length of a minimum path between these two same nodes in the spanner.

Traversal: An exploration technique for visiting every node (or link) of a network.

UDG (Unit Disk Graph): A graph consisting of wireless hosts with identical transmission range.

Chapter 7
The Four Facets of Multimedia Streaming

Florence Agboma
University of Essex, UK

Antonio Liotta
Eindhoven University of Technology, The Netherlands

ABSTRACT

Recent advances in media coding techniques and network access technologies have made multimedia streaming practicable and affordable in both fixed and mobile environments. Multimedia streaming services from anywhere and at anytime is fast becoming a reality. This chapter provides a snapshot of the facets involved in delivering multimedia streaming contents to the end user's terminal. The facets discussed in the chapter extend from the streaming paradigms, media coding techniques, the network support for multimedia streaming services to the techniques used to evaluate the user's quality of experience for multimedia streaming applications. Each facet is also discussed for both fixed and mobile environments.

INTRODUCTION

The term multimedia streaming refers to the instantaneous delivery of media contents (combination of voice, video and data) from a server application to a user's device. Recent advances in media coding techniques and network access technologies have made multimedia streaming practicable and affordable in both fixed and mobile environments. Examples of the emerging multimedia streaming services include Video-on-Demand (VoD), Peer-to-peer (P2P) television, distance learning applications, digital video libraries, and Internet Protocol Television (IPTV). The beneficiaries of this service penetration include terminal manufacturers who will see increased sales in compatible terminals, service providers and network operators who will see a rise in their average revenue per user (ARPU), and content providers and aggregators who will be able to reach a wider audience. The other beneficiaries will be the consumers, who will be able to enjoy a wider range multimedia streaming services.

The service quality, service assurance (reliability and availability), and the user experience when using these services, will determine the

DOI: 10.4018/978-1-60566-250-3.ch007

evolution of multimedia streaming. It is therefore important to provide a positive user quality of experience (QoE) in order to attract new customers and retain existing ones. QoE is defined as the perception of the user about the quality of a particular service or network (David, 2006, p. 3). The user's experience will depend on the type of network environment (i.e. fixed or mobile) that the user's terminal is operating in. In this chapter, the term "fixed environment" refers to a stationary terminal that is used from a fixed location within a network, whilst the term "mobile environment" refers to a moving terminal used across cell or network boundaries. Multimedia streaming in the mobile environment introduces new challenges such as the low transmission bandwidths, high error rates, poor network coverage, low device processing power and small-sized terminals. These shortcomings lead to unimpressive viewing experiences. This chapter discusses the facets of delivering multimedia streaming contents from four orthogonal angles. In the first facet, we review the streaming paradigms currently in use, paying attention to the key differences between real-time and on-demand streaming and between client-server streaming and peer-to-peer streaming architectures. The second facet introduces the media coding standards and discusses the techniques available to perform media adaptation, illustrating how media streams can adapt to the dynamic variation of network conditions. In the third facet, we discuss the various ways, in which the current networks support streaming services, starting off with Best-Effort networks and then proceeding to QoS-enabled networks. We later explain how and why real-time streaming services benefit from traffic engineering. Finally, the fourth facet deals with the evaluation of the multimedia streaming applications. We present the various methods used to assess the user's QoE in the context of multimedia streaming services. The chapter concludes with our views on the evolution of streaming techniques. We identify potential areas of research that focus on three themes: ubiquitous streaming; peer-to-peer streaming; and streaming based on quality of experience management.

FACET 1: STREAMING PARADIGMS

Streaming multimedia services can generally be grouped under two main categories, real time and on-demand. Within the real time paradigm, the multimedia contents are usually broadcast in real time and are centrally controlled. There is an increased requirement to maintaining constant service availability. For example, a disruption in a broadcast session, means that a subscriber cannot go back to replay or continue from the missing session because content viewing is according to the broadcast timelines. The on-demand paradigm offers a more personalized service, where individuals can control the type of contents they want to watch. It allows the playback of an archived video and also facilitates the new growing demand of user-generated contents where users upload their personal video contents to a specific online community. Examples can be found in YouTube.com, myspace.com, etc. With a large number of users, the server capacity, storage and bandwidth requirements for on-demand services become challenging because the multimedia streams are unicast. The service availability requirements are not as stringent as that of the real time paradigm because of the degree of control that is given to the subscriber. A subscriber can interact with the multimedia streams using the start, stop, pause and replay functions on the device. This on the other hand, makes the client viewing behaviour more flexible. The on-demand paradigm enables service providers to differentiate their services, which could further lead to increase ARPU.

The traditional approach of client-server streaming and content distribution networks (CDNs) is gradually moving towards the area of (P2P) streaming. In the client-server architecture, a client hosts requests and receives service from an always-on server. The addition of new clients to the network always degrades the overall network performance. This centralized platform of client-server architectures acts as a single point of failure because if the server goes down, clients experience a total denial of service. An alternate solution is to use replicate servers (CDNs) to deliver contents to users, thus reducing the effect of network congestion. With the client-server architecture, application tools can be easily implemented to enable rate control adaptation, security, route selection, path computation and end-to-end error correction of the multimedia streams (Jurca, 2007).

The P2P streaming architecture is a fundamentally different paradigm to the client-server architecture, as it does not rely on the use of dedicated servers but on the users (also known as peers) and the availability of multimedia contents on the network. In P2P streaming architectures, the users' terminals collaborate to form an application layer overlay network with the purpose of making the stream distribution process more efficient, thereby reducing the number of client-server interactions to a minimum. One of the key features in a P2P streaming system is that each user contributes resources including bandwidth, storage space, and computing power. Thus, the total system capacity can actually increase as more users join the system. P2P systems promise to solve the bandwidth scalability issues, provide more robustness to central failures, and to be more cost effective in deployment. Despite these improvements offered by P2P streaming, there are still other issues such as, the instability of users joining or leaving the network, interoperability among devices, security, the issue of digital rights management and end-to-end delay that inhibit its successful deployment.

Multimedia streaming applications have been deployed in both fixed and mobile environments using mainly the client-server architectures. However, the scalability issues and the overhead costs of the client-server architecture are very high especially taking into consideration the high costs of the network resources. Due to this limitation, P2P architectures have gained more interests from both the telecommunication industry and the research community in providing an inexpensive service platform. P2P streaming applications are generally well adopted in fixed networks. The architecture and performance analysis of P2P streaming systems in fixed environments can be found in (Sentinelli, 2007; Tang, 2007). The experimental results from (Sentinelli, 2007) highlight the issue of end-to-end delay on P2P streaming services. There are studies in the literature on multimedia streaming applications in mobile environments (Diaz, 2007; Leung, 2007). In a mobile environment, the user's terminal moves around quite often and handovers can occur fairly frequently, i.e. the user's terminal detaches itself from one access network to another. Seamless mobility across different access networks will pave the way towards a ubiquitous multimedia streaming environment.

FACET 2: MULTIMEDIA CODING TECHNIQUES

Multimedia streaming is bandwidth intensive and requires a large amount of storage space on the network. The International Standardisation Organisation (ISO) and the International Telecommunications Union (ITU) are standardization bodies that develop international standards for image and video coding algorithms. Examples of the coding standards are MPEG-1, H.261, MPEG-2, H.263, MPEG-4 and H.264. These different standards are targeted for different applications. H.264 is the latest standard designed for a wide range of applications ranging from mobile video to HDTV.

The performance of the various coding standards has been compared in (Wiegand, 2003) using the Peak Signal-to-Noise Ratio (PSNR) and subjective testing techniques. The results in (Wiegand, 2003) show that H.264 outperforms previous standards in terms of average bit rate savings and rate distortion performance. H.264

coding algorithm is used in a number of multimedia services like mobile TV, IPTV, video conferencing, broadcast TV and Internet video to significantly reduce the amount of data to be transmitted, without causing much distortion to the output of the video quality. The key features of H.264 include compression efficiency, transmission efficiency - with a number of built-in features to support reliable, robust transmission over a range of channels and networks (Richardson, 2003).

The various coding standards use several techniques to adapt the media streams to varying network conditions. The media streams that are produced by the source encoder can either be at a constant bit rate (CBR) or variable bit rate (VBR). In CBR encoding, the quantization step size (quality level) varies within the media stream in order to guarantee a constant bit rate. This is useful for constant rate links such as the Integrated Services Digital Network (ISDN). The quality of the compressed media stream depends mainly on the complexity[1] of the content, as long as the complexity of the scene is more or less constant; CBR works fine (Balaouras, 2005). An example of an application requiring a CBR encoding is video conferencing. VBR encoding allows for bit rate variation and consequently, is more suitable for dynamic bandwidth allocation. An example of an application requiring a VBR encoding is one-way live video streaming (Balaouras, 2005). To deliver an acceptable service quality for multimedia streaming, several mechanisms can be applied to adapt to bandwidth fluctuation. These adaptive mechanisms can either be implemented at the source or receiver-end. Some examples are:

- Buffer-based rate control
- Feed-forward rate control
- Improved quality rate control using the Region Of Interest (ROI) coding
- Rate control using internal feedback loop
- Reduced resolution rate control
- Rate control using multi-layer coding
- Fine granular scaleability

Detailed descriptions of these adaptive mechanisms can be found in (Abdul, 2002; Ghanbari, 2003). In multimedia streaming applications, the media quality can be affected by degradations caused by the media coding, and also by the network condition parameters, such as delay, jitter, bandwidth and packet loss. The effect of jitter and end-to-end delay in multimedia streaming services can to some extent be mitigated by adding a playout buffer at the decoder. The playout buffer stores the received video packets until its playout time. Depending on the size of the buffer, the playout delay is typically between 5 and 15 seconds, which is acceptable for most video streaming applications (Balaouras, 2005). The use of these playout buffers also makes the retransmission of lost packets possible. However, the use of playout buffers is not well suited for real time conversational video applications because of their high delay constraint.

Multimedia streaming applications in a fixed environment present significant technical challenges, further exacerbated in mobile environments, which comprises dynamic network characteristics, delays due to handoffs, high error rates and packet losses. Packet loss or corrupted media streams visibly affects a user's viewing experience. In order to reduce the effects that packet loss and bit errors may have on perceptual video quality, error control mechanisms are employed to increase robustness and resilience against transmission errors. Three types of error control mechanisms can be identified based on their implementation. They are error concealment techniques, error resilience techniques and transport-based error control techniques (Abdul, 2002). Some examples of these techniques are

- Motion vectors data recovery
- Data partitioning
- Unequal Error Propagation (UEP)
- Forward Error Correction (FEC)
- Video Redundancy Coding (VRC)
- INTRA refresh
- Adaptive INTRA refresh

More examples and detailed description on these error control mechanisms can be found in (Abdul, 2002; Ghanbari, 2003).

FACET 3: NETWORK SUPPORT FOR STREAMING SERVICES

Different types of traffic demand different network-level QoS (NQoS) requirements and guarantees from the underlying network. Real-time streaming application in which the underlying network must always provide responses within a specified window of time, and require very stringent guarantees of its network traffic compared to Email and Internet browsing applications, which do not require such guarantees. Traffic prioritization and guaranteeing of QoS requirements becomes extremely important. Connectionless networks (e.g. the Internet) provide services on a Best-Effort basis and cannot reserve bandwidth for a specific flow[2]. They deliver data packets using hop-by-hop forwarding techniques to the destination address that is encapsulated in the packet header. They do not provide delivery guarantees. The Internet Engineering Task Force (IETF) has proposed QoS frameworks to meet the demand for Internet QoS. Examples of these QoS-enabled frameworks are Integrated Services (IntServ), Differentiated Services (DiffServ), Multiprotocol Label Switch (MPLS), Generalized Multiprotocol Label Switch (GMPLS) and Traffic Engineering (TE) (Farrel, 2004)

IntServ's goal is to augment the existing best-effort Internet with a range of end-to-end services for real-time streaming and interactive applications. IntServ is based on resource reservation and supports per-flow QoS guarantees. It uses a signaling ReSerVation Protocol (RSVP) to set-up paths and reserve resources prior to data transmission. The use of RSVP with IntServ is specified in (RFC 2210). IntServ QoS framework includes two types of services in addition to best-effort service, namely: 1) Guaranteed Services (GS) (RFC 2212) for real-time applications requiring strict bandwidth and latency requirements; and 2) Controlled Load (CL) (RFC 2211) for traditional applications requiring enhanced best-effort service. Although, IntServ provides hard QoS guarantees on a per-flow basis, it is not scalable due to its per-flow end-to-end resource reservation (Bai, 2004).

DiffServ is based on a class-based approach. It operates using traffic classification, where data packets are marked and placed into a limited number of traffic classes. Packets in different classes of traffic, receive different levels of service on a per-hop basis. There are three standard service classes defined for DiffServ: the default Best-Effort (BE), Expedited Forwarding (EF) (RFC 3246) and Assured Forwarding (AF) (RFC 2597). EF minimises delay, jitter and loss, hence making it suitable for real-time services e.g. interactive voice, video etc. On the other hand, AF allows more flexible and dynamic sharing of network resources. DiffServ does not provide guarantees to QoS, it simply prioritizes traffic. However, when a large proportion of the traffic within a network is classified as high-priority, the DiffServ priority scheme becomes unable to handle the requirements adequately and all high-priority data flows are equally degraded (Farrel, 2004)

MPLS (RFC 3031) is a connection-oriented form of IP network and it also supports conventional IP QoS protocols (e.g. IntServ, DiffServ). It creates a virtual path called Label Switched Path (LSP) between the ingress and egress of an MPLS enabled-network. It uses labels to forward data packets through these predetermined paths. Using this technique, routing can be done very quickly and any necessary resources can be reserved along the path. This feature also provides traffic engineering capabilities to IP networks to optimize the distribution of QoS-sensitive and Best Effort traffic around their networks. Traffic engineering is the process in which data is routed through the network according to a management view of the availability of resources, and the current and expected traffic loads (Farrel, 2004, p. 399). (RFC 2707) provides requirements for traffic engineering over MPLS. RSVP-TE is used for managing MPLS paths. GMPLS (RFC 3945) extends MPLS traffic engineering signaling protocols and traffic engineering routing protocols to provide a common, standardized suite for controlling core networks. The interested reader is referred to (Farrel, 2004) for more information on these QoS-enabled frameworks.

The RSVP protocol designed for fixed networks assumes stable bandwidth availability in the network. However, mobile environments on the other hand are characterized by heterogeneous network capabilities where the bandwidth availability and network characteristics are highly dynamic and unpredictable. Therefore, this makes RSVP inappropriate for resource reservation in mobile environments.

Real-time streaming services such as voice or video are sensitive to packet loss, delay, jitter and bandwidth variations. In the event of network failure (this can be anything from link/path failure to the network being congested), end-to-end restoration may take several seconds – which is unacceptable for real-time applications (Aslam, 2005). The network should be able to provide robustness to such failures with minimal disruptions to its service quality. In MPLS-enabled networks, restoration and path protection are two mechanisms of providing network availability and reliability (Iannaccone, 2004). Restoration is based on finding a new path on demand to restore connectivity once a failure has occurred, whilst path protection is based on providing a fixed and pre-determined failure recovery path. In case of network failure, the pre-determined path is used to forward the traffic (Iannaccone, 2004). Restoration mechanisms provide a cost effective approach since additional resources are allocated only after a failure has occurred, unlike the path protection mechanism which pre-establishes a recovery path or path segments in case a failure is going to occur. On the downside, restoration mechanism offers slower recovery because of the delay involved in finding these resources after the failure has been detected (Iannaccone, 2004). Based on this, path protection mechanism is more efficient for real-time applications in the event of network failures. There are two distinct path protection solutions. They are one-to-one backup approach and one-to-many backup approach (Aslam, 2005). Both restoration and path protection mechanisms perform re-routing to recover from network failures. Re-routing of traffic paths often leads to packet losses; this on the other hand, degrades the QoE of real-time streaming applications.

FACET 4: QOE ASSESSMENT METHODS FOR MULTIMEDIA SERVICES

The multimedia streams can be affected by impairments due to conversions between analog and digital formats, coding and transmission channels errors. Therefore, methods are required to assess the user's QoE of multimedia streaming applications. There are two assessment methods namely: subjective and objective assessments methods. The International Telecommunication Union (ITU) has recommendations on how to carry out subjective assessments for television picture quality (ITU-R, BT500) and multimedia application applications (ITU-T, P.911 & P.910). The standards for objective perceptual metrics for digital cable television and digital broadcast television can be found in (ITU-R, BT.1683; ITU-T, J.144).

Subjective Assessment

Subjective assessments use human viewers (subjects), where subjects watch a video clip and rate its quality using quality scores. Quality scores are then averaged to produce the Mean Opinion Score (MOS). Depending on the purpose of the application that is been evaluated examples of subjective assessment methods are:

Single Stimulus Continuous Quality Evaluation (SSCQE), Absolute Category Rating (ACR), Double Stimulus Continuous Quality Evaluation (DSCQE), Double Stimulus Impairment Scale (DSIS), Double Stimulus Comparison Scale (DSCS), and Subjective Assessment Methodology for VIdeo Quality SAMVIQ (which is currently being standardized). SAMVIQ has been designed specifically for multimedia content, taking into account codec types, image format, bitrates, packet loss, etc. (EBU BPN 056:SAMVIQ2003)

Pros and Cons of Subjective Assessment

Subjective assessments provide accurate results because they produce quality ratings as perceived by the viewer. The results are also used as a reference in building and evaluating the performance of Video Quality Metrics (VQMs) used in objective assessments. However, subjective assessments require immense planning and design so as to produce reliable subjective quality ratings, are time consuming and expensive to carry out. Based on these weaknesses, the use of objective quality metrics is often desirable.

Objective Assessment

Objective assessments employ the use of metrics (algorithm) to evaluate the quality of the video automatically. PSNR and Mean Square Error (MSE) are the most widely used objective quality metrics because it is simple to implement and has clear physical meanings. This is based on the assumption that a high PSNR value usually indicates high quality and a low PSNR value specifies low quality. It is often used to evaluate and compare the performance of codecs. This precise computation does not necessarily correlate with subjective assessments (Richardson, 2003) because it does not take into account the region of interest that might be less sensitive to the viewer. This shortcoming has led to the development of more complex measures based on the Human Visual System (HVS) to produce a prediction that is close to the human perception. The performance of HVS-based objective video quality assessment outperforms that of PSNR and MSE in terms of correlating with subjective ratings but has high computational complexity. Standards for objective perceptual metrics for digital cable television and digital broadcast television can be found in (ITU-R, BT.1683; ITU-T, J.144) As yet, there is no standardized objective perceptual model specifically for mobile terminals.

Subjective Versus Objective Assessment

We have briefly discussed the assessment methods used in evaluating the quality of multimedia applications. The objective metrics discussed herein eliminate the importance of the audio counterpart that usually comes with the video. Thus, these metrics produce less accurate results compared to subjective assessment. Previous studies e.g., (Joly, 2001) have suggested that good audio quality tends to increase subjective ratings of video quality. Also, these objective metrics do not indicate the impact of different monitor types (e.g. mobile video phone, PDA, etc.). Subjective assessments remains the most accurate and reliable way of evaluating the user's QoE because it produces quality ratings as perceived by the viewer.

FUTURE TRENDS

Multimedia streaming is still an evolving technology. New network technologies, media coding techniques and enabled devices will emerge to provide seamless user experience. The traditional approach of client-server streaming is gradually paving the way for P2P streaming where large-scale, cost effective multimedia streaming can be observed (Tang, 2007). As different networks merge with the aim of providing more services and better user experience (an example is seen in triple-play services for IPTV), there is an increased requirement to implement QoE management. Network and service providers must manage, monitor and optimize their service provisioning efficiently. In this context, research on streaming based on QoE management needs to be conducted in order to understand the user's requirement and expectations, and apply this knowledge in providing acceptable service quality and user experience.

CONCLUSION

In this chapter, a snapshot of state-of-art streaming technologies from four orthogonal angles has been presented. Starting from the streaming paradigms to the techniques used in assessing the user's QoE, to ensuring end-to-end user satisfaction for multimedia streaming. The different facets of multimedia streaming need to be studied, understood, evaluated and integrated as a whole unit rather than in separate entities.

REFERENCES

Abdul, H. S. (2002). *Compressed Video Communications*. John Wiley & Sons, Ltd.

Aslam, F., Raza, S., Dogar, E. R., Ahmad, I. U., & Uzmi, Z. A. (2005). NPP: A Facility Based Computation Framework for Restoration Routing Using Aggregate Link Usage Information. In *Lecture Notes in Computer Science, Quality of Service in Multiservice IP Networks* (Vol. 3375, pp. 150-163). Springer Berlin.

Bai, Y., & ITO, R.M. (2004). QoS Control for Video and Audio Communication in Conventional and Active Networks: Approaches and Comparison. *IEEE Communications Surveys, 6*(1), 42-49.

Balaouras, P., & Stavrakakis, I. (2005). Multimedia Transport Protocols for Wireless Networks. In Salkintzis, A. S., & Passas, N. (Eds.), *Emerging Wireless Multimedia Services and Technologies*. John Wiley & Sons, Ltd.

David, S., Man, L., & Renaud, C. (Eds.). (2006). *QoS and QoE Management in UMTS Cellular Systems*. Wiley.

Diaz, A., Merino, P., & Panizo, L. (2007). Experimental Analysis of Peer-to-Peer Streaming in Cellular Networks. *IEEE International Conference on Advanced Information Netowrking and Applications* (pp. 784-791).

EBU. (2003). BPN 056:SAMVIQ - Subjective Assessment Methodology for Video Quality.

Farrel, A. (2004). *The Internet and its Protocol: A Comparative Approach*. Morgan Kaufmann.

Ghanbari, M. (2003). *Standard Codecs: Image Compression to Advanced Video Coding*. Institution of Electrical Engineers.

Iannaccone, G., Chuah, C.-N., Bhattacharyya, S., & Diot, C. (2004). Feasibility of IP Restoration in a Tier-1 Backbone. *IEEE Network, 18*(2), 13–19. doi:10.1109/MNET.2004.1276606

ITU-R. (2002). *Recommendation BT500*. Methodology for the Subjective Assessment of the Quality of Television Pictures.

ITU-R. BT.1683 (2004). Objective Perceptual Video Quality Measurement Techniques for Standard Definition Broadcast Television in the Presence of a Full Reference.

ITU-T. (1998). *Recommendation* (p. 911). Subjective Audiovisual Quality Assessment Methods for Multimedia Applications.

ITU-T. (1999). *Recommendation* (p. 910). Subjective Video Quality Assessment Methods for Multimedia Applications.

ITU-T. J.144 (2004). Objective Perceptual Video Quality Measurement Techniques for Digital Cable Television in the Presence of a Full Reference.

Joly, A., Nathalie, M., & Marcel, B. (2001). *Audio-Visual Quality and Interactions between Television Audio and Video*. Paper presented at the International Symposium on Signal Processing and its Applications, Malaysia.

Jurca, D., Chakareski, J., Wagner, J. P., & Frossard, P. (2007). Enabling Adaptive Video Streaming in P2P Systems. *IEEE Communications Magazine*, *45*(6), 108–114. doi:10.1109/MCOM.2007.374427

Leung, M. F., & Chan, G. (2007). Broadcast-Based Peer-to-Peer Collaborative Video Streaming among Mobiles. *IEEE Transactions on Broadcasting*, *53*(1), 350–361. doi:10.1109/TBC.2006.889093

RFC 2210 Wroclawski, J. (1997). The use of RSVP With IETF Integrated Services.

RFC 2211 Wroclawski, J. (1997). Specification of the Controlled-Load Network Element Service.

RFC 2212 Shenker, S., Partridge,C., & Guerin, R. (1997). Specification of Guaranteed Quality of Service.

RFC 2597 Heinanen, J., Baker, F., Weiss, W., & Wroclawski, J. (1999). Assured Forwarding PHB Group.

RFC 2707 Bergman, R., Hastings, T., Isaacson, S., & Lewis, H. (1999). Job Monitoring MIB-V1.

RFC 3031 Rosen, E., Viswanathan, A., & Callon, R. (2001). Multiprotocol Label Switching Architecture.

RFC 3246 Davie, B., Charny, A., Bennett, J.C.R., Benson, K., Le Boudec, J.Y., Courtney, W., Davari, S., Firoiu, V., & Stiliadis, D. (2002). An Expedited Forwarding PHB (Per-Hop Behaviour).

RFC 3945 Mannie, E (2004). Generalized Multi-Protocol Label Switching (GMPLS) Architecture.

Richardson, I. E. G. (2003). *H.264 and MPEG-4 Video Compression: Video Coding for Next Generation Multimedia*. Wiley. doi:10.1002/0470869615

Sentinelli, A., Marfia, G., Gerla, M., & Kleinrock, L. (2007). Will IPTV Ride the Peer-to-Peer Stream? *IEEE Communications Magazine*, *45*(6), 86–92. doi:10.1109/MCOM.2007.374424

Tang, Y., Luo, J.-G., Zhang, Q., Zhang, M., & Yang, S.-Q. (2007). Deploying P2P Networks for Large-Scale Live Video Streaming Service. *IEEE Communications Magazine*, *45*(6), 100–106. doi:10.1109/MCOM.2007.374426

Wiegand, T., Schwarz, H., Joch, A., Kossentini, F., & Sullivan, G. J. (2003). Rate-Constrained Coder Control and Comparison of Video Coding Standards. *IEEE Transactions on Circuits and Systems for Video Technology*, *13*(7), 688–703. doi:10.1109/TCSVT.2003.815168

KEY TERMS AND DEFINITIONS

Multimedia Streaming: refers to the instantaneous delivery of media (combination of voice, video and data) files from a server application to a user's device.

Client-Server Streaming Architecture: refers to the use of dedicated servers to deliver media files to the users (clients).

Peer-to-Peer Streaming Architecture: is based on the concept of equal peers in a network. It allows users to distribute media files instantaneously to each other.

QoE: refers to the perception of a user about the overall quality of a particular service or network.

QoS: is a metric used to represent the capability of a network to provide guarantees to selected network traffic.

Connection-Oriented Networks: provides QoS guarantees by setting up virtual paths between the sender and the receiver before transmission.

Traffic Engineering: is a process of arranging how traffic flows through the network.

Objective Video Quality Assessment: uses a metric / algorithm to evaluate the quality of a video.

Subjective Video Quality Assessment: employs the use of human viewers to evaluate the quality of a video.

ENDNOTES

1 The complexity of the content is typically determined by the overall amount of motion and also by the level of detail in each particular picture (Balaouras, 2005).

2 A flow can be defined as any data packet from a certain application

Chapter 8
Real-Time Communications in Wireless Sensor Networks

Isabelle Augé-Blum
Université de Lyon, France

Fei Yang
Université de Lyon, France

Thomas Watteyne
Université de Lyon, Orange Labs, France

ABSTRACT

This chapter presents the state-of-the-art of real-time communication in the challenging topic of Wireless Sensor Networks (WSNs). In real-time communication, the duration between the event which initiates the sending of a message, and the instant this message is received must be smaller than a known delay. Because topologies are extremely dynamic and not known priori, this type of constraint is very hard to meet in WSNs. In this chapter, the different communication protocols proposed in the literatures, together with their respective advantages and drawbacks, are discussed. We focus on MAC and routing because they are key layers in real-time communication. As most existing protocols are not suitable under realistic constraints where sensor nodes and wireless links are unreliable, we give, at the end of this chapter, some insights about future trends in designing real-time protocols. We hope to give the reader an overview of recent research works in this complex topic which we consider to be essential in critical applications.

INTRODUCTION

Wireless Sensor Networks (WSNs) have experienced a tremendous upsurge in recent years, both in the academy and the industry. These networks are composed of a potentially (very) large number of battery-powered nodes capable of measuring physical parameters, processing them, and communicating wirelessly. Communication is done in a multi-hop ad-hoc manner without requiring a pre-installed fixed infrastructure. Data gathered by the sensors is collected by a sink node which is generally more powerful and less energy-constrained than the low-end sensors nodes. WSNs are foreseen to be used in a wide range of applications, generally classified in five domains: military, environment, health, intelligent homes and other commercial solutions (Akyildiz

DOI: 10.4018/978-1-60566-250-3.ch008

et al., 2002). As an example, a company called Coronis Systems (Dugas, 2005) based in Southern France has deployed a 25,000-node network in Les Sables d'Olonne, France. Each sensor node is attached to a home's water meter and reports daily consumption to the local water provider.

Because of the plethora of applications, new constraints emerge, including timeliness constraints. These emerge from critical application such as a fire detection. Sensor nodes are scattered throughout a forest to monitor temperature. Whenever they sense an abnormal raise in temperature, an alarm message is sent to the sink node, which is directly connected to the fire brigade. In this type of application, the fire brigade needs to be warned within three minutes from the detection of the fire. If the alarm message has not reached the sink node within those three minutes, the surveillance system is considered to have failed. In this case, the impact of the fire on the environment or even on human lives can be dramatic.

Industrial applications with similar constraints exist, and industrial wired network solutions have been successfully rolled out for more than two decades (Rodd et al., 1998). Such networks are embedded in planes, rockets, cars. They are also used for nuclear power plants surveillance or production chain monitoring. Although this technology is ripe, it has never been used outside local and small areas.

WSN applications are different for mainly two reasons: the use of the wireless link, and a deployment over larger area. For an application to cope with timeliness constraints, the underlying communication network protocols need to be real-time. This chapter deals with the new field of real-time communication in the context of WSNs.

BACKGROUND

Applications with timeliness constraints are called "real-time". When an event happens, the application should deal with it within a known and bounded delay. The application taking more time can have a potentially (very) negative impact. Real-time applications are classified in two categories. Hard real-time applications are such that the bounded delay absolutely needs to be met. Examples include forest fire detection. Soft real-time applications are looser in that they can tolerate not to meet a deadline. Examples include multimedia applications. Applications can be classified in four categories, according to how data exchange is initiated. In event-driven application, communication starts when some event happens. In query-driven applications, communication only starts when a specific node sends a query. This specific node can be the sink node in the case of WSNs. In time-triggered applications, communication happens at predefined instants. Finally, some applications such as live video transmission require data to continuously flow through the network. Each one of these application categories has specific time-related constraints.

Distributed real-time applications have emerged in the 1980's, and many solutions exist today. For an application to be real-time, the underlying network must be able to deal with real-time constraints. So far, only local area networks were considered, as they needed to be deployed only over small areas. The size of such an area is typically a building (e.g. nuclear power plant surveillance network) or smaller (network embedded in planes, cars or train). As a result, all communicating parts are connected to the same (wired) network and it is unnecessary to implement the complete OSI stack (Tanenbaum, 2002). Only physical, medium access control (MAC) and application layers are required. In such a simplified setting, the MAC protocol is central, and determines whether the medium access delay is bounded or not. If this delay is deterministic, known and bounded, the system is considered real-time. A "real-time" protocol is not per se "fast". It only means it guarantees the transmission of data within a time frame. Such protocols can be classified in the following categories:

- contention-based or event-triggered protocols (e.g. CAN, 802.3 DCR)
- contention-free protocols which can be event-triggered (e.g. FIP, TTP/C). Alternatively, the right to access the medium can be attributed by a central entity (in a master/slave setting) or be decentralized (e.g. Token Ring).

Studies have formally validated these real-time protocols. Because of the space limitation, we cannot detail each of these protocols, yet the interested reader can refer to (Thomesse, 1998).

As WSN technology becomes ripe, new applications emerge. Thousands of nodes are deployed over large areas. The wireless medium allows for nodes to move, appear and disappear in the network. Deployment is also much faster and less expensive. An extreme deployment scenario is wireless nodes dropped from a helicopter in the to-be-monitored area. An interesting example of a logical architecture for WSN-based tracking solutions is given by (He et al., 2007). The authors identify six phases to be used in such solutions: first event detection, filtering out the false-positives, verifying close sensor nodes have detected the same event, data aggregation, transmission to the sink and processing. For the resulting communication architecture to be real-time, each of these six phases needs to meet real-time constraints. This architecture is called VigilNet, and (He et al., 2007) shows the complexity of real-time applications.

In this document, we address the communication problem. As communication is done in a multi-hop ad-hoc manner, the following challenges appear when designing real-time communication systems:

- **Routing.** It is essential to know the worst case number of hops, and the worst case delay of a single hop.
- **Medium access.** Not all nodes can hear each other, and hidden/exposed node problems yield collisions. To be real-time, collision resolution needs to be deterministic. This is however not the case with popular wireless MAC protocols such as IEEE802.11.
- **Radio link dynamicity.** Transmission ranges and packet error rate vary over time, making the radio link unreliable.

Besides these challenges, a real-time communication protocol for WSNs needs to be energy-efficient to increase network lifetime. Energy-efficiency includes turning off a node's radio as frequently as possible. There is a trade-off between a highly energy-efficient network (where nodes have their radio off most of the time) and a network with low transmission delay (where radios are on most of the time to relay messages quickly).

REAL-TIME COMMUNICATION PROTOCOLS

Real-time communication protocols for WSNs are scientifically at the crossroads of two distinct worlds: distributed real-time applications and wireless networking. Their design is thus a challenge. On one hand, a number of real-time MAC protocols for WSNs are largely inspired by industrial wired networks. Although they are real-time, run-time constraints of WSNs and wired LAN settings are very different. Their strict and sometime unrealistic constraints severely limit them to be used in WSNs directly. On the other hand, routing solutions emerging from the world of wireless networking are functional, yet they present only soft real-time characteristics. In this section, we will give an overview of the existing solutions for each category, starting with MAC protocols.

Real-Time MAC Protocols

Several MAC protocols borrow ideas from existing industrial LAN solutions. The wireless BlackBurst protocol (Sobrinho & Krishnakumar, 1999) is close to its wired counterpart CAN (ISO 11898, 1993). It uses the Carrier-Sense Multiple Access (CSMA) principle. When a node wants to transmit a message, it sends out noise for a duration proportional to the message's priority. Only the message with highest priority is eventually sent as other potential transmitters hear a noise when listening before transmitting. To avoid collisions and ensure all nodes can access the medium, other protocols ensure medium access time to all nodes. This can be done either by a central authority, e.g. IEEE802.15.4 star topology (IEEE Std.802.15.4, 2003) or in a distributed way, e.g. Wireless Token Ring (Ergen et al., 2004). Although valid, these solutions only poorly apply to a WSN setting as they assume nodes are fully meshed. Moreover, these solutions are not energy-efficient.

Solutions specifically designed for WSNs are needed Energy-efficiency has been the most active research goal for WSNs. S-MAC (Wei et al., 2004) and its extensions - T-MAC (Dam & Langendoen, 2003), DS-MAC (Lin et al., 2004) - are probably the best known WSN MAC protocol family. Such contention-based protocols are suitable for large scale WSNs because (1) nodes do not need a precise synchronization and (2) nodes can (dis)appear without affecting the network. Yet, contention based protocols face the problem of collision. Collisions are inevitable, and solutions to resolve them are not deterministic. As a result, the time to access the medium is not bounded and the protocol is not real-time. Moreover, as nodes are sleeping most of the time, multi-hop transmission delay can be very high.

To our knowledge, F-MAC (Roedig et al., 2006) is the only contention-based MAC protocol guaranteeing bounded bandwidth and delay without requiring synchronization. In this protocol, each message is transmitted using small data packets called framelets. The frequency at which framelets are transmitted varies from node to node. With its neighbors having different frequencies, a given node will be able to successfully receive at least one framelet from a source even when all its neighbors are transmitting. Moreover, nodes can be grouped into clusters. All the nodes of the same cluster can communicate with one another at a given frequency, different from neighbor clusters. The biggest drawback of this approach is its poor bandwidth utilization as the same information is sent in multiple framelets. Furthermore, the worst case delay will increase exponentially with the number of nodes within the same collision domain. This protocol is not suitable for large scale and dense sensor networks.

A second class of protocols are contention-free. These avoid collisions by using Time Division Medium Access (TDMA) or Frequency Division Medium Access (FDMA) techniques. Some of them guarantee bounded medium access delay. The Implicit Earliest Deadline First (I-EDF) protocol (Caccamo et al., 2002) assumes the network is organized in regular hexagonal cells, with a router node in the middle of each cell. FDMA is used in inter-cell communication to avoid conflicts between cells, and TDMA is adopted in intra-cell transmission. The nodes inside each cell are assumed to be fully connected, which means a node can transmit messages to any other nodes in a single hop. Although it guarantees bounded end-to-end delay, its main drawback is that the hexagonal cell structure is extremely constraining. Moreover, the protocol is not energy-efficient, and assumes all nodes can communicate using 7 different frequencies. Another solution is to assume the sink node knows the full network topology, learnt for example during an initialization phase. With this information, it is able to build a network-wide TDMA schedule. PEDAMACS (Ergen & Varaiya, 2006) assumes the sink has a transmission power strong enough for its message to be heard by any node in the network. The sink node uses this link to send the global TDMA schedule determined after learning the topology. Messages sent by the sensor nodes follow a multi-hop path to the sink. TSMP (Pister, 2008) is TDMA-based and employs FDMA and frequency hopping. All the nodes are synchronized, and the sink (which has the list of the nodes, their neighbors and their requirements in terms of traffic generation) builds a scheduling table in both time and frequency. Experimental results report a delivery ratio of over 99.999%. RT-LINK (Mangharam et al., 2007) is TDMA-based and applicable to networks which require predictability in throughput, latency and energy consumption. Hardware-based global time synchronization is used. Two phases, namely topology-gathering and scheduling, are included in RT-LINK. A cycle is defined as the duration between two synchronization pulses, and consists of a large number of frames divided into two parts: scheduled and contention slots. Each node that wants to transmit data periodically sends HELLO messages by randomly selecting a slot within the contention slots. A HELLO message is transmitted in a multi-hop manner to the sink node, which is responsible for network-wide slot assignment. The node will then be active in the assigned scheduled slot. PR-MAC (Chen et al., 2007) is used for monitoring applications where data is sent periodically. A sensor node starts by sending a message to the sink using a (non real-time) contention-based MAC protocol. This message contains a description of the sensed value, and the path taken by the message. Using the reverse path, the sink node sends a series of control messages to the relaying nodes, which indicate the periodicity of the subsequent messages and act as resource reservation messages. Once all relaying nodes are contacted, the path is set up and the sink node can expect data message to reach it in a real-time fashion. BitMAC (Ringwald & Römer, 2005) constructs a MAC protocol on top of On-Off Keying modulation. Under this assumption, and when nodes are tightly synchronized, the channel performs a logical "OR" on overlapping signals. BitMAC uses this feature to arbitrate medium access in a deterministic way; deterministic bounds are given on the execution time of all protocol elements. Although I-EDF, TSMP, PEDAMACS, TR-LINK, PR-MAC and BitMAC allow for real-time communication, they are based on a central controller.

To our knowledge, only DUAL-MODE (Watteyne et al., 2006) creates a schedule in a distributed way. It is designed for linear networks. The protocol switches between two modes: unprotected (contention based) and protected (contention-free). Alarm messages are transmitted in "unprotected mode" as long as there are no collisions. Each node relays a message when a backoff time proportional to its distance to the sink elapses (and no other message is heard). In case of collision, the protocol switches to protected mode, which avoids collision by channel reservation. Because the unprotected mode allows for faster transmission, the protocol switches back to this mode when possible. This protocol does not require synchronization and constructs a schedule in a fully distributed manner. Yet, besides it only focuses on linear networks, it is not energy-efficient.

Real-Time Routing Protocols

We will now focus on real-time routing protocols. In theory, if the underlying protocol is TDMA based (i.e. nodes follow a schedule to access the medium) and a central authority decides upon this schedule, it is possible to integrate the multi-hop routes in the schedule, yielding a fully hard real-time solution. A solution along these lines is, however, very unpractical as the central authority will need to know the complete topology (i.e. scalability is an issue) and the inherent dynamicity of WSNs cannot be integrated easily.

A more general real-time solution would require a protocol to bound the time taken for a single-hop transmission, and to know what the worst-case number of hops is. Two solutions are possible. The first one is to group nodes into clusters, each having a clusterhead, and different clusterheads communicate through a virtual backbone (Théoleyre & Valois, 2005). Paths between clusters can be known in advance and real-time communication is possible. This principle is used in I-EDF (described above). The problem here is that creating and maintaining such a structure is complex and costly. A single node appearing or disappearing may disturb the whole network, and large routing table need to be maintained, limiting the scalability.

The second one is to use geographic routing protocols (Frey et al., 2006). With high-density networks, it is fair to assume geographic distance and hop count between nodes are related. Such routing protocols are based on the notion of velocity: distance over end-to-end delay. Their advantage is that they are scalable as a node only needs to maintain a (small) neighborhood table.

In RAP (Lu et al., 2002), a priority is attached to each message depending on the velocity required by a message to reach destination before its deadline. The required velocity can be determined once and for all at the source, or at each hop. RAP requires a MAC protocol which handles message priority. SPEED (He et al., 2003) provides end-to-end soft real-time guarantees, using geographic forwarding. Each node maintains a neighbor table which contains distance and time to reach its neighbors. It can thus calculate the velocity needed by each packet, and choose a neighbor which can offer this velocity. SPEED also offers congestion avoidance. In both protocols, if a message cannot keep up with its required velocity, it is discarded on the route. These are thus soft real-time protocols. FT-SPEED (Zhao et al., 2007), MMSPEED (Felemban et al., 2006) and RPAR (Chipara et al., 2006) are recent improvements of SPEED. FT-SPEED solves the "void problem" (if the node density is low, geographic forwarding can fail because a current node has no neighbors closer to the sink than itself, although a path exist which goes around this void). MMSPEED allows for service differentiation, guaranteeing both timeliness and reliability. RPAR introduces transmission power adjustment, resulting in a routing scheme which adapts to the traffic load in order to balance energy-efficiency and low multi-hop transmission delay. POWER-SPEED avoids control packets and estimates the delay of the downstream path using the upstream path delay. Transmission power is tuned as a function of deadline in a way similar to RPAR.

Some proposals, such as ARP (Han et al., 2007), use velocity and the nodes' remaining energy to elect the next hop. Note that ARP handles both real-time and non real-time data. RTLD (Ali et al., 2008) elects the next hop according to a metric which mixes velocity, link quality and remaining energy.

CED (Yuan et al., 2007) selects as next hop the node with the best delay-energy trade-off. In order to do so, the next hop node is not only closer to the sink, but also further away from the sender from which it receives the packet. Note that, strictly speaking, this protocol cannot be considered real-time as deadlines are not taken into account. (Mahapatra et al., 2006) is also based on geographic forwarding, but selects the best two nodes (according delay and remaining energy) as next hop to increase the reliability of data delivery.

A solution close to geographic routing is gradient-based routing (Ye et al., 2005). During an initialization phase, each node learns its "height", i.e. the number of hops separating it from the sink node. A message is sent to the neighbor node with smallest height, "rolling down" to gradient until reaching the sink node. As each node knows how many hops are needed for a message to reach the sink, transmission can guarantee hard real-time constraints. The main problem is that this type of routing is very static, and topological changes are hard to deal with.

FUTURE TRENDS

The main problems faced by real-time WSN protocols are collisions and interference. We believe that a MAC protocol should switch between an event-triggered mode (when traffic is low, thus low collision probability) and a time-triggered mode (high traffic, avoiding collisions). The main problem is that schedules need to be determined in an decentralized way: each node needs to determine when it can access the medium using only local information. To our knowledge, current solutions for two-dimensional networks are based on schedules created by the sink. This MAC protocol should be coupled with data aggregation techniques to avoid that multiple sensors send distinct messages when sensing the same event. Data aggregation needs to lower the number of data flows while allowing for sufficient robustness in the multi-hop transmission. As any other protocol, the data aggregation should offer real-time characteristics. We believe that geographic routing protocols are very suitable for WSNs because of their scalability and energy-efficiency. Provided one can know the worst case number of hops, they can be used as a basis for real-time routing.

Recent studies are proposing communication protocols allowing for entirely autonomous WSNs. These networks would self-organize without human intervention, and be extremely robust to network dynamicity (Watteyne et al., 2008). Similarly, it would be interesting for those networks to determine what timeliness constraints they can offer, as a function of network size, density, radio conditions, etc. This information could be reported to the sink node. In case the network changes (nodes appearing/disappearing), the sink would receive an update on the network's timeliness constraints. Based on these constraints, the network administrator could determine what applications to deploy over the network, whether or not to deploy new nodes, etc.

CONCLUSION

This chapter presents the state-of-the-art on real-time communications in WSNs. This research domain can currently be considered a "hot topic" in research, with numerous recent proposals. Nevertheless, no current solution is entirely satisfactory, as they often make strong and sometimes unrealistic assumptions. These assumptions are specific to a given application, and can thus not be applied to other applications with similar constraints. Solutions are needed which apply to different application types (event-driven, query-driven, time-triggered or continuous). We believe a hybrid solution (contention-based under low traffic, contention-free under high traffic) could offer better performances. The most recent proposals take into account the energy-efficiency, which is essential in the context of WSNs. This is all the more thru because there is a trade-off between energy-efficiency and delay. To be energy-efficient, nodes need to sleep most of the time, which increases the communication delay. One of the biggest challenges is link dynamicity. Even with MAC and routing layers meeting hard real-time constraints, it is not feasible to have a physical layer guaranteeing message delivery. As a consequence, the notions of reliability and real-time communication should be considered jointly in the context of WSNs.

REFERENCES

Akyildiz, I. F., Su, W., Sankarasubramaniam, Y., & Cayirci, E. (2002). Wireless Sensor Networks: A Survey. *Computer Networks*, *38*(2), 393–422. doi:10.1016/S1389-1286(01)00302-4

Ali, A., & Fisal, N. (2008). A real-time routing protocol with load distribution in wireless sensor networks. *Computer Communications*, *31*(14), 3190–3203. doi:10.1016/j.comcom.2008.04.030

Caccamo, M., Zhang, L. Y., Sha, L., & Buttazzo, G. (2002). An Implicit Prioritized Access Protocol for Wireless Sensor Networks. *In Proceedings of the 23rd IEEE Real-Time Systems Symposium (RTSS'2002)* (pp. 39-48). Austin.

Chen, J., Zhu, P., & Qi, Z. (2007). PR-MAC: Path-Oriented Real-Time MAC Protocol for Wireless Sensor Network. *In Proceedings of the 2007 International Conference on Embedded Software and System (ICESS'2007)* (pp.530-539). Daegu.

Chipara, O., He, Z., Xing, G., Chen, Q., Wang, X., Lu, C., et al. (2006). Real-time Power-Aware Routing in Sensor Networks. *In Proceedings of the 14th International Workshop on Quality of Service (IWQoS'2006),* (pp. 83-92). New Haven.

Dam, T. V., & Langendoen, K. (2003). An Adaptive Energy-Efficient MAC Protocol for Wireless Sensor Networks. *In Proceedings of the first ACM Conference on Embedded Networked Sensor Systems (SenSys'2003)* (pp. 171-180). Los Angeles.

Dugas, C. (2005). Configuring and managing a large-scale monitoring network: solving real world challenges for ultra-low-powered and long-range wireless mesh networks. *International Journal of Network Management, 15*(4), 269–282. doi:10.1002/nem.573

Ergen, M., Lee, D., Sengupta, R., & Varaiya, P. (2004). WTRP - Wireless token ring protocol. *IEEE Transactions on Vehicular Technology, 53*(6), 1863–1881. doi:10.1109/TVT.2004.836928

Ergen, S. C., & Varaiya, P. (2006). PEDAMACS: Power Efficient and Delay Aware Medium Access Protocol for Sensor Networks. *IEEE Transactions on Mobile Computing, 7*(5), 920–930. doi:10.1109/TMC.2006.100

Felemban, E., Lee, C. G., & Ekici, E. (2006). MMSPEED: Multipath Multi-SPEED Protocol for QoS Guarantee of Reliability and Timeliness in Wireless Sensor Networks. *IEEE Transactions on Mobile Computing, 5*(6), 738–754. doi:10.1109/TMC.2006 79

Frey, H., & Stojmenovic, I. (2006). On Delivery Guarantees of Face and Combined Greedy-FaceRouting in Ad Hoc and Sensor Networks. *In Proceedings of the Twelfth ACM Annual International Conference on Mobile Computing and Networking (MobiCom'2006)* (pp.390-401), Los Angeles.

Han, P., Zhou, X., Li, Y., Chen, X., & Gao, C. S. (2007). An Adaptive Real-Time Routing Scheme for Wireless Sensor Networks. *In Proceedings of the 21st International Conference on Advanced Information Networking and Applications Workshops (AINAW'2007)*, (pp. 918-922). Ontario.

He, T., Stankovic, J. A., Lu, C., & Abdelzaher, T. F. (2003). SPEED: a stateless protocol forreal-time communication in sensor networks. *In Proceedings of the International Conference on Distribued Computing Systems* (ICDCS'2003). Rhode Island.

He, T., Vicaire, P., Yan, T., Luo, L., Gu, L., Zhou, G., et al. (2007). Achieving Real-Time Target Tracking Using Wireless Sensor Networks. *In Proceedings of the 12th IEEE Real-Time and Embedded Technology and Applications Symposium* (RTAS'2007), (pp.37-48). Washington, DC.

ISO 11898:1993 (1993), Road Vehicles – Interchange of Digital Information – Controller Area Network (CAN) for High-Speed Communication.

Lin, P., Qiao, C., & Wang, X. (2004). Medium Access Control with A Dynamic Duty Cycle for Sensor Networks. *In Proceedings of the IEEE Wireless Communications and Networking Conference* (WCNC'2004) (pp. 1534-1539). Atlanta.

Lu, C., Blum, B. M., Abdelzaher, T. F., Stankovic, J. A., & He, T. (2002). Rap: A real-time communication architecture for large-scale wireless sensor networks. *In Proceedings of the 8th IEEE Real-Time and Embedded Technology and Applications Symposium* (RTAS'2002). San Jose.

Mahapatra, A., Anand, K., & Agrawal, D. P. (2006). QoS and energy aware routing for real-time traffic in wireless sensor networks. *Computer Communications*, *29*(4), 437–445. doi:10.1016/j.comcom.2004.12.028

Mangharam, R., Rowe, A., & Rajkumar, R. (2007). FireFly: A Cross-Layer Platform for Real-Time Embedded Wireless Networks. *Real-Time Systems*, *37*(2), 183–231. doi:10.1007/s11241-007-9028-z

Pister, K. (2008). Technical Overview of Time Synchronized Mesh Protocol (TSMP) [White paper]. Retrieved from http://www.dustnetworks.com

Ringwald, M., & Römer, K. (2005). Bitmac: a deterministic, collision-free, and robust mac protocol for sensor networks. *In Proceedings of the second European Conference on Wireless Sensor Networks* (EWSN'2005) (pp. 57-69). Istanbul.

Rodd, M. G., Dimyati, K., & Motus, L. (1998). The design and analysis of low-cost real-time fieldbus systems. *Control Engineering Practice*, *6*(1), 83–91. doi:10.1016/S0967-0661(97)10053-3

Roedig, U., Barroso, A., & Sreenan, C. J. (2006). f-MAC: A Deterministic Media Access Control Protocol without Time Synchronization. *In Proceedings of the 3rd European Conference on Wireless Sensor Networks (EWSN'2006)* (pp. 276-291). Zurich.

Sobrinho, J., & Krishnakumar, A. (1999). Quality-of-Service in Ad Hoc Carrier Sense Multiple Access Wireless Networks. *IEEE Journal on Selected Areas in Communications*, *17*(8), 1353–1368. doi:10.1109/49.779919

IEEE Std.802.15.4(2003) Wireless Medium Access Control(MAC) and Physical Layer(PHY) Specifications for Low-Rate Wireless Personal Area Networks(LR-WPANs).

Tanenbaum, A. S. (Ed.) (2002). *Computer Networks* (4th ed.). Prentice Hall.

Théoleyre, F., & Valois, F. (2005). Virtual structure routing in ad hoc networks. *In Proceedings of the IEEE International Conference on Communications (ICC'2005)* (pp.3078-3082), Seoul.

Thomesse, J. P. (1998). A review of the fieldbuses. *Annual Reviews in Control*, *22*, 35–45. doi:10.1016/S1367-5788(98)00003-0

Watteyne, T., Augé-Blum, I., & Ubeda, S. (2006). Dual-Mode Real-Time MAC Protocol for Wireless Sensor Networks: A Validation/Simulation Approach. *In Proceedings of the First International Conference on Integrated Internet Ad hoc and Sensor Networks (InterSense'2006)* (pp. 2-8). Nice.

Watteyne, T., Dohler, M., & Augé-Blum, I. (2009). Beyond Localization: Communicating Using Virtual Coordinates. In Mao, G., & Fidan, B. (Eds.), *Localization Algorithms and Strategies for Wireless Sensor Networks: Monitoring and Surveillance Techniques for Target Tracking* (pp. 446–467). Hershey, PA: IGI Global.

Ye, F., Zhong, G., Lu, S., & Zhang, L. (2005). GRAdient Broadcast: A robust data delivery protocol for large scale sensor networks. *Wireless Networks*, *11*(3), 285–298. doi:10.1007/s11276-005-6612-9

Ye, W., Heidemann, J., & Estrin, D. (2004). Medium Access Control with Coordinated Adaptive Sleeping for Wireless Sensor Networks. *IEEE/ACM Transactions on Networking*, *3*(12), 493-506.

Yuan, L. F., Cheng, W. Q., & Du, X. (2007). An energy-efficient real-time routing protocol for sensor networks. *Computer Communications*, *30*(10), 2274–2283. doi:10.1016/j.comcom.2007.06.002

Zhao, L., Kan, B. Q., Xu, Y. J., & Li, X. W. (2007). FT-SPEED: A Fault-Tolerant, Real-Time Routing Protocol for Wireless Sensor Networks. *In Proceedings of the International Conference on Wireless Communications, Networking and Mobile Computing (WiCom'2007)* (pp. 2531-2534). ShangHai.

Zhou, Y. F. Ngai, Edith C.-H., Lyu, Michael R., & Liu, J. (2007). POWER-SPEED: A Power-Controlled Real-Time Data Transport Protocol for Wireless Sensor-Actuator Networks. *In Proceedings of the IEEE Wireless Communications and Networking Conference (WCNC'2007)* (pp. 3739-3743), Hong Kong.

KEY TERMS AND DEFINITIONS

Critical Application: application which requires timeliness constraints to be met. These constraints not being met could have a catastrophic impact on costs and/or human lives.

Hard Real-Time: a deadline needs to be met under all circumstances. If not, the system could have a catastrophic impact on costs and/or human lives.

Multi-Hop Ad-Hoc Communication: communication mode which does not require a fixed infrastructure, in which all nodes may serve as a relay to relay a message between source and destination node.

Real-Time System: the maximum duration between an event happening and this event being taken into account by the system should be known and bounded. This duration depends on the application dynamics. Note that "real-time" and "fast" are not synonymous, real-time meaning "before a given deadline".

Sensor Node: electronic device capable of doing three complementary tasks: measuring a physical value, processing that value and communicating wirelessly. A sensor node is very constrained, especially in terms of energy.

Soft Real-Time: a deadline may not be met under certain circumstances, without impacting the application.

Wireless Sensor Network: a WSN is composed of a large number of sensor nodes (1,000+) which self-organize in order to be able to communicate. The goal of a WSN is to collect information from the environment they are deployed in, and to transmit this information to a specific collecting node called sink node.

Chapter 9
A Survey of IP Layer Mobility Management Protocols in Next Generation Wireless Networks

Li Jun Zhang
Geninov Inc., Canada

Liyan Zhang
Dalian Jiaotong University, China

Laurent Marchand
Ericsson Canada Inc., Canada

Samuel Pierre
Ecole Polytechnique de Montreal, Canada

ABSTRACT

This chapter provides a survey of IP layer mobility management protocols in next generation wireless networks. In all-IP-based next generation wireless networks, mobile nodes freely change their points of attachment to the network while communicating with correspondent nodes. Hence, mobility management consists of a critical issue, which is to track mobile users' current location and to efficiently route packets to them. This chapter elaborates the mobility management procedure for the protocols such as mobility support in IPv6 (MIPv6), hierarchical mobile IPv6 mobility management (HMIPv6), fast handovers for mobile IPv6 (FMIPv6), fast handover for hierarchical mobile IPv6 (F-HMIPv6) and proxy mobile IPv6 (PMIPv6). Furthermore, future trends about mobility management are described as well. He authors hope that understanding these mobility management protocols can not only help the researchers to find more advanced solutions in this field, but also provide a training toolkit within the mobile operators.

INTRODUCTION

The commercial proliferation of multimode wireless terminals (such as laptops, Personal Digital Assistants (PDAs), smart-phones), the explosion of mobile data communications, the emergence of multi-technology environments with disparate capabilities, the integration of such diverse wireless environments and the wide variety of traditional

DOI: 10.4018/978-1-60566-250-3.ch009

and value-added services offered to end-users have placed extra requirements for mobile communications and computing. On the other hand, traditional voice, fax, email and paging services are increasingly being substituted by real-time multimedia applications, such as video-conferencing, image transfer and audio-streaming. However, the provision of such multimedia services imposes strict Quality of Service (QoS) requirements on the networks. Furthermore, since next-generation wireless networks (NGWNs) must not only provide integrated services, but also capabilities for dynamically relocating mobile terminals (Akyildiz et al., 1999), new challenges have sparked research laboratories to realize an advanced level of tetherless, seamless multimedia services for mobile users roaming between different domains or systems. Given such circumstances, mobility management becomes an important issue to enable telecommunication networks to locate roaming terminals and deliver calls or data to them while these terminals moving into a new service area.

Currently, assorted wireless technologies and networks exist to meet the various needs of mobile users. For example, wireless local area networks (WLANs) enable the delivery of high data-rate services with small radio coverage, cellular networks provide voice and data services with a relatively lower data-rate, yet large coverage, and satellite networks enable global roaming with worldwide coverage at drastic costs. Since these networks are designed for specific service requirements and vary significantly in terms of bandwidth, delay, coverage area, costs and QoS provisioning (Akyildiz et al., 2005), they tend to complement one another, and their integration enables mobile users to be "always best connected" (Gustafsson & Jonsson, 2003) to the most appropriate network.

On other hand, NGWNs introduce a variety of heterogeneities in terms of radio access technologies, network architectures, network protocols and service demands, and such inherent heterogeneities require a common infrastructure to interconnect multiple access systems (Akyildiz et al., 2004). Hence, using all-IP-based infrastructure to support ubiquitous communication appears to be very promising. First of all, IP-based wireless networks are better suited to support the rapidly growing mobile data and multimedia applications (Chen & Zhang, 2004). This is confirmed by the fact that IPv6 (Deering & Hinden, 1998) is designated as the only IP version supported for IP Multimedia Subsystem (IMS) within the Third Generation Partnership Project (3GPP) (Arkko et al., 2003). Secondly, IP-based wireless networks have already brought global success to Internet services and they will also prove to be a successful platform to foster future mobile services (Chen & Zhang, 2004). Last but not least, IP-based wireless networks are independent of the underlying radio access technologies, making it possible and feasible to maintain seamless connectivity over different radio technologies, while offering global roaming capabilities (Chen & Zhang, 2004). Therefore, NGWNs are designed to take advantage of IP-based technologies to achieve global roaming amongst a variety of access technologies (Akyildiz et al., 2004).

This chapter provides a survey of IP-layer mobility management protocols designed within the working groups of the Internet Engineering Task Force (IETF) for IPv6-based NGWNs. The remainder of the chapter is organized as follows. First, the challenges for mobility management in NGWNs are outlined in order to provide a global view of the research background. Typical IP-layer mobility management protocols such as mobile IPv6, hierarchical mobile IPv6, and fast handovers for mobile IPv6, fast handover for hierarchical mobile IPv6 and proxy mobile IPv6 are described in detail. Future trends in the design of intelligent mobility management protocols are then presented. A conclusion is drawn in the last section.

BACKGROUND

Generally, mobility management comprises two components: location management and handoff management (Quintero et al., 2004). In NGWNs, mobile users perform two types of movements: intrasystem and intersystem roaming. Intrasystem roaming refers to movements between the cells within a system, and its mobility management solutions are based on similar network interfaces and protocols. On the other hand, intersystem roaming refers to movements between different technologies, protocols, backbones or service providers. Based on intra- or intersystem roaming, the corresponding location management and handoff management can be further classified into intra- and intersystem location management and handoff management (Akyildiz et al., 2004).

Location management is the process that allows the network to locate the access point (AP) of the mobile node for calls or data delivery. This process can be further split into two steps: location update and call (or data) delivery. The former requires mobile nodes to provide the network with their location information, while the latter indicates that the network is queried for the location information of mobile users in order to deliver calls (or data) to them. The challenges in the design of intersystem location management protocols include (Akyildiz et al., 2004):

- How to reduce signaling loads and latencies that pertain to service delivery.
- How to guarantee on-demand QoS in different wireless networks (or systems).
- How to select the most suitable network for mobile users to perform location registrations (or updates) when the service areas of heterogeneous wireless networks fully overlap.
- Where and how to store the mobile user's updated location information when the service areas of heterogeneous wireless networks fully overlap.
- How to determine mobile users' exact location within a specific time constraint when the service areas of heterogeneous wireless networks fully overlap.

Typically, handoff (or handover) refers to the process of transferring signals from one AP to another. Handoff management aims to maintain network connectivity as mobile users change their points of attachment to the network. Obviously, handoff protocols need to preserve connectivity as mobile users move about, while simultaneously curtailing disturbance, or disruption from ongoing call (or data sessions) transfers. Therefore, minimal handoff disruption and session continuity are the primary goals of handoff management (Dimopoulou et al., 2005). Generally handoffs require certain key features such as low latency, minimal packet loss, minimal jitter for multimedia streaming, high reliability, sustainable scalability to large networks, etc. Here are some of the challenges encountered when designing handoff management protocols (Akyildiz et al., 2004):

- Minimizing signaling overheads and power consumption related to handoff management;
- Making efficient use of network resources during handoff;
- Improving network scalability, reliability and robustness;
- Guaranteeing on-demand QoS during handoff: (1) reducing intra- and intersystem handoff latency which is composed of signaling message processing time, resources allocation and route setup delay, format transformation time, etc. (2) alleviating user-perceptible service degradation (3) decreasing handoff failure to a near-zero level and (4) mitigating packet loss rate to a near-zero level to achieve seamless mobility.

IP LAYER MOBILITY MANAGEMENT PROTOCOLS

In all-IP-based NGWNs, mobile nodes (MNs) freely change their APs while communicating with correspondent nodes (CNs). Accordingly, mobility management consists of a critical issue, which is to track mobile users' current location and to efficiently route packets to them. Several typical mobility management protocols are designed within the IETF working groups, such as mobile IPv6, hierarchical mobile IPv6, fast handovers for mobile IPv6, fast handover for hierarchical mobile IPv6 and proxy mobile IPv6. Moreover, a number of working groups are striving to improve the performance of such specifications.

Generally, IP mobility includes macro- and micro-mobility. Macro-mobility designates mobility over a large area; this refers to situations where MNs move between IP domains (Manner & Kojo, 2004). Typically, Mobile IPv4 (MIPv4) (Perkins, 2002, 2007) and mobile IPv6 (Johnson et al., 2004) are best suited for macromobility management. Nevertheless, this chapter only addresses mobile IPv6. Micro-mobility refers to mobility over a small area, i.e. within an IP domain. Usually, micro-mobility protocols confine movements related signaling within a local domain without propagating to the mobile user' home network and those hosting their communicating peers. Location updates to a distant home agent (HA) and CNs are eliminated as long as MNs remain inside their local domain. Hence, micro-mobility protocols yield better performance than macro-mobility solutions for intra-domain roaming. This section presents macromobility protocol, like mobile IPv6 and micro-mobility protocols, such as hierarchical mobile IPv6, fast handovers for mobile IPv6, fast handover for hierarchical mobile IPv6 and proxy mobile IPv6.

Mobile IPv6

Mobile IPv6 (MIPv6), IETF's mip6 work group's baseline, was designed for macromobility management protocols for future all-IP wireless networks (Johnson et al., 2004). It enables MNs to remain connected while moving around within the Internet topology. An MN is always identified by its home address, regardless of its current location on the Internet. While away from the home network, the MN is associated with a care-of-address (CoA). To route packets to its current location in an efficient manner, binding between an MN's home address and the CoA is managed by an HA, which is actually a router on the MN's home link. Packets destined for the MN are intercepted by the HA and tunneled to the MN's current CoA. Consequently, given this tunnel mode, triangular routing problems become unavoidable. Furthermore, HAs also create bottlenecks in the network. To avoid triangular routing, MIPv6 defines route optimization that enables an MN to register its current location with a CN. As a result, the CN can directly send packets to the MN without passing though the HA.

MIPv6 mobility management procedure consists of movement detection, new CoA configuration, duplicate address detection (DAD) and binding registration (or update) with the HA and all CN with route optimization mode. Movement detection aims to determine whether the MN has moved to a new network or not. Such detection is important since, in situations where the MN has moved, the addresses configured in a previous network become invalid, requiring additional configuration to establish or maintain upper layer connectivity. Generally, this is accomplished by assessing the reachability of the current (or default) router, the validity of the configured addresses and finding a new available router on the network. To accelerate movement detection, MNs can multicast Router Solicitation (RS) messages requesting immediate Router Advertisements (RAs) from access routers, rather than wait for the subsequent periodic advertisements to arrive (Narten et al., 2007).

Based on the RA it received, the MN then configures a CoA on the new link. The address can be configured in two different manners. First, using a stateless way, by combining the MN's respective interface identifier with the network prefix found in the RA (Thomson et al., 2007), or it can occur via a stateful way during which a server assigns an address using the dynamic host configuration protocol (DHCP) (Droms et al., 2003).

When the new address is added to the specific interface, detection on duplicate address is performed in order to preserve the uniqueness of the new address. Hence, the MN sends a multicasting Neighbor Solicitation (NS) to its neighbors and this procedure requires waiting at least one second to listen to responses from other nodes. If no responses are received before the time limit expires, the new IPv6 address is considered available (Narten et al., 2007). To alleviate this lengthy delay, optimistic DAD procedure (Moore, 2006) is preferred for fast address configuration.

Home registration involves the exchange of signaling messages (Binding Update (BU) and Binding Acknowledgment (BA)) between the MN and the HA to enable the creation and maintenance of a binding between the MN's home address and its CoA. Accordingly, packets destined for the MN will be intercepted by the HA and tunneled to its CoA. Such registration is authorized by the use of IPsec (Arkko et al., 2004).

In case of Route Optimization (RO) that follows a successful home registration, correspondent registration is executed via a BU message sent by the MN in order to create the same binding for the CN. However, as empowering nodes with the ability to redirect packets from one IP address to another raises security concerns, return routability is performed prior to the correspondent registration.

The return routability procedure consists of home-address test and CoA-test. During the home-address test, the MN sends a Home Test Init message to the CN via the HA. The CN responds with a Home Test message containing a secret Home Keygen Token. Addressed to the MN's home address, the Home Test is forwarded to the MN's current CoA by the HA. While CoA-test is in progress, the MN sends a Care-of Test Init message to the CN, which returns a Care-of Test message that contains a secret Care-of Keygen Token. The Care-of Test Init and Care-of Test are directly routed between the MN and the CN, bypassing the HA.

The MN needs both the Home Keygen Token and the Care-of Keygen Token to validate the ensuing correspondent registration. The former token proves the MN to be the legitimate owner of its home address, while the latter token confirms that the MN is currently present at the new CoA. The return routability procedure also enables the CN to be reasonably confident that the MN can be reached at both its CoA and its home address. Afterwards, the MN sends BU to the CN through a direct path. Accordingly, the CN sends packets directly to the MN's CoA without involving the HA. Figure 1 illustrates the mobility management procedure in MIPv6.

In a nutshell, MIPv6 defines an IP-layer mobility management scheme to provide MNs with continuous Internet access while they move from one domain or AP to another. As mentioned above, the process of changing APs is known as handoff. During handoff, there is a period during which MNs are unable to send or receive packets due to delays resulting from both link switching and IP protocol operations. Eventually, handoff latency results in packet loss and degrades network performance, which is unacceptable and detrimental for real-time traffic with user-perceptible service deterioration (Kempf et al., 2003). As reducing handoff latency could be beneficial to real-time applications, non-real-time applications and throughput-sensitive applications as well, new enhancements are necessary for optimal performance, especially in situations where mobile users perform local movements and change APs

Figure 1. Mobility Management Procedure in MIPv6

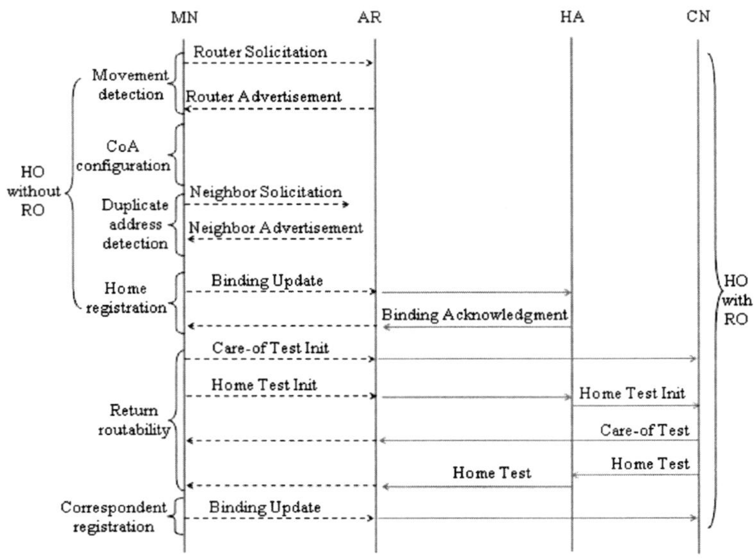

frequently. Moreover, in order to keep bindings updated, an MN periodically informs the HA and CNs about its CoA if and when its location changes or the binding lifetime expires. Under such circumstances, the network is burdened with a huge amount of signaling overhead. Hence, minimizing signaling overhead is also important to improve the network bandwidth usage and extend the battery lifespan of MNs.

Hierarchical Mobile IPv6

Even though MIPv6 provides appropriate macro-mobility management features, certain pitfalls pertaining to mobility management procedures remain, such as high signaling overhead, long handoff latency and unacceptable packet loss rate. Hence, improvements are required to enhance the performance of MIPv6. In this context, Hierarchical Mobile IPv6 (HMIPv6) was designed to reduce signaling overhead and location update delays in MIPv6 using hierarchical mobility agents called Mobility Anchor Points (MAPs). Such entities control the local movements of MNs (Soliman et al., 2005, 2007).

When an MN enters a new MAP domain, it receives RAs from access routers. Based on the received information, the MN configures two IP addresses: a Regional CoA (RCoA) on a specific MAP subnet and an on-link CoA (LCoA). After performing a DAD for the LCoA, the MN sends a local BU to the MAP in order to establish binding between the RCoA and the LCoA. Upon receiving this message, the MAP performs a DAD for the MN's RCoA and returns a BA to the MN.

Once the MN moves into the local MAP domain, it only needs to inform the MAP of its new LCoA without further updating the binding at the HA and all CNs. However, if the MN performs inter-domain movements, i.e. moving from one MAP domain to another, the MN first needs to register new binding at the new MAP. After receiving the BA from the new MAP, the MN sends a BU to the HA to establish binding between its home address and the new RCoA. With route optimization, the MN also needs to carry out return routability and correspondent registration procedures in order to create (or update) the identical binding for every CN. Packets addressed to the MN are intercepted by the MAP and tunneled to the MN's current LCoA. Figure

Figure 2. Mobility Management Procedure in HMIPv6

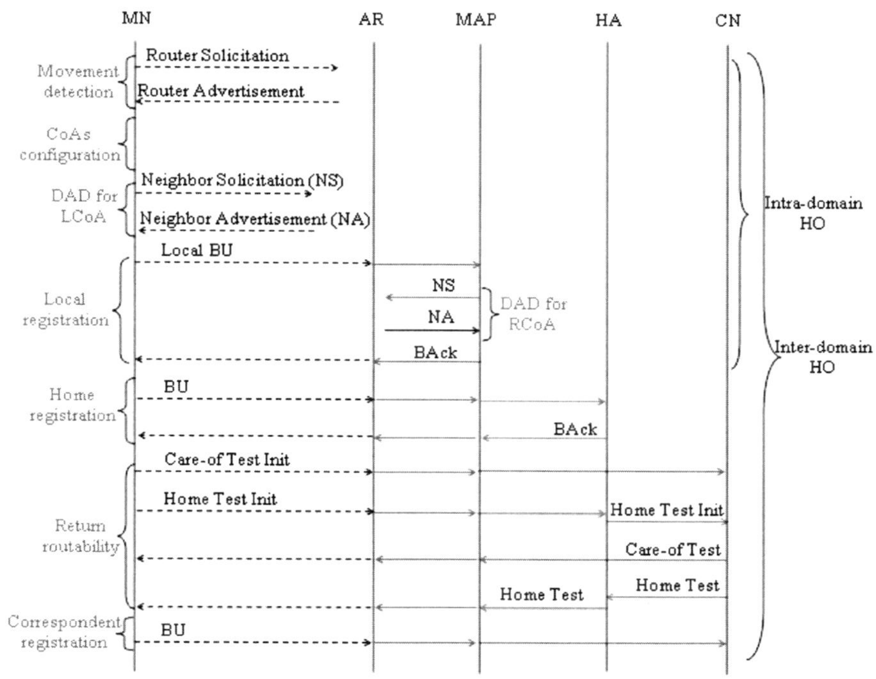

2 illustrates the mobility management procedure in HMIPv6.

On the whole, the HMIPv6 protocol confines BUs to a local MAP for intra-domain movements, rather than propagates them back to the HA and all CNs, thus improving the MIPv6 performance. However, this approach still needs further improvements to support real-time applications as HMIPv6 is mainly concerned with BU delays pertaining to intra-domain roaming, which does not improve the latencies caused by movement detection, CoA configuration as well as CoA verifications (Jung et al., 2005).

Fast Handovers for Mobile IPv6

Fast handovers for Mobile IPv6 (FMIPv6) enables MNs to rapidly detect their movements and formulate a prospective CoA while still connected to their current subnets (Koodli, 2005, 2007). It also provides MNs with the opportunity to use available link-layer event notifications (or triggers) to accelerate network-layer handoffs. Consequently, delays due to network prefix discovery and the generation of new IP addresses are completely eliminated during handoff. Moreover, a bidirectional tunnel is setup between the Previous Access Router (PAR) and the New Access Router (NAR) to avoid packet drops during handoff. In addition, the PAR maintains binding between the MN's Previous CoA (PCoA) and the New CoA (NCoA). Hence, packets addressed to an MN are intercepted by the PAR and tunneled to the MN's NCoA. Such an approach avoids BUs for the HA and all CNs in the event of route optimization.

FMIPv6 supports two operation modes: the predictive and the reactive modes. The former implies that MNs receive Fast Binding Acknowledgement (FBack) messages from their previous links. As for the latter, MNs do not receive FBack messages from their attached PARs.

After discovering one or more nearby APs, an MN sends a Router Solicitation for Proxy Advertisement (RtSolPr) message to the PAR (the

Figure 3. Predictive Mobility Management Procedure in FMIPv6

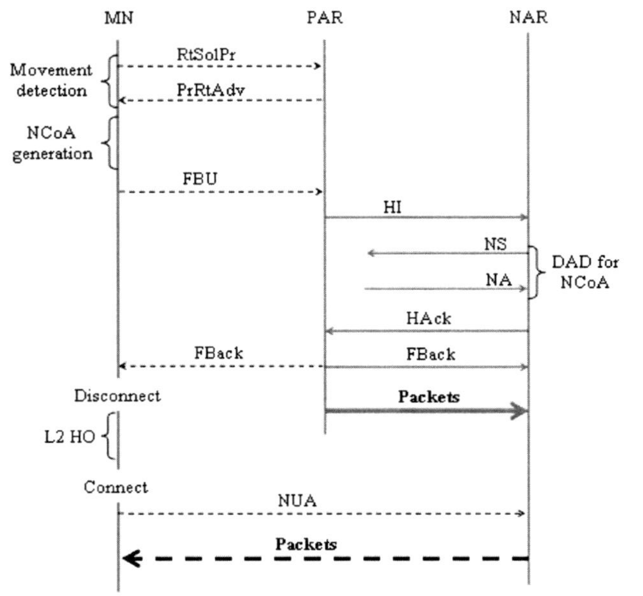

MN's default router prior to handoff) to resolve AP identifiers to subnet router information. In response to the RtSolPr, the PAR sends a Proxy Router Advertisement (PrRtAdv) to the MN to indicate whether or not it holds information regarding the NAR. Upon receiving the PrRtAdv, the MN formulates a prospective CoA and sends a Fast Binding Update (FBU) to the PAR.

The PAR sends the NAR a Handover Initiate (HI) message to set up a bidirectional tunnel. In return, the NAR validates the proposed NCoA via DAD. If the NCoA is invalid, the NAR allocates an NCoA for the MN and sends the PAR a Handover Acknowledge (HAck). The PAR then binds the MN's PCoA with the NCoA, and forwards an FBack to the MN. Afterwards, the PAR intercepts packets destined for the MN's PCoA and tunnels them to the NCoA. The NAR then buffers these packets. Upon receiving the FBack, the MN disconnects from the PAR and initiates link-layer switching procedures. Once attached to the new link, the MN sends an Unsolicited Neighbor Advertisement (UNA) to the NAR, so that both arriving and buffered packets can be forwarded immediately to the MN. Figure 3 shows the predictive mobility management procedure in FMIPv6.

Predictive mobility management enables MNs to receive FBack on their previous link. Furthermore, receiving FBack from the PAR means that packet tunneling is already in progress when MNs handover to the NAR. If MNs do not receive the FBack message on the previous link, they must follow the reactive mobility management procedures.

Performing reactive mobility management can occur for two reasons: either the MN did not send the FBU, or the MN left the link after sending the FBU, which was possibly lost, but before receiving the FBack from the PAR. Without receiving an FBack, the MN cannot be sure that the PAR successfully processed the FBU or that the tunnels are ready to use when it handovers to the NAR. As a consequence, the MN sends a UNA to the NAR immediately after attaching to the new link and launches a DAD probe for its NCoA. Upon receiving the UNA, the NAR can detect that the designated NCoA is in use

Figure 4. Reactive Mobility Management Procedure in FMIPv6

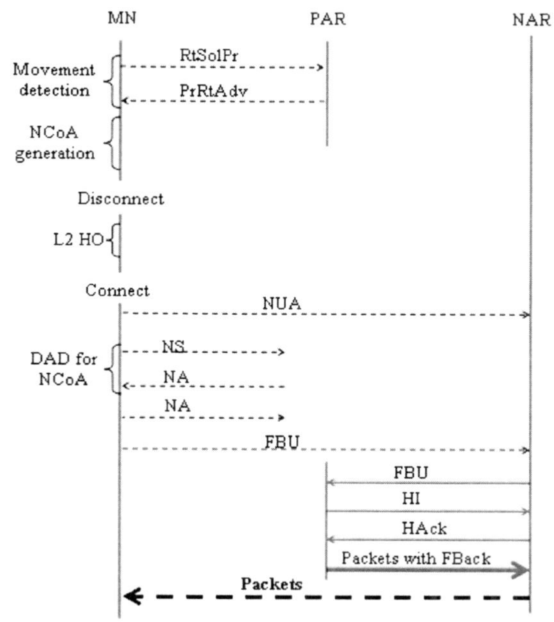

when processing the UNA. In this case, it sends the MN a RA with a Neighbour Advertisement Acknowledge (NAACK) option where the NAR assigns an alternate IP address for the MN to use.

Once the MN confirms the validity of its NCoA either via DAD or provided by the NAR, it sends a Neighbor Advertisement (NA) to the all-nodes-multicast address to join the multicast group. The MN then sends an FBU to the PAR via the NAR. Upon receiving the FBU, the PAR binds the MN's PCoA and the NCoA. If necessary, HI and HAck messages are exchanged to establish a bidirectional tunnel between the PAR and the NAR. Accordingly, packets destined to the MN's PCoA are intercepted by the PAR and tunneled to the NCoA via the NAR. Moreover, the FBack message can also be piggybacked with these packets. Figure 4 illustrates the reactive mobility management procedures.

Basically, FMIPv6 addresses the following issues: how can mobile users send packets as soon as they detect a new subnet link? How can they receive packets as soon as the NAR detects their attachment? As the FMIPv6 protocol utilizes pre-handover triggers, its performance, in terms of number of lost packets, depends dramatically on the pre-handoff trigger time, thus becoming unreliable when the pre-handoff trigger is delivered too closely to the actual link switch (Kempf et al., 2003).

Fast Handover for Hierarchical Mobile IPv6

In their own ways, both FMIPv6 and HMIPv6 are designed to improve MIPv6 performance in terms of handover delay and signaling overhead, thus making it necessary to amalgamate these two schemes. However, simply superposing FMIPv6 over HMIPv6 induces unnecessary processing overhead for PAR re-tunneling, as well as inefficient use of network bandwidth (Jung et al., 2005). Hence, an effective integration called Fast handover for Hierarchical MIPv6 (F-HMIPv6) is designed to enable an MN to exchange handoff related signaling messages with a local MAP,

Figure 5. Intra-Domain Mobility Management Procedure in F-HMIPv6

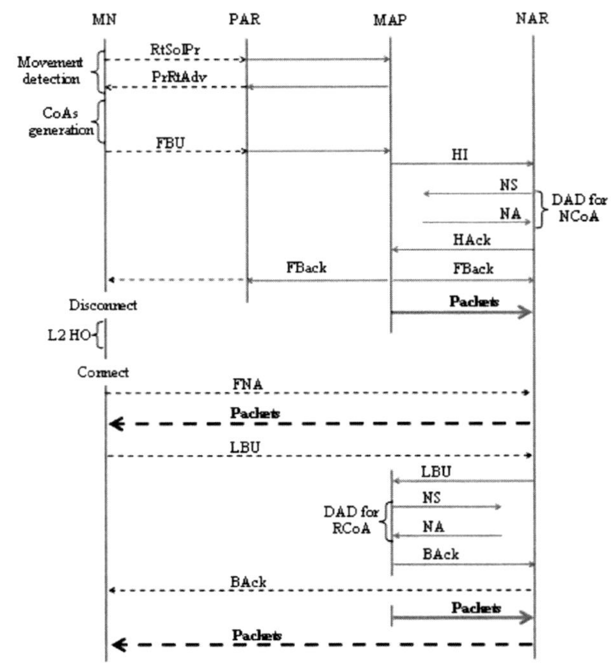

and a bidirectional tunnel is established between the MAP and the NAR, instead of between the PAR and the NAR. Figure 5 depicts the mobility management procedure in F-HMIPv6 for intra-domain roaming.

Even though F-HMIPv6 allows mobile users to benefit from both FMIPv6 and HMIPv6, the handoff latency for intra-domain roaming lasts about 90 ms whereas the handoff delay for inter-domain roaming rises to about 240 ms (Jung et al., 2005), making this protocol unsuitable for multimedia streaming traffic.

Note that all of the above-mentioned mobility management protocols prevent knowing the exact moment when mobile users move or attach to NARs. Furthermore, as packets are transmitted too early or too late in many cases, this dilemma results in tremendous quantities of lost packets. Simultaneous binding could be an interesting solution for this challenge (Haseeb and Ismail, 2007). Furthermore, MIPv6, HMIPv6, FMIPv6 and F-HMIPv6 protocols consist of host-centric solutions for IP mobility, generating problems when MNs cannot signal their mobility information to the HAs and their correspondents.

Proxy Mobile IPv6

MIPv6 requires certain IPv6 client functionalities for MNs, i.e. MNs must be capable to manage mobility by themselves. However, as certain MNs lack such functionalities, Proxy Mobile IPv6 (PMIPv6) is designed to enable network-based mobility management for MNs without their participation in mobility related signaling activities (Gundavelli et al., 2007).

PMIPv6 extends MIPv6 signaling messages and reuses the functionality of HA to support mobility for MNs without host involvement. In the network, mobility entities are introduced to track the movements of MNs, initiate mobility signaling on behalf of MNs and setup the routing state required.

When an MN attaches to a new access link connected to a Mobile Access Gateway (MAG), the MAG authenticates and authorizes the MN before providing PMIPv6 services. While access is authenticated or network attachment events are notified, the MAG acquires the MN's identifier and uses it to access the MN's policy store and retrieves the address of the Local Mobility Anchor (LMA) that serves as the MN's HA.

The MAG also configures a proxy CoA for the MN, and then sends a Proxy Binding Update (PBU) message to the LMA. In return, the LMA updates its Binding Cache Entry (BCE) for that MN and replies with a Proxy Binding Acknowledgment (PBA) that contains the MN's home network prefix assigned by the LMA. Upon receiving the PBA, the MAG establishes a bidirectional tunnel between its proxy-CoA and the LMA address. Then, the MAG periodically sends RA messages to the MN on the access link advertising the MN's home network prefix as the hosted on-link-prefix. Or, the MN can send an RS to request for an immediate RA from the MAG. In return, the MAG replies with an RA equipped with the MN's home network prefix. Hence, the MN can configure a home address using either stateful or stateless mode depending on the modes permitted on that access link. As a result, the MN only uses its home address for all its communication and the proxy-CoA is invisible to the MN. Packets sent to the MN are intercepted by the LMA, tunneled to the MAG which removes the tunnel header before forwarding them to the MN. Likewise, packets sent by the MN are intercepted by the MAG and tunneled to the LMA which then removes the outer header and forwards them to the CN.

When an MN handoffs from a Previous MAG (P-MAG) to a New MAG (N-MAG), the MN detached event is detected by the P-MAG, which sends a PBU to the LMA in order to update the MN's binding and routing states. After receiving the PBU, the LMA waits for a certain *MinDelayBeforeBCEDelete* amount of time before deleting the MN's entry from its BCE. Then, the LMA sends a PBA to the P-MAG.

When the MN attaches to the new link, the N-MAG obtains its identifier either via access authentication or notified network event. Then, the N-MAG accesses the MN's profile and obtains the LMA address. The N-MAG also configures a new proxy-CoA for the MN on one of its interfaces. Subsequently, it sends a PBU to the LMA which assigns a home network prefix to the MN and sends a PBA to the N-MAG that contains this home network prefix. Upon receiving the PBA, the N-MAG sets up a bi-directional tunnel with the LMA before configuring a data path for the MN's traffic. Finally, it emulates the MN's home link by sending RA messages to the MN that contains the MN's home network prefix. Figure 6 illustrates the handoff management procedure in PMIPv6.

As PMIPv6 was only recently drafted by the IETF community, certain shortcomings remain to be addressed, such as triangular routing problems without route optimization support, security concerns, etc. Furthermore, since detecting MNs' detachment and attachment events remains difficult in many wireless networks, in-fly packets will certainly be dropped at N-MAGs. Aside from that, a link-layer access authentication protocol is required to enable the MAG to obtain MNs' identifiers. However, such proposals have not yet been developed.

FUTURE TRENDS

Aside from the IETF's standardization activities, one of the well-known activities is that the IMS -based solution for service provisioning and handover between fixed and mobile operators is to be used in NGWNs. Within the 3GPP and the 3GPP2, research is well underway to provide handoff capabilities between disparate technologies and different domains. This 3GPP activity falls under the projects undertaken by System

Figure 6. Handoff Management Procedure in PMIPv6

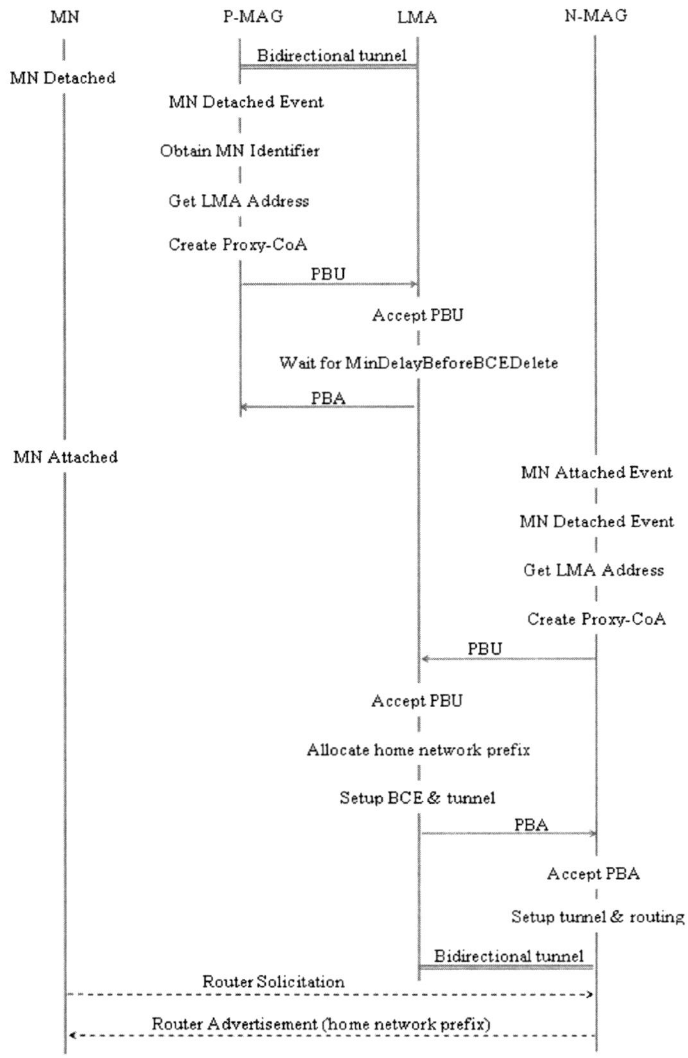

Architecture Evolution (SAE) and Long Term Evolution (LTE).

The IEEE 802.21 working group delves into the Media-Independent Handover (MIH) project while IEEE 802.11u addresses inter-technology handovers. The IEEE 802.11r task group is currently attempting to enhance handoff performance within IEEE 802.11. Additionally, a number of R & D activities are ongoing by academia, research institutes and industries simultaneously. To date, numerous research projects have focused on certain facets pertaining to mobility management, such as security, QoS provisioning and routing. Nevertheless, it is essential to develop an end-to-end mobility management solution that covers the entire span of the network, from radio access to packet delivery with guaranteed QoS. In the near future, mobile users carrying an integrated terminal can use a wide range of applications provided by multiple wireless networks (Hui & Yeung, 2003).

Fourth-Generation (4G) wireless networks are expected to integrate a large number of heterogeneous wireless technologies towards universal seamless access and omnipresent computing through seamless mobility that enables users to access sought services anywhere, at any time (Zhang et al., 2003). One of the main research challenges for seamless mobility consists of providing available and reliable intra- and intersystem handoff solutions (Nasser et al., 2006). This implies the design of new and efficient mobility management schemes to optimize QoS while providing flawless mobility.

Moreover, as 4G wireless networks tend to revolve around Universal Mobile Access (UMA) and ubiquitous computing via seamless mobility, another major challenge for seamless mobility lies in the design of an effective and efficient vertical handoff protocol for MNs that move between different types of networks. As traditional operations for handoff detection policies, decision metrics and radio link transfers cannot adapt to dynamic handoff criteria, nor deliver context-aware services or ensure network interoperability, it is obvious that new techniques are required to manage intersystem mobility (McNair & Zhu, 2004).

CONCLUSION

This chapter presents some basic concepts and background related to IP-layer mobility management protocols proposed by IETF's working groups. MIPv6, the baseline protocol, made significant contributions to enable MNs access to the Internet while roaming around. However, the MIPv6 mobility management procedure leads to impaired handoff performance in terms of signaling load, handoff delays and packet losses. Hence, improvements are crucial to enhance network performance. In this context, HMIPv6, FMIPv6, F-HMIPv6 and PMIPv6 were designed for localized mobility management. All of these improvements have their own merits and shortcomings. Furthermore, several standardization activities are underway to create an intelligent mobility management scheme that enables mobile users to benefit from its cost-effectiveness, enhanced features, location-independence and seamless mobility across different systems via multimode radio interfaces.

REFERENCES

Akyildiz, I. F., McNair, J., Ho, J. S. M., Uzunalioglu, H., & Wang, W. (1999). Mobility Management in Next-Generation Wireless Systems. *Proceedings of the IEEE*, *87*(8), 1347–1384. doi:10.1109/5.775420

Akyildiz, I. F., Mohanty, S., & Xie, J. (2005). A Ubiquitous Mobile Communication Architecture for Next-Generation Heterogeneous Wireless Systems. *IEEE Communications Magazine*, *43*(6), 29–36. doi:10.1109/MCOM.2005.1452832

Akyildiz, I. F., Xie, J., & Mohanty, S. (2004). A Survey of Mobility Management in Next-Generation All-IP-Based Wireless Systems. *IEEE Wireless Communications*, *11*(4), 16–28. doi:10.1109/MWC.2004.1325888

Arkko, J., Devarapalli, V., & Dupont, F. (2004). Using IPsec to Protect Mobile IPv6 Signaling Between Mobile Nodes and Home Agents. IETF RFC 3776.

Arkko, J., Kuijpers, G., Soliman, H., Loughney, J., & Wiljakka, J. (2003). Internet Protocol Version 6 (IPv6) for Some Second and Third Generation Cellular Hosts. IETF RFC 3316.

Chen, J.-C., & Zhang, T. (2004). *IP-Based Next-Generation Wireless Networks: Systems, Architectures, and Protocols*. New York: Wiley-Interscience. doi:10.1002/0471478253

Deering, S., & Hinden, R. (1998). Internet Protocol, Version 6 (IPv6) Specification. IETF RFC 2460.

Dimopoulou, L., Leoleis, G., & Venieris, I. S. (2005). Fast Handover Support in a WLAN Environment: Challenges and Perspectives. *IEEE Network*, *19*(3), 14–20. doi:10.1109/MNET.2005.1453394

Droms, R., Bound, J., Volz, B., Lemon, T., Perkins, C., & Carney, M. (2003). Dynamic Host Configuration Protocol for IPv6 (DHCPv6). IETF RFC 3315.

Gundavelli, S., Leung, K., Devarapalli, V., Chowdhury, K., & Patil, B. (2007). Proxy Mobile IPv6. IETF Draft. draft-ietf-netlmm-proxymip6-07.txt.

Gustafsson, E., & Jonsson, A. (2003). Always Best Connected. *IEEE Wireless Communications*, *10*(1), 49–55. doi:10.1109/MWC.2003.1182111

Haseeb, S., & Ismail, A. F. (2007). Handoff Latency Analysis of Mobile IPv6 Protocol Variations. *Computer Communications*, *30*(4), 849–855. doi:10.1016/j.comcom.2006.10.010

Hsieh, R., Zhou, Z. G., & Seneviratne, A. (2003). S-MIP: A Seamless Handoff Architecture for Mobile IP. *22nd Annual Joint Conference of the IEEE Computer and Communications Societies (INFOCOM 2003)*, (pp. 1774-1784). San Francisco, CA, USA.

Hui, S. Y., & Yeung, K. H. (2003). Challenges in the Migration to 4G Mobile Systems. *IEEE Communications Magazine*, *41*(12), 54–59. doi:10.1109/MCOM.2003.1252799

Johnson, D., Perkins, C., & Arkko, J. (2004). Mobility Support in IPv6. IETF RFC-3775.

Jung, H. Y., Kim, E. A., Yi, J. W., & Lee, H. H. (2005). A Scheme for Supporting Fast Handover in Hierarchical Mobile IPv6 Networks. *ETRI Journal*, *27*(6), 798–801. doi:10.4218/etrij.05.0205.0009

Kempf, J., Wood, J., & Fu, G. (2003). Fast Mobile IPv6 Handover Packet Loss Performance: Measurements for Emulated Real Time Traffic. *IEEE Wireless Communications and Networking Conference (WCNC 2003)* (pp. 1230-1235). New Orleans, LA, USA.

Koodli, R. (2005). Fast Handovers for Mobile IPv6. IETF RFC 4068.

Koodli, R. (2007). Mobile IPv6 Fast Handovers. IETF Draft, draft-ietf-mipshop-fmipv6-rfc4068bis-03.txt.

Manner, J., & Kojo, M. (2004). Mobility Related Terminology. IETF RFC-3753.

McNair, J., & Zhu, F. (2004). Vertical Handoffs in Fourth-Generation Multinetwork Environments. *IEEE Wireless Communications*, *11*(3), 8–15. doi:10.1109/MWC.2004.1308935

Moore, N. (2006). Optimistic Duplicate Address Detection (DAD) for IPv6. IETF RFC 4429.

Narten, T., Nordmark, E., Simpson, W., & Soliman, H. (2007). Neighbor Discovery for IP version 6 (IPv6). IETF RFC 4861.

Nasser, N., Hasswa, A., & Hassanein, H. (2006). Handoffs in Fourth Generation Heterogeneous Networks. *IEEE Communications Magazine*, *44*(10), 96–103. doi:10.1109/MCOM.2006.1710420

Perkins, C. (2002). IP Mobility Support for IPv4. IETF RFC-3344.

Perkins, C. (2007). IP Mobility Support for IPv4, Revised. IETF Draft. draft-ietf-mip4-rfc3344bis-05.

Quintero, A., Garcia, O., & Pierre, S. (2004). An Alternative Strategy for Location Update and Paging in Mobile Networks. *Computer Communications*, *27*(15), 1509–1523. doi:10.1016/j.comcom.2004.06.001

Soliman, H., Castelluccia, C., El Malki, K., & Bellier, L. (2005). Hierarchical Mobile IPv6 Mobility Management (HMIPv6). IETF RFC 4140.

Soliman, H., Castelluccia, C., El Malki, K., & Bellier, L. (2007). Hierarchical Mobile IPv6 Mobility Management (HMIPv6). IETF Draft, draft-ietf-mipshop-4140bis-00.txt.

Thomson, S., Narten, T., & Jinmei, T. (2007). IPv6 Stateless Address Autoconfiguration. IETF RFC 4862.

Zhang, Q., Guo, C., Guo, Z., & Zhu, W. (2003). Efficient Mobility Management for Vertical Handoff between WWAN and WLAN. *IEEE Communications Magazine*, *41*(11), 102–108. doi:10.1109/MCOM.2003.1244929

KEY TERMS AND DEFINITIONS

Handoff Management: prevents interruptions of ongoing data sessions or calls when mobile nodes change channels or move from one subnet to another.

Horizontal (Intra-System) Handoff: also called intra-technology handoff. A handoff process where a mobile node moves between base stations or access points that support the same network technology, e.g. a mobile node changes the associated access point in wireless local area networks.

Inter-Domain Handoff: a handoff process where a mobile node transfers between different networks (supporting the same or different types of network interfaces) managed by different administrative domains.

Intra-Domain Handoff: a handoff process where a mobile node transfers between different networks (supporting the same or different types of network interfaces) which are managed by the same administrative domain.

Location Management: tracks and locates mobile nodes in order to deliver incoming calls or data sessions.

Location Update: process performed by a mobile node away from its home network to explicitly notify its home agent or correspondent nodes of its new location (e.g. care-of-address on the new attached link) and to store changes to its user location profile (e.g. to update the corresponding binding cache).

Mobility Management: enables telecommunication networks to locate roaming mobile nodes for data or call delivery and to maintain a connection with mobile nodes as they change their network point of attachment (or access point). Mobility management includes two components: location management and handoff management.

Seamless Handoff: a handoff process where a mobile node moves between networks without causing disruption to its ongoing sessions, i.e. with low latency, minimal packet loss, guaranteed QoS with continuous and reliable connectivity.

Vertical (Inter-System) Handoff: also referred to as inter-technology handoff. A handoff process where a mobile node moves among access points that support different network technologies, e.g. the changeover of signal transmission from an IEEE 802.11b access point to an overlaid cellular network.

Chapter 10
Auditory Feedback for Network Intrusion Detection

M.A. Garcia-Ruiz
University of Colima, Mexico

M. Vargas Martin
University of Ontario Institute of Technology, Canada

R. Aquino-Santos
University of Colima, Mexico

O. Alvarez-Cardenas
University of Colima, Mexico

A. Edwards
University of Colima, Mexico

M.G. Mayoral-Baldivia
University of Colima, Mexico

ABSTRACT

Network intrusion (the unauthorized access to a computer system perpetrated by persons or a piece of software) is a major concern of network administrators. Carrying out network intrusion detection (NID) manually is a tiring, complex and time-consuming task, often overloading the visual sensory channel. To overcome this, real-time auditory alarms and auditory information provide immediate feedback, helping to identify trends or patterns of the attacks in network logs and in real-time network accesses, allowing professionals make rapid decisions, thus representing an interesting alternative to conventional NID. The purpose of this chapter is to describe literature research on the benefits and challenges of Auditory Display (the use of non-speech sound to represent meaningful information at a computer interface) applied to wireless, wired, and mobile network intrusion detection. This chapter also highlights further work on Auditory Display and multimodal interfaces to support intrusion detection in wireless mesh networks.

DOI: 10.4018/978-1-60566-250-3.ch010

INTRODUCTION

The increasing threat of cybernetic attacks in wireless and traditional networks has become one of the major concerns of network equipment designers and administrators. An intrusion is defined as an unauthorized access to a computer system that violates some security policy. One of the main problems caused by intruders is that they consume or hijack resources (e.g., bandwidth, processing power, services) and compromise vulnerable systems. Network administrators cannot always assume that a network intrusion detection system (NIDS) can discern between malicious and non-malicious traffic. Even after an intrusion is discovered, considerable time is often spent deciding on what course of action to take, or whether or not it is necessary to disconnect or shut down services, which is not always a viable option. Carrying out the intrusion detection manually is a tiring, complex and time-consuming task, often overloading visual sensory channel. To overcome this, sound parameters have been used to represent information about network attacks. Real time auditory alarms provide immediate feedback, identify and register trends or patterns of the attacks in network logs, and help professionals make rapid decisions, thus representing an interesting alternative to conventional NIDS.

This chapter highlights usefulness of auditory feedback, in conjunction with other types of sensory information regarding intrusion detection in wireless networks, as well as technological advantages and challenges of its application, are also discussed in this chapter.

The remainder of the chapter describes further trends and work on Auditory Display and multimodal interfaces to support intrusion detection in wireless mesh networks.

Section 2 (Background) describes definitions on Auditory Display, sonification and other information regarding to the application of parameters of sound to human-computer interfaces.

Section 3 (Network Intrusion Detection Systems) outlines a general description on network intrusion detection systems and its placement in a network.

Section 4 (Wireless Networks and their Challenges on Network Intrusion Detection) refers to a list and a description of intrusion vulnerabilities of wireless networks.

Section 5 (Auditory Feedback applied to Computer Networks and NIDS) looks at a literature review on visual and auditory techniques applied to NIDS. This section also includes technical aspects on the design of Auditory Display for NIDS. In it, useful auditory feedback design guidelines are explained, as well as the challenges present in auditory feedback construction.

Section 6 (Future Trends) deals with the future trends on Auditory Display applications, for example in mesh networks. In addition, the use of multimodal information for NID carrying out visual, auditory information, and integrating other sensory channels, is a promising area of research that needs to be explored.

The objective of this chapter is to describe literature research on the benefits and challenges of Auditory Display (according to Kramer (1994), the use of non-speech sound to represent meaningful information at a computer interface) applied to wireless, wired, and mobile network intrusion detection.

BACKGROUND

Auditory Display can be defined as an area of computer science that studies how non-speech sounds are designed and applied to represent meaningful information in a computer interface (Kramer, 1994). Sounds can be very effective in Human-Computer Interaction (HCI) for discriminating data patterns, identifying trends in information, mapping complex data, complementing or supplementing other sensory channels, and representing alarms, among other applications, as long as the sounds are adequately designed and adapted to the computer interface. This can be done by following well-established auditory and HCI guidelines and principles (Brewster et al., 1995). However, sounds can be unpleasant if they are played too loud, or distract persons who are nearby (Shneiderman & Plaisant, 2004). Sonification (also called auralization) is the action of mapping data onto parameters of non-verbal sound, such as volume, pitch, timbre, duration, frequency, amplitude, and rhythm in a computer interface (Kramer et al., 1997). Several experimental and empirical studies have demonstrated the effectiveness and efficacy of auditory feedback (AF) in computer interfaces. For example, Brewster (1997) found that sound is effective to solve the problem of visualization overload, where AF acted as complementary information for the visual sensory channel.

There are two main types of non-speech sounds in Auditory Display: Earcons and auditory icons. Blattner et al. (1992) defined an earcon as a group of one or more musical notes, called motives, which make use of specific auditory parameters, such as changes in duration, timbre and loudness. In addition, their timbre is made of sounds of wind, stringed or percussion instruments. Earcons are associated with either objects or actions presented in a computer interface. Because earcons make abstract associations with data, users must learn them in an initial training process. Earcons can be arranged into families to map groups of data, and can inherit their particular properties. Moreover, auditory icons are composed of everyday sounds, or sound effects, that should have a direct association with objects or events in a computer interface (Gaver, 1986). For instance, a sound of splashing water played at different tempo can be associated with a computer simulation of a working water pump. The user should identify easily and rapidly this association because there is a semantic relationship between the sounds and the objects or actions. However, it is often difficult to find ambient sounds for designing useful auditory icons with an effective semantic relationship about NIDS.

Sonification is the action of mapping scientific and other types of abstract data onto attributes of sounds, mainly doing this by modifying their amplitude, frequency and duration, commonly occurring in real time. Sonification has been employed to uncover patterns or trends in those data, as well as to serve as alarms, among other applications (Kramer, 1994). Scientific sonification is the equivalent to scientific visualization, and can make use of auditory icons and earcons.

NETWORK INTRUSION DETECTION SYSTEMS

An intrusion is defined as an unauthorized access to a computer system that violates some security policy compromising confidentiality, integrity, or availability. Intrusion detection can be dealt with at different levels of layered systems. In particular, network intrusion detection consists of a set of actions aimed at detecting intrusions by analyzing network traffic. Network administrators would generally monitor the traffic visually, making it a tedious task. Several proactive and reactive NIDS approaches have been proposed, although administrators still have to extensively visualize their results. Some of these are signature and anomaly-based, and some of them use self-learning techniques, ranging from probabilistic analysis (Venkataraman et al., 2005) to neural networks (Hofmann et al., 2002). The reaction techniques can vary from raising an alarm or delaying traffic to complex auto-configurable mechanisms (Singh et al., 2003) or automatic generation of signatures (Newsome et al., 2005). Typical NIDSs rely on the presence of the characteristics of common attacks such as performing many similar actions in a short period of time (van Oorschot et al., 2006), spoofing IP addresses (Park, K., & Lee, H., 2001), attempting connections to or from non-existing hosts or services (Twycross & Williamson, 2003),

Figure 1. NIDS placement in a typical network system

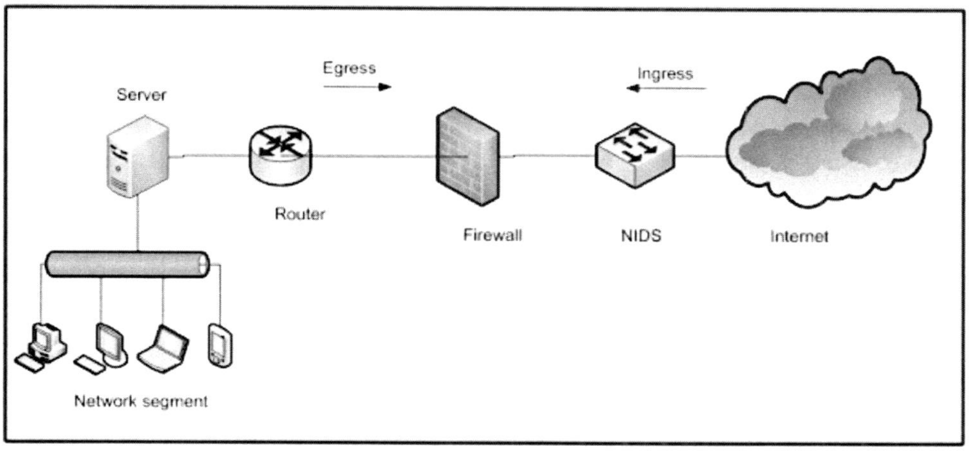

among others. Some existing mitigation systems have been sonificated (Qi et al. 2007, and references therein).

A NIDS may be placed in a number of physical locations and devices, as long as it can see and analyze network traffic (ingress and egress) for the protected network segments. A NIDS can be implemented as hardware or software inside a switch or router, or as a separate dedicated device. Now, depending on the physical location, the NIDS may have access to different kinds of traffic. For example, if a NIDS is located outside the firewall perimeter (Figure 1), the NIDS is able to analyze ingress traffic (i.e., traffic generated outside the protected network), however the NIDS may not have access to a full spectrum of egress traffic because the firewall may block some of it. Conversely, if the NIDS were placed inside the perimeter, it would have access to egress traffic, but ingress traffic may be limited according to the firewall rules. The physical placement of a NIDS ultimately depends on the needs and intentions of the system's architecture. As per core network infrastructures, a NIDS can be an integral component built-in within a high performance router, thus being able to analyze, and possible correlate, traffic to and from a number of networks.

WIRELESS NETWORKS AND THEIR CHALLENGES ON NETWORK INTRUSION DETECTION

A wireless local area network (WLAN) is the linking of two or more computers without using wires of any type, via electromagnetic microwaves (O´Hara and Petrick, 1999). During the nineties, the IEEE 802.11 Working Group established WLAN standards, described in detail in: http://www.ieee802.oorg/11/. Often the terms "Wi-Fi" and IEEE 802.11 have been used interchangeably. The WLAN technology has several sets of standards, being the 802.11a, 802.11b. 802.11g, and 802.11n the more used to date (Earle, 2006). Wireless networks can also be an extension of wired networks forming bridges. This can be used for many applications, such as connecting networks of buildings nearby, and for Internet source providers (ISP) supplying last mile services (Balachandran et al., 2005). Since WLANs can be an extension of wired networks, it is not surprising that attacks of persons and malicious software can go beyond wired LANs, compromising resources of both types of networks.

Anyone with basic WLAN skills and specialized pieces of software for detecting WLAN characteristics (popularly termed as "sniffers") can get access to WLANs from any place at any time, within the range of a wireless transmission. A previous step for a hacker to trespass a WLAN is by carrying out "wardriving", which is just the search and detection of private WLANs at the street level using a mobile computer using programs like Netstumbler (a controversial practice that has presumably been banned in some countries, although for some, wardriving does not contravene any laws). Interestingly, a version of Netstumbler was developed for personal digital assistants (PDAs). The unauthorized use of private wireless communications and its resources is called "piggybacking" (Butler et al., 2003). This is a type of intrusion that will probably not be detected by an intrusion detection system, where a typical NID assumes that the underlying computing architecture is functioning properly. The intrusion detection in wireless networks is considered the art of detecting inappropriate and illegal anomalous activity (Yang et al., 2004).

The literature has reported the following intrusion vulnerabilities of wireless networks:

- Network Discovery Attacks are employed several times by wireless LAN discovery tools such as Netstumbler. It aims at discovering as much useful information about the network as possible (Tsakountakis et al., 2007). The derived information will be exploited later on for launching a real attack against the network employing wardriving.
- Eavesdropping/Traffic Analysis is an attack in which an unauthorized person gains access to a wireless transmission and does not modify its content (Karygiannis & Owens, 2002).
- Masquerading/Impersonation Attacks are when the attacker impersonates an authorized user and thereby gains certain unauthorized privileges (Karygiannis & Owens, 2002).
- Man-in-the-middle (MITM) Attacks are, in the scope of a WLAN a technique where an attacker is able to redirect all traffic between two hosts of that same WLAN for packet sniffing or data manipulation, without the end hosts being aware of it (Belenger, 2007).
- Denial-of-Service (DoS) Attacks aim to disrupt a legitimate user's access to the network, servers, or applications by either targeting a server directly or by consuming network bandwidth (El-Moussa et al., 2007).

AUDITORY FEEDBACK APPLIED TO COMPUTER NETWORKS AND NIDS

Auditory feedback has been researched in a number of fields, such as for identifying and tracking satellites in an aerospace control room (Albers, 1994), and for positively analyzing 3D properties of molecular structures (Hansen et al., 1999). Contrary to information visualization of network intrusion detection, the application of sound for NIDS analysis has not been extensively researched and developed, although a number of important AF research projects and applications about computer networks have emerged. It is believed that one of the main technical reasons why auditory feedback for NIDS is not widely developed and applied nowadays is because its difficulty to design effective auditory icons and earcons, especially because more formal studies are needed about designing the mapping of abstract and complex information of attacks onto sound parameters. Nevertheless, general earcon and auditory icon design guidelines have been developed (Kramer, 1994; Blattner et al, 1989; Brewster, 1995; Gaver, 1989). Garcia-Ruiz et al. (2007) conducted an empirical usability study with 29 computer science engineering students, which consisted of selecting earcons and auditory icons to be applied to NIDS, working as alarms for network attacks. It was found in the study that students suggested making auditory icons with sounds of violent nature (i.e. a sound of cars crashing), and most of the learners preferred using auditory icons instead of earcons for further developments of sonification of network intrusion detection.

With respect to Auditory Display applications to computer networks, Barra et al. (2001) and Gilfix and Couch (2000) used sound parameters to effectively represent Web server status, working as alarms to inform the network administrator about Web malfunctions, high load, email spam, and network traffic. Kimoto and Ohno (2002) explored the use of pitch, melody and chords of musical notes made of MIDI (Musical Instrument Digital Interface, a standard protocol for communicating electronic musical devices and computers and to play digital music) sounds and open source software to represent network traffic, particularly the number of packets per second. Preliminary tests indicated that sound helped people perceive network traffic variations and marked changes in network function. Malandrino et al. (2003) successfully used MIDI tracks for representing network components, to monitor the behavior of distributed systems with information obtained from the Simple Network Management Protocol (SNMP).

Moreover, auditory feedback has been researched specifically for conducting intrusion detection analysis. Varner and Knight (2001) proposed an audio/visual and agent-based system for monitoring the network in real time to identify malicious hacker attacks. Gopinath (2004) carried out a study where he mapped data from network logs to signaling malicious attacks by identifying false positives (an alarm unnecessarily generated under the absence of any intrusion) and denial of service (DoS).

Technical Aspects on the Design of Auditory Display for NIDS

Designing useful auditory alarms and auditory parameters to map information for NIDS is not a trivial task. One of the main challenges is to map adequate sound effects (auditory icons) and musical sounds (earcons) onto information on network attacks, to allow network administrators to easily identify and remember those mappings. One strategy is to let the network administrator to choose his/her own sounds for NIDS from a repertoire. A selection of auditory icons and earcons for NIDS can be seen in Garcia-Ruiz et al. (2007) (shown in Table 1). Earcons or auditory icons like those shown in Figure 1 can be used as auditory warnings to represent network anomalies, for example, mapping unusual port accesses that may be generated by impersonating attacks or a denial of service (DoS). The advantage of using sound effects (auditory icons) or musical sounds (earcons) is that they can be pleasant for the network administrator to listen to, as well as being a powerful "attention grabber".

A number of general guidelines on acoustics for sound design in Auditory Display have been experimentally developed, for example see Table 2, which need to be addressed when desiging Auditory Display for NIDS.

Table 1. A selection of musical instruments and sound effects to be used on Auditory Display for NIDS (Garcia-Ruiz et al., 2007).

Proposed musical instruments for making earcons	Proposed sound effects for making auditory icons
• Guitar • Electric guitar • Drums • Trumpets • Violin • Bass • Cello • Piano • Flute	• Breaking glass • Breaking china (plates) • Water stream • Thunder • Nature sounds • Rain • Beach waves • Cars crashing • Screams • Dog barking

Table 2. A summary of Auditory Display design guidelines.

Sanders and McCormick(1987) general guidelines	Brewster et al. (1995) earcon guidelines
• Use frequencies of between 0.2 and kHz • Use frequencies below kHz if the signal must travel a considerable distance • Use a frequency below 0.5 KHz if there are obstacles • Use a modulated signal • Use signals with frequencies different from the background noise • Make warnings discernible from each other • Use a separate communication system for warnings	• If listeners are to make absolute judgments of earcons then pitch/register should not be used. • The maximum pitch used should be no higher than 5 kHz (four octaves above C3) and no lower than 125 Hz-150 Hz (the octave of C4) • Make rhythms as different as possible. Putting different numbers of notes in each rhythm is very effective. • Great care must be taken over the use of intensity (loudness) because it is the main cause of annoyance due to sound. • Stereo position or sound spatialization may be useful for differentiating sounds played in parallel. • Use rhythm or pitch (perhaps combined with lower intensity) to capture user attention.

Technical and usability guidelines for designing sonifications with MIDI sounds are found in Brown et al. (2003). In addition, important guidelines for designing auditory icons (sound effects) are described in Gaver (1986) and in Kramer (1994).

Two key aspects to improve the usability of the sounds are to let the users program and choose their own sounds in the monitoring/security software, and to serve as redundant information feedback to visual processes.

Current Applications of Auditory Alarms and Warnings in Network Security and Monitoring Software

A number of commercial and open source programs have been developed for analyzing network traffic (firewalls, antivirus, NIDS, network sniffers, and others), which include the option of playing auditory alarms and warnings, where their main purpose is to grab network users' and administrators' attention on network anomalies, like unauthorized accesses to network ports, warnings on malicious software attacks, and suspicious programs that try to install and run by themselves in a computer, among others. Another purpose is to complement visual information on the network traffic, since most of the network monitoring programs show colored text warnings on the computer screen, or even send emails to alert the network administrator (Walker, 2001). Table 3 shows a list of examples of types of network applications that make use of auditory alarms and warnings, with respect to network security.

From Table 3 and from the literature review carried out in this chapter, it can be seen that it is technically possible to use MIDI sounds working as alarms. Those MIDI sounds are generally played by the pc sound card. An advantage of MIDI files over pre-recorded sounds is that the former are really small in size. Most of the computational process for playing the MIDI sound is done by the pc sound card, thus freeing the PC CPU of processing the sound. In addition, an external MIDI device can be used for decoding and playing MIDI sounds.

FUTURE TRENDS

The IEEE 802.11 Task Group has recently started research on mesh networks, under the name of 802.11s, investigating different mesh configurations with wireless devices carrying out a Network layer routing protocol. This new forming standard can be applied to build a network mesh within an enterprise, or for a broader and extended number of users, such as a neighborhood, and wide-area networks (WANs) (Hiertz et al., 2007). Wireless

Table 3. Types of network monitoring and security programs that include auditory alarms, warnings, and other types of Auditory Display.

Type of network program	Examples	Types of sounds used	Purpose of the sound(s)
Firewall	• Zone Alarm (tm)	Pre-recorded musical sounds and sound effects, user configurable	An alarm is played when inbound or outbound programs are executed, attempts of port attacks, etc.
Antivirus	• Avast (tm)	Pre-recorded soundeffects and musical sounds, user configurable	An alarm is played when a virus is found, when a task is finished, or when the virus database is updated, among other purposes.
Sniffer/Network monitoring	• Netstumbler (tm) • Netgong (tm)	• MIDI sounds (musical tones), user configurable • Pre-recorded musical sounds	• Sonification of an access point signal strength, according to signal-to-noise ratio (SNR) • An audible alarm is played when Netgong detects an absent or failing connection in a local network

Mesh Networks (WMN) are a group of static mesh routers which employ multipoint and multi hop forwarding to provide wireless Internet connectivity. Current routing protocols in WMNs assume nodes to be non-malicious, therefore a future research work will consider integrate security and auditory feedback mechanisms for routing algorithms in WMN.

The use of various sensory information channels at the computer interface, based on HCI techniques (this is, the mapping of various types of network information onto visual, auditory or tactile sensory channels in a multimodal interface), occurring either in a desktop or a mobile computer, seems a viable option for carrying out intrusion detection with effectiveness and efficacy, since information from one sensory channel can complement or supplement information from another channel. In addition, 3D audio-visual NIDSs may ease the identification of network attacks (Papadopoulos et al., 2004).

CONCLUSION

This chapter described an overview of Auditory Display and network intrusion detection, applied to wireless, wired and mobile network intrusion detection. Usefulness of auditory feedback, in conjunction with other types of sensory information of NIDS, as well as technological advantages and challenges of its application, were also discussed in this chapter. It was also reviewed that wireless networks pose special types of attacks that need to be carefully addressed. Experimental studies are needed about the application of auditory feedback in NIDS, particularly the design of the mapping of abstract information of attacks onto sound parameters.

The increased density of wireless networks with different physical and operative functions makes network attacks an increasingly frequent reality. The great variety of network attacks and abusive practices makes refining network intrusion techniques more important than ever. Present practices are long, tedious, and insufficient, as they require intense visual scrutiny of network logs and real time network monitoring. Employing other human sensory channels to detect network intrusion can contribute to improve monitoring of network activity. Specifically, Auditory Display permits real time detection employing the auditory channel to provide preliminary feedback that can be used to then visually refer to log activity, and more promptly address network abuse.

REFERENCES

Albers, M. C. (1994). The Varese system, hybrid auditory interfaces, and satellite-ground control: Auditory icons and sonification in a complex, dynamic system. In *Proceedings of ICAD '94 - The Second International Conference on Auditory Display, Santa Fe, NM. Santa Fe Institute.*

Balachandran, A., Voelker, G. M., & Bahl, P. (2005). Wireless hotspots: current challenges and future directions. *Mobile Networks and Applications*, *10*(3).

Barra, M., & Cillo, T. De Santis, A., Petrillo, U.F., Negro, A., & Scarano, V. (2001). Personal Web-Melody: Customized sonification of web servers. In *Proceedings of the International Conference on Auditory Display (ICAD),* Espoo, Finland.

Belenger, J., & Calafate, C. (2007). A low-cost embedded IDS to monitor and prevent Man-in-the-Middle attack on wired LAN environments. In *Proceedings of t international conference on emerging security information and technologies.*

Blattner, M. M., Sumikawa, D. A., & Greenberg, R. M. (1989). Earcons and icons: their structure and common design principles. *Human-Computer Interaction*, *4*(1), 11–44. doi:10.1207/s15327051hci0401_1

Brewster, S. A. (1997). Using non-speech sound to overcome information overload. *Displays, Special issue on multimedia displays*, 17.

Brewster, S. A., Wright, P. C., & Edwards, A. D. N. (1995). Experimentally derived guidelines for the creation of earcons. *In Adjunct Proceedings of HCI'95*, Huddersfield, UK.

Brown, L. M., Brewster, S. A., & Ramlol, R. (2003). Design guidelines for audio presentation of graphs and tables. In *Proceedings of the 2003 International Conference on Auditory Display, Boston, MA, USA, July 6-9.*

Butler, J., Undercoffer, J., & Pinkston, J. (2003). Hidden processes: the implication for intrusion detection. In *Proceedings of Workshop on Information Assurance, Systems, Man and Cybernetics Society.*

Earle, A. E. (2006). *Wireless security handbook.* Boca Raton, FL: Auerbach Publications.

El-Moussa, F., Linge, N., & Hope, N. (2007). *Active router approach to defeating denial-of-service attacks in networks.* IET Communications white paper.

Garcia-Ruiz, M. A., Edwards, A., Aquino-Santos, R., Vargas Martin, M., & Mendoza Gonzalez, R. (2007). Using sonification to teach network intrusion detection: A preliminary usability study. In *Proceedings of ED-Media*, Vancouver, Canada.

Gaver, W. W. (1986). Auditory icons: Using sound in computer interfaces. *Human-Computer Interaction*, *2*, 167–177. doi:10.1207/s15327051hci0202_3

Gilfix, M., & Couch, A. (2000). Peep (The network auralizer): Monitoring your network with sound. In *Proceedings of 14th System Administration Conference*, New Orleans, USA.

Gopinath, M. C. (2004). *Auralization of intrusion detection system using Jlisten.* Unpublished thesis, Birla Institute of Technology and Science, India.

Hansen, M. D., Charp, E., Lodha, S., Meads, D., & Pang, A. (1999). PROMUSE: A system for multi-media data presentation of protein structural alignments. In *Proceedings of the Pacific Symposium on Biocomputing*, Hawaii, EUA.

Hiertz, G. R., Sebastian, M., Rui, Z., Denteneer, D., & Berlemann, L. (2007). Principles of IEEE 802.11s. In *Proceedings of 16th International Conference on Computer Communications and Networks*, ICCCN 2007.

Hofmann, A., Horeis, T., & Sick, B. (2002). Feature selection for intrusion detection: An evolutionary approach. In *Proceedings of the IEEE International Joint Conference on Neural Networks*, Budapest, Hungary (Vol. 2).

Karygiannis, T., & Owens, L. (2002). *Wireless network security: 802.11, Bluetooth, and Handheld Devices*. National Institute of Standards and Technology white paper.

Kimoto, M., & Ohno, H. (2002). Design and implementation of Stetho - network sonification system. In *Proceeding of ICMC2002* (pp 273-279).

Kramer, G. (Ed.). (1994). *Auditory display: Sonification, audification, and auditory interfaces. Santa Fe Institute Studies in the Sciences of Complexity, Proc (Vol. XVIII)*. Reading, MA: Addison-Wesley.

Kramer, G., Walker, B., Bonebright, T., Cook, P., Flowers, J. H., Miner, N., et al. (1997). *Sonification report: status of the field and research agenda.* Available at: http://www.icad.org/websiteV2.0/References/nsf.html

Malandrino, D., Mea, D., Negro, A., Palmieri, G., & Scarano, V. (2003). Nemos: Network monitoring with sound. In *Proceedings of the 2003 International Conference on Auditory Display (ICAD)*, Boston, MA, USA, July 6-9.

Newsome, J., Karp, B., & Song, D. (2005). Polygraph: Automatically generating signatures for polymorphic worms. In *Proceedings of the 2005 IEEE Symposium on Security and Privacy*, Oakland, USA.

O'Hara, B., & Petrick, A. (1999). *The IEEE 802.11 handbook: A designer's companion. Standards Information Network.* Los Alamitos, CA: IEEE Press.

Papadopoulos, C., Kyriakakis, C., Sawchuk, A., & He, X. (2004). CyberSeer: 3D audio-visual immersion for network security and management. In *Proceedings of the 2004 ACM Workshop on Visualization and Data Mining For Computer Security,* Washington DC, USA.

Park, K., & Lee, H. (2001). On the effectiveness of route-based packet filtering for distributed DoS attack prevention in power-law internet. In *Proceedings of Special Interest Group on Data Communication.* San Diego, USA: SIGCOMM.

Qi, L., Vargas Martin, M., Kapralos, B., Green, M., & Garcia-Ruiz, M. A. (2007). Toward sound-assisted intrusion detection systems. In *Proceedings of the Information Security Symposium. LNCS 4804*, Algarve, Portugal.

Sanders, M. S., & McCormick, E. J. (1987). *Human factors in engineering and design.* New York: McGraw-Hill.

Shneiderman, B., & Plaisant, C. (2004). *Designing the user interface: Strategies for effective human-computer interaction* (4th ed.). Boston, MA: Addison-Wesley.

Singh, S., Estan, C., Varghese, G., & Savage, S. (2003). *The EarlyBird system for realtime detection of unknown worms* (Technical Report CS2003-0761). University of California, San Diego, USA.

Tsakountakis, A., Kambourakis, G., & Gritzalis, S. (2007). Towards effective wireless intrusion detection in IEEE 802.11i. In *Proceedings of Third International Workshop on Security Privacy and Trust in Pervasive and Ubiquitous Computing*, IEEE.

Twycross, J., & Williamson, M. M. (2003). Implementing and testing a virus throttle. In *Proceedings of the 12th USENIX Security Symposium*, Washington, USA.

Van Oorschot, P. C., Robert, J.-M., & Vargas Martin, M. (2006). A monitoring system for detecting repeated packets with applications to computer worms. *International Journal of Information Security*, *5*(3), 186–199. doi:10.1007/s10207-005-0081-8

Varner, P. E., & Knight, J. C. (2001). Security monitoring, visualization, and system survivability. In *Proceedings of IEEE/SEI. Information Survivability Workshop (ISW).*

Walker, J. (2001). *Security event correlation: Where are we now? NetIQ whitepaper.* Retrieved from http://www.net-. iq.com/products/sm/whitepapers.asp

Yang, H., Xie, L., & Sun, J. (2004). Intrusion detection solution to WLANs. In *Proceedings of IEEE Circuit and System Symposium on Emerging Technologies: Frontiers of Mobile and Wireless Communications*. IEEE.

KEY TERMS AND DEFINITIONS

Auditory Display: branch of Computer Science that studies the application of sound to convey meaningful information at the computer interface.

Auditory Feedback: non-speech and verbal sounds that provide meaningful information to computer users in response of user actions and computer events.

Human-Computer Interaction: discipline concerned with the design, evaluation and implementation of easy-to-use, productive, safe and interactive computer interfaces for human use, and the study of its context of use.

Network Intrusion Detection (NID): the act of searching and detecting malicious activities of software and persons in computer networks, such as port scans, denial of service, and unauthorized network traffic monitoring done by third persons.

Piggybacking: the unauthorized use of private wireless communications and its resources.

Wardriving: the search and detection of private WLANs at the street level, using a mobile computer and specialized software.

Wireless Intrusion Detection System (WIDS): a system that monitors the radio spectrum for the presence of unauthorized access points, and can assist to proactively and reactively protect potential attacks in wireless networks.

Wireless Local Area Network (WLAN): the linking of two or more computers without using wires, via electromagnetic microwaves.

Chapter 11
Peer-to-Peer over Mobile Ad-Hoc Networks

Nadia N Qadri
University of Essex, UK

Antonio Liotta
Eindhoven University of Technology, The Netherlands

ABSTRACT

In this chapter we review various approaches for the convergence of Peer-to-Peer (P2P) and Mobile Ad hoc Networks (MANETs), identifying strengths and weaknesses, and putting things in perspective. P2P and MANETs are among the most active research topics in pervasive computing. The convergence of P2P networks and MANETs would allow existing P2P applications such as P2P file sharing and P2P streaming to benefit from the ubiquitous connectivity of ad-hoc. A P2P network over an ad-hoc infrastructure is a powerful combination that provides users with means to access different kinds of information anytime and anywhere. Realizing such a system is, however, not straightforward.

INTRODUCTION

The goal of mobile computing is to provide access to information and resources at anytime, anywhere. The traditional client-server architecture is insufficient to provide these needs, due to various issues like scalability (as the number of nodes grows the server can be overloaded) and link failure (which may lead to server unreachable). To cope with these issues there is a need for an architecture which allows both decentralized and spontaneous connectivity. Peer-to-Peer (P2P) systems offer the means to realize decentralized networks, which can be used to share resources over the internet. On the other hand, a Mobile Ad-hoc Network (MANETs) is a spontaneous network made of mobile nodes connected wirelessly but without relying on a specific infrastructure network. A P2P network over an ad-hoc infrastructure is a powerful combination that provides users with means to access different kinds of information anytime and anywhere. P2P systems and MANETs have been developed by different communities in order to address entirely different requirements. Despite being fundamentally different, MANETs and P2P networks share the common vision of

DOI: 10.4018/978-1-60566-250-3.ch011

load distribution based on the user's willingness to share resources in a dynamic and decentralized manner.

The P2P network concept has drawn increasing attention, and has been widely deployed on the Internet for various purposes, including distributed data storage, file sharing, collaborative computing, and Internet telephony (Wu, 2006). The P2P system is popular for being scalable, fault-tolerant, and self-organized. Nowadays P2P has turned out to be one the most active research areas of distributed computing and networking. P2P was originally used to describe the communication between two peers. Recently, it refers to a system that enables two or more peers to share resources (based on peer's willingness) by using some appropriate communication system, without relying on central coordination. P2P applications offer a way to take advantage of different computer resources on the internet.

The term P2P became popular in 1999 with the emergence of Napster (Napster, 1999). Napster changed the way files were shared on the internet because it let peers download digital materials from each other, rather than from dedicated servers, increasing the system capacity and scalability. Napster, though, was still a centralized P2P application since it used a centralized indexing server.

By contrast, a second generation of P2P applications exemplified by Gnutella (Gnutella, 2000), Freenet (Clarke, 2000), KaZaA (KaZaA, 2002) and BitTorrent (BitTorrent, 2001) provided a truly decentralized architecture.

On the other side, MANET technology also draws great attention of worldwide researchers and scientists. Since the first appearance of wireless ad-hoc networks in the DARPA packet radio networks in the 1970s (Jubin, 1987), it became an interesting research object in the computer industry. In the 1990s, the concept of commercial ad-hoc networks arrived with notebook computers and other viable wireless communications equipment. At the same time, the idea of a collection of mobile nodes was proposed at several research conferences. The IEEE 802.11 subcommittee (IETF-manet) had adopted the term "ad-hoc networks" and the research community had started to look into the possibility of deploying ad-hoc networks in other areas of application. During the last couple of years tremendous improvements have been made in the research of ad hoc networks. Due to their ability to create and organize a network without any central management, MANETs are characterized as the art of networking without a network (Jiang, 2004). A MANET is a special kind of wireless ad-hoc network where there is no fixed backbone infrastructure. Due to this, the MANET can be flexible and rapidly deployed (Goldsmith, 2002).

MANETs and P2P networks are decentralized, autonomous and highly dynamic in a fairly similar way. In both cases, network nodes contribute to the overall system performance intermittently and unpredictably. Both rely on mechanisms to publish, share, index, cache, discover, and reorganize resources dynamically. They both aim at supporting efficient data distribution mechanisms such as multicasting. The strong functional similarities between MANETs and P2P networks, has recently sparked a new research thread based on a fundamental question: is it possible and beneficial to envision a convergence between P2P and MANETs? Such convergence would be appealing as it would allow distributed applications to run over spontaneous, infrastructure-less networks, adding a new dimension to the current Internet.

In this chapter we review MANETs and P2P systems from this new perspective. First we look at each technology independently, considering their commonalities and differences. Then we survey existing literature on the topic of P2P over MANETs. A critical discussion of the state-of-the-art will lead us to the identification of the key areas that require further research.

Table 1. Classification of P2P Systems

Centralized	Decentralized		Hybrid
	Unstructured	Structured	
Napster	Freenet	Chord	KazaA
SETI@Home	Gnutella	CAN	Morpheus
Bit Torrent		Pastry	Overnet/eDonkey2000
		Tapastry	
		Kademlia	

BACKGROUND

Peer-to-Peer (P2P) Systems

P2P can be broadly defined as a self organizing, decentralized distributed system that consists of potentially un-trusted, unreliable nodes with symmetric roles (Borg, 2003; Wu, 2006). P2P systems are designed for the sharing of resources such as CPU cycles, storage, files and data directly without any centralized control or hierarchical organization. P2P architectures are characterized by their ability to adapt to failures, handling and accommodating large number of nodes while maintaining acceptable connectivity and performance (A-Theotokis, 2004). P2P networks are mainly based on application-level overlays (P2P overlays) built over the physical network, enabling a range of P2P applications such as P2P file sharing and P2P streaming. They have their own routing protocols that permit computing devices to share information and resources directly, without dedicated servers (A-Theotokis, 2004). In P2P networks, each peer can act as server or as a client. Each peer can connect to one or more peer at any time to provide resource sharing.

The P2P paradigm has largely adopted a layered approach. A P2P overlay network built on top of the Internet provides a general purpose substrate that provides many common properties desired by distributed applications, such as self-organization, decentralization, diversity and redundancy. Such an overlay shields distributed application designers from the complexities of organizing and maintaining secure overlays, tolerating node failures, load balancing, and locating application objects (Wu, 2006).

P2P network architectures can be classified in three main categories: centralized, decentralized and hybrid (Franciscani, 2003; Minar, 2001; Wu, 2006). Table 1 shows some categories, derived from (Minar, 2001), whereas Table 2 shows the advantages and disadvantages of the P2P categories considering different properties. In a centralized P2P network, the topology is based on a central indexing server (or servers) that maintains a directory of the resources available on the peers and coordinates the interaction between peers. However, after receiving the information from the central server, peers communicate directly. The advantage of server based P2P systems is that they can be more easily managed and therefore fewer security issues are likely to be involved. However they are less fault-tolerant as some of the information is only held by the central server. If anything goes wrong with the server, the whole system goes down. The extensibility and scalability of these systems depends on the capacity of the server which is usually limited by its processing power capabilities.

In decentralized P2P networks, all peers act as both server and client equally. The communication is made through multiple unicasts, where peers forward messages on behalf of others peer. The decentralized topology is further divided into two categories i.e. Unstructured P2P overlays, is

Table 2. Advantages and Disadvantages of P2P Systems based on different properties

	Centralized	Decentralized		Hybrid
		Unstructured	Structured	
Manageable	Yes	No	Yes	No
Coherent	Yes	No	No	partially
Extensible	No	Yes	Yes	Yes
Fault-Tolerant	No	Yes	Yes	Yes
Secure	Yes	No	No	No
Lawsuit-Proof	No	Yes	Yes	Yes
Scalable	Depends on server	May be	Yes	Depends on super peers

formed when the overlay links are established arbitrarily by peers. Such networks can be easily constructed as a new peer that wants to join the network can copy existing links of another node and then form its own links over time and structured P2P overlays, which is precisely controlled and determined by some algorithm and were developed to improve the search mechanism to discover data. The advantage of decentralized systems is that they can be more easily extended and are more fault-tolerant than centralized P2P. However, security is a big issue in these systems.

Decentralized Structured P2P system can be managed more easily and are more scalable than their unstructured counterpart. For instance, DHT may be used to manage the P2P overlay (Dubnicki, 2004; Lua, 2005). In unstructured P2P systems scalability is difficult to measure due to the lack of deterministic overlay management mechanism. Increasing number of nodes supports the resource sharing system but also increases signaling overheads as most of unstructured P2P protocols periodically broadcast messages.

In hybrid (centralized + decentralized) P2P networks, there are some super-nodes or super-peers that play a more important role than others. Peers forward their queries to super-peers which communicate with each other in a decentralized manner. Hybrid P2P systems have same advantages as decentralized P2P systems but have difficulty in managing the peers and in this case the scalability depends on the actual number of super peers and their capabilities.

Mobile Ad-Hoc Networks (MANETs)

A MANET can be defined as a self organizing and autonomous system of mobile nodes that communicate over bandwidth-constrained wireless links. Since the nodes are mobile, the network topology may change rapidly and unpredictably over time. The network is decentralized, where all network activity including discovering the topology, routing functionality and message delivering is executed by the nodes themselves (Giordano, 2001; IETF-manet). MANETs represent an alternative to the traditional infrastructure model of mobile communication: they do not require a fixed communication infrastructure, but instead rely on wireless nodes to act as mobile routers.

As the network topology in such a network may keep changing randomly, the routing protocols used in traditional wired networks cannot be directly applied in MANETs due to their highly dynamic topology, absence of established infrastructure for centralized administration (e.g base station or access point), bandwidth constrained wireless links, and resource (energy-constrained nodes).

A variety of routing protocols for MANET have been proposed in the literature (Royer, 1999). These protocols can be classified into several

Figure 1. Classification of MANET routing protocols

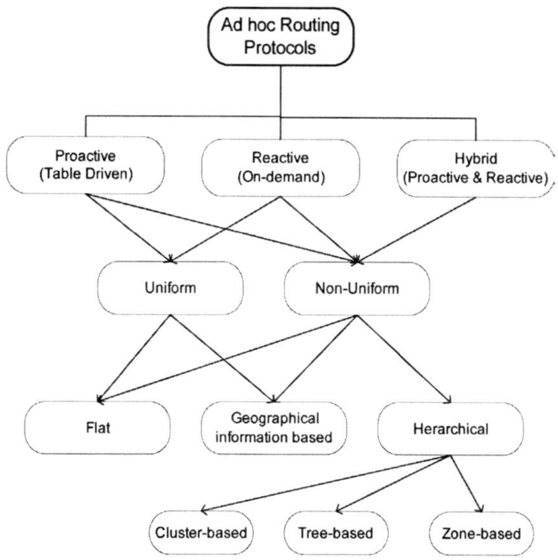

types based on different criteria. Ad hoc network routing protocols can be classified into three major categories based on the routing information update mechanism.

1. Proactive routing protocols (table-driven)
2. Reactive routing protocols (on-demand)
3. Hybrid routing protocols (proactive + reactive)

A detailed classification based on different criteria can be found in (Murthy, 2004). A general classification is shown in Figure 1.

P2P OVER MANET

Although conceptually and practically appealing, deploying P2P over MANETs is not straightforward since many of the classic assumptions that can be made for conventional P2P over wired networks do not hold any longer. In fact, the ad-hoc networks flexibility and convenience comes with new issues (Goldsmith, 2002). MANETs not only inherit the traditional problems of wireless networking such as bandwidth optimization, power control, and transmission quality enhancement (Giordano, 2001; IEEE-draft, 1999) but, due to the lack of a fixed infrastructure suffer from a number of complexities and design constraints that are specific to ad-hoc networks (Chiasserini, 1999; Corson, 1999). Examples are: dynamically changing network topology (Chlamtac, 1998), energy constraints, signaling overheads, variation in link and node capabilities (Vaidya, 2004), misbehaving nodes, and unreliable links.

A lot of research in the field of MANETs addresses the crucial problem of packet routing and service location issues - this is for example reflected by the IETF MANET working group. However, little effort has been made to investigate other services required by higher-layer protocols and applications. There is a magnitude of potentially useful applications for mobile ad-hoc networks, but application development for mobile ad-hoc networks is not easy. This issue is exacerbated in the case of P2P applications, which are highly distributed and generate traffic surges from and to unpredictable locations.

Comparison between P2P and MANET

Recently, the synergy between MANETs and P2P networks was recognized (Borg, 2003; Ding, 2004; Franciscani, 2003; Hu, 2003; Klemm, 2003; L. B. Oliveira, 2005; Schollmeier, 2002). By analyzing the synergy of MANET and P2P networks, we summarize the similarities between them as follows.

- *Dynamic network topology.* Nodes in P2P networks may randomly join and leave the network without any signs, which causes the network topology to change frequently with time. It is the same with MANETs. Furthermore, the mobility character of mobile nodes in MANETs can make network topology changes to be even more frequent.

- *Multi-hop connection.* Nodes in both types of networks connect with each other via multi-hop routing. Hop-to-hop connections in P2P network are typically via TCP links with physically unlimited range, while hop-to-hop connections in MANETs are via wireless links, which are limited by the radio transmission range.
- *Decentralized control (infrastructure less network).* Both P2P networks and MANETs have a decentralized structure. And there is no central administration point in the network
- *Node's multi-identity.* Nodes in P2P networks and MANETs act both as client and a server, as well as implementing routing functionality. Each node can route and transfer messages independently and their nodes have equivalent functionalities and capabilities. They can also provide resource downloading and communicate with each other directly.

Besides, unstructured P2P networks such as Gnutella share additional similarities with MANET, including (i) flooding-based routing protocols and (ii) limited scalability due to bandwidth consuming traffic from flooding.

Despite these similarities there are some major differences between both the networks that are described in Table 3. A detailed discussion on similarities and differences can be found in (Schollmeier, 2002).

Challenges in Converging P2P and MANETs

The similarities and differences between P2P and MANETs discussed in previous sub section lead to interesting although challenging research issues on P2P systems over MANETs. Such as,

- *Lack of Infrastructure.* There are some P2P protocols that rely on some infrastructure components in their designs. For example, CAN (Content Addressable Network) assigns node identifiers based on locations determined from static landmarks (Ratnasamy, 2001). These protocols cannot be adapted in ad hoc scenarios due to lack of infrastructure.
- *Frequent topology changes.* In MANETs due to node mobility and limited transmission range, network topology changes rapidly and unpredictably, which is not the case in internet. Therefore P2P Protocols integrated in MANETs need to update their overlay topology frequently which results in more overhead to maintain the overlay.
- *Node Churn.* The P2P protocol implemented in MANETs could suffer from the frequent node joining and leaving the network due to mobility or limited transmission range and may result in poor performance.
- *Bandwidth limitation.* The limited bandwidth in MANETs may limit the usability of P2P applications which are usually bandwidth-intensive.
- *Limited Battery Power.* Many P2P applications on the internet are not designed to use minimum message transmissions. In MANETs it is necessary for nodes to reduce the number of message transmissions while keeping the performance acceptable to conserve battery power.

- *Scalability.* In structured P2P routing protocols, nodes can scale up to millions. MANETs are likely to be much smaller in scale and have limited capacity than P2P overlays in the internet. Therefore it is difficult to say that such a large number of nodes can ever effectively be organized into one infrastructure-less network.

Table 3. Differences between P2P and MANETs

	P2P	MANET
Layer level	Application layer overlay	Network layer
Reason of creating a network	To provides service over logical infrastructure	To provides Connectivity over physical infrastructure
Purpose of communication b/w nodes	Search data do not necessarily communicate	Communicate with other users
Communication pattern	Multiple unicast with virtual broadcasting	Physical broadcasting
Nodes behavior	Static	Mobile
Communication link	Direct	Indirect
Relay nodes	Not required	Required intermediate nodes
Connection medium	Fixed wired	Wireless
Connection establishment	Hop by Hop via TCP links, Whereas the single hop path length is not physically limited	Hop by Hop via radio links, which are thus limited by the radio transmission
Connection maintenance	Comparatively Easy	Difficult
Node location	Anywhere on internet	Limited area
Network Structure	Logical structure apart from physical structure	Logical structure corresponds to physical Structure
Physical position of the node	Difficult to find	Can be roughly estimated
Routing	Reactive only (reliable algorithms not implemented yet) stops when TTL field is 0	Proactive, Reactive and Hybrid (reliable algorithms exist), Stops when destination is found
Network Topology	Changes less frequently	Changes more frequently due to mobility
Network Size	Large networks (millions of nodes)	Usually small networks (few nodes)
Network connectivity	Not effected by nodes joining or leaving	Effected by nodes joining and leaving
Bandwidth	Large bandwidth	Small bandwidth
Available Resources	Practical unlimited	limited

- *Addressing.* No specific addressing architecture has been standardized for MANETs. Due to node churn in MANETs, it is reasonable to assume that nodes will have changing IP addresses over the time. This may be very challenging for structured P2P protocols that store logical to physical addresses. In some cases node Identifiers are assigned as hashed IP addressed and therefore node identifiers will be continuously changing leading to inconsistency issues.
- *Flooding.* In MANETs many routing protocols are based on a flooding approach. In the case of Unstructured P2P protocols, which also rely on information flooding, performance degradation will result from signaling overheads.

Both P2P and MANETs have recently become popular research areas due to the wide deployment of P2P applications over the Internet and rapid progress of wireless communication, but little research has been done on the convergence of these two overlay network technologies. Many decentralized applications have been built upon P2P overlays and these same ideas may be useful in developing higher level services in an ad hoc setting.

A number of publications have been devoted to bringing P2P protocols into ad hoc networks. Many of those publications are mainly conceptual, presenting architectural proposals but not evaluating them (Conti, 2004; Datta, 2003; Ding, 2004; Eberspacher, 2004; Hu, 2003; Yan, 2004). A first approach for implementing P2P file sharing for MANETs constitutes the protocol 7DS (Papadopouli, 2001), which focus on single-hop settings. 7DS uses local broadcast transmissions for sharing Web documents among peers in order to enable online web browsing without connecting to the Internet.

PROEM (Kortuem, 2001) aspires to provide a platform to develop P2P applications for MANETs. Its messages are based on TCP, UDP or HTTP and employ XML for the representation of the messages. However it also does not provide any mechanism to adapt its virtual to the physical topology. Thus its feasibility in larger MANET scenarios is doubtful.

Lawrence proposes an approach to use JADE-LEAP in an ad hoc environment (Lawrence, 2002). The author proposes the removal of some mandatory components and adds a Discovery Agent to support ad hoc networks. Unfortunately this does not address the question of routing, which is one of the most critical questions in a MANET scenario. Later on, the concept of Passive Distributed Indexing (PDI) (Lindemann, 2002) and Optimized Routing Independent Overlay Network (ORION) were introduced (Klemm, 2003). Both of them focused on building a file sharing system on MANET, based on flooding mechanisms to discover files on demand. A protocol named MPP emphasized on communication between application layer and network layer to deploy P2P systems effectively in mobile environments (Schollmeier, 2003). This protocol establishes an intermediate layer which helps passing queries from application layer to network layer. Although this protocol is an important step towards an efficient P2P application on MANET, it still relies on flooding the network with query lookup. The use of flooding quickly becomes the performance bottleneck as the size of MANET grows.

To deploy a P2P system on MANET, maintaining the overlay connectivity is the most challenging issue due to node mobility. If we apply existing Internet P2P solutions to MANET, the static overlay structure will cause extremely high connection overheads and inaccurate search for a file because of node mobility. In addition, flat static addressing mechanisms are assumed in MANET. As a consequence, routing is done on demand by flooding or, otherwise, routing tables have to be maintained dynamically. Intuitively, routing or searching a file by flooding yields extremely high cost. Therefore, the flooding approach is only applicable to small-scale MANET. Scalability is a critical challenge for deploying a P2P system to large MANET with thousands of nodes(Cheng, 2005). In next section we will be discussing the various design approaches for the convergence of P2P and MANET.

Figure 2. Taxonomy of P2P over MANET design Approaches

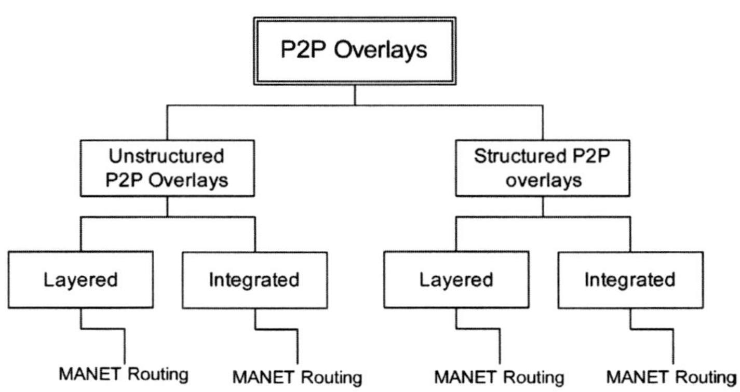

P2P OVER MANET DESIGN APPROACHES

P2P can be mapped on the top of MANET in different ways. Various design approaches can be adopted, based on the classification of decentralized P2P overlays, as shown in Figure 2.

As discussed previously, P2P overlays are classified into unstructured P2P and structured P2P overlays. In addition, the nodes in MANETs have to act as end hosts and routers and are need to involve in supporting the P2P overlay. The P2P overlay can be implemented either at the network layer or at application layer. If a P2P protocol is layered on top of existing MANETs routing protocol the corresponding design approach is termed layered approach. Figure 3 shows a general scenario of P2P application layered on the top of a MANETs.

If a protocol is embedded with a MANETs routing protocol at the network layer then this approach is called integrated or cross-layer approach, which are used interchangeably. Figure 4 shows a simple example of P2P application integrated with network layer of MANET.

UNSTRUCTURED P2P OVERLAY OVER MANET

In (L.B. Oliveira, 2003), the performance of ad hoc routing protocol under an unstructured P2P application in MANETs is observed. The authors studied the performance of three existing MANETs routing protocols, i.e. AODV (Perkins, 1999), DSR (Johnson, 1996) and DSDV (Perkins, 1994), when Gnutella-like (Gnutella, 2000) application was layered on top. Their results show that each protocol analyzed, performed well under some conditions and according to some metrics but had drawbacks in others; particularly the delivery rate was quite good but with more number of hops and latency. Furthermore, it was observed that the performance of these protocols was different from those observed in previous studies (Broch, 1998; Das, 2000) for ad hoc network. Later on, different unstructured P2P protocols (Franciscani,

Figure 3. A P2P application Layered over MANET(L. B. Oliveira, 2005)

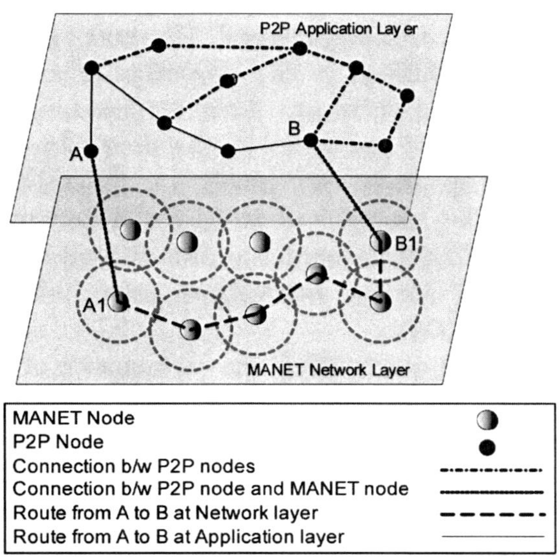

Figure 4. A P2P application integrated with MANET(L. B. Oliveira, 2005)

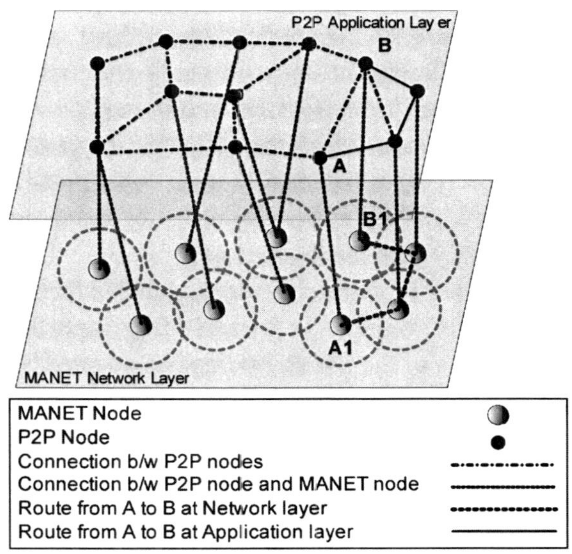

2005; Heer. T, 2006; Lau. G, 2005) were introduced but their performance over MANET has not yet been evaluated.

Optimized Routing Independent Overlay Network (ORION) (Klemm et al., 2003) is also based on the integrated approach. ORION proposes a P2P file sharing system for MANET. The basis of ORION is AODV (Perkins, 1999) and the Simple Multicast and Broadcast protocol for MANET. ORION combines application-layer query processing with the network layer process of the route discovery from AODV. The results show that ORION significantly increases the search accuracy as the number of nodes grows and reduces overheads from searching. In layered approach the search accuracy was always lower than that of integrated approach due to more overheads at large network size.

In (Conti, 2005) a cross-layer Gnutella-like P2P application is proposed for MANET and its performance over MANET is investigated. The comparison with a similar layered design (L.B. Oliveira, 2003) is made. Results show that a straightforward implementation of the protocol is not satisfactory under the point of view of signaling overheads and average overlay connectivity.

A similar work is done in (Rajagopalan, 2006) who propose a cross-layer decentralized BitTorrent for MANET. A comparison is made with a modified version of BitTorrent (BitTorrent, 2001) over MANET. The results observed were analogous to those obtained in the cross-layered Gnutella study mentioned above, which shows that the integrated approach is more resistant to node mobility and network partitions and performs better and more consistently than the layered one.

In summary, the key factor responsible for the worst performance of the layered design is not the inefficiency of the search mechanism adapted by unstructured P2P protocol but due to the overheads generated in maintaining the overlay topology whose neighbor connections are not adapted at the dynamic physical topology.

The layered design not only the limited scalability of unstructured systems but also the limited scalability of flooding based ad hoc routing protocols as both system strongly rely on information flooding. Additionally, this approach does not take cross-layer optimization techniques into consideration so both layers flood the network without being aware of each other (Olaf, 2006). The results of the study indicate that the integrated unstructured design has significantly lower overhead compared to the layered design while achieving better performance according to application-specific metrics.

Structured P2P Overlay Over MANET

In (Cramer, 2006) a Structured P2P protocol, chord (I.Stoica, 2001) is layered on the top of three existing MANET routing protocols, AODV, DSR, and OLSR (Clausen, 2003). In the majority of all tested scenarios, which varied in network size, node mobility and application load, it was found that Chord's ability to consistently resolve lookups was significantly impaired. The results show that chord is unable to perform well under high mobility.

Similar results were observed in (Hu, 2003; Pucha, 2004) when slightly modified version of pastry (A.Rowstron, 2001) was layered on the DSR ad-hoc routing Protocol. The results show that it is difficult to take advantage of many optimization techniques from the interactions between DHT and ad hoc routing protocol with layered approach. For example, it is difficult for the routing structures of the DHT and the route cache of DSR to coordinate with each other to optimally discover and maintain source routes (Pucha, 2004).

In (Delmastro, 2005) the performance of a structured P2P system (Freepastry), running on a real ad hoc network, is evaluated. The results observed were almost same as in (Cramer, 2006; Pucha, 2004). From the implementation of FreePastry (Freepastry) it was observed that carrying out the overlay management through a high number of remote connections, introduces a heavy overhead on the ad hoc networks, reducing the overall system performances. In order to improve the system, each Pastry node should become aware of the network topology, maintaining locally a correspondence between physical and logical address spaces.

Hu and others (Hu, 2003) proposed DPSR, a network layer unicast routing protocol, which uses a cross-layer approach for implementing the structured P2P overlay Pastry as a network-layer routing protocol, and integrates it with the MANET routing protocol DSR. No performance evaluation of DPSR is available though. Ekta, which is based on DPSR, is a DHT for MANETs. Ekta and DPSR share the essence of tightly integrating Pastry and DSR at the network layer. However, unlike DPSR, Ekta does not use DHT for unicast routing. By the optimizing the cross-layer techniques, integrating route cache of DSR and DHT protocol into one structure can help to prevent overheads and increase the overall efficiency of the system in Mobile ad hoc environment. IP addresses are directly mapped to pastry node IDs, enabling efficient routing to particular nodes. Ekta uses eavesdropping to learn about paths to close nodes. This knowledge is used to find close nodes within certain ID spaces for proximity routing and thus optimize the performance of Pastry. However this work does not address the issues like network splits, mergers, and the existence of multiple DHTs and mostly focuses on the methods to improve the efficiency of the routing techniques. The simulation scenario which was chosen makes network splits impossible, and therefore does not allow making statements about the performance of Ekta in these extreme but common situations (Heer, 2006).

CrossROAD (Delmastro, 2005) is also based on Pastry. In CrossROAD, the performance of routing protocols and middleware platforms have been evaluated in terms of overhead and reconfiguration delays, in case of mobility scenarios and network topology changes. In addition, a real prototype of an optimized P2P system (CrossROAD) that exploits the cross-layer interactions with a proactive routing protocol (OLSR) has been presented. However, the evaluation is made on a network of only eight nodes, so it is still doubtful that this will work on large MANETs.

In (Cramer, 2005b) the performance of Proximity Neighbor Selection (PNS) for DHT (Distributed Hash Table) over MANET have been evaluated. The authors used Chord as their DHT for simulations. It was observed that the performance of PNS is strongly dependent on network density, degrading with increasing density. A different strategy than PNS for adapting a virtual network to its underlying network is proximity identifier selection (PIS) (Gummadi, 2003) where a node's overlay address is chosen according to the node's current location (Du, 2004; Zahn, 2004).

Iterative Successor Pointer Rewiring Protocol (ISPRP) (Cramer, 2005a, 2005c) is a P2P routing protocol for MANET based on Chord. ISPRP is a self-stabilizing protocol that creates the basic ring structure of a Chord overlay for overlay-based routing at the network layer of an ad-hoc network. It does not employ flooding for discovering the routes to the successor nodes. However, it does not handle node mobility at present. Recently, Zupeng Li and others in (Zupeng Li, 2006) have proposed a new routing protocol named Peer Computing based Dynamic Source Routing (PDSR) for MANET which integrates P2P computing with MANETs routing algorithms. PDSR adopts a node naming mechanism similar to Chord. The performance of PDSR over MANET is not yet evaluated.

In summary, virtually in every scenario the integrated architecture appear to perform better than the layered one. The integration of the P2P structured layer and ad hoc routing layers is crucial to achieve efficiency, as it significantly reduces the duplication of routing information and maintenance efforts. On the other hand, unstructured P2P does not have precise control over the overlay topology. As a result of lacking structure, the query protocols in unstructured P2P networks in general revert to flooding that limits scalability. A summary of the advantages and disadvantages of the approaches discussed above is given in Table 4.

FUTURE TRENDS

This chapter focuses on the interesting developments surrounding P2P protocols over MANETs. It is observed that most work involves Gnutella as an unstructured P2P overlay and CHORD and Pastry as a structured P2P overlay, in both layered and integrated approaches over MANETs. The performance of these protocols were observed along with few new P2P protocols for MANETs but it would be interesting to see how other structured P2Pand unstructured P2P protocols will perform over MANETs. On the other hand if we look at the literature we observe that only few of the available MANET routing protocols have been investigated in relation to P2P, including mainly AODV, DSR, DSDV and OLSR. The feasibility of other MANET routing protocols under P2P applications is still not known. Along with that, the protocols mentioned in previous sections are only applicable in some specific scenarios or are feasible under some metrics. It is unclear which of the P2P overlay, structured or unstructured, would be more efficient for supporting distributed application over MANET. Which design approach would support scalable unicast in MANETs? Would it will be possible to apply to the case of MANETs all the incentive-based and security techniques which are used when P2P systems run over internet?

CONCLUSION

In this chapter we reviewed the convergence of P2P over MANETs. First we discussed each technology separately, providing background as to where and when they emerged. We also discussed the classification of P2P overlays based different topologies along with examples. Then we made the comparison between those topologies in respect of different properties, giving an insight on which topology works well under which scenarios. Furthermore we discussed the similarities and dissimilarities between P2P and MANET technologies from various perspectives. It was interesting to see that although both work under almost the same scenarios, but have many differences as well. From this point it will be easier to understand that in order to implement P2P protocols over MANET we need to mitigate those differences to get an optimal system for MANET. Our review identifies areas that require further investigation, indicating that a promising way to tackle the problem is to build P2P-aware MANETs, rather than attempting a layered approach.

Table 4. Summarized advantages and disadvantages of different architectures

	Architectures			
	Unstructured		Structured	
	Layered	*Integrated*	*Layered*	*Integrated*
Manageable	No	No	partial	Yes
Scalable	No	Medium	Limited	Yes
Handle mobility	No	Yes	At low mobility	Yes
Overheads	High	Low	Medium	Low
Implementation	Easy	Partially difficult	Partially difficult	Difficult

REFERENCES

A-Theotokis, S., & Spinellis, D. (2004). A Survey of Peer-to-Peer Content Distribution Technologies. *ACM Computing Surveys*, *36*(4), 335–371. doi:10.1145/1041680.1041681

BitTorrent. (2001). BitTorrent webpage. Retrieved from http://bitconjurer.org/BitTorrent

Borg, J. (2003). *A comparative study of ad hoc & peer to peer networks*. Unpublished Master's thesis, University College London.

Broch, J., Maltz, D. A., Johnson, D. B., Hu, Y. C., & Jetcheva, J. (1998). *A performance comparison of multi-hop wireless ad hoc network routing protocols*. Paper presented at the 4th annual ACM/IEEE international conference on Mobile computing and networking Dallas, Texas, United States.

Cheng, J., Wang, C.-Y., Hoh, C. C., & Hwan, R.-H. (2005). *A Scalable Location-Aware Peer-to-Peer System for MANET.* Paper presented at the Mobile Technology, Applications and Systems, 2005 2nd International Conference.

Chiasserini, C. F., & Rao, R. R. (1999). *Pulsed battery discharge in communication devices*. Paper presented at the 5th annual ACM/IEEE international conference on Mobile computing and networking.

Chlamtac, I., & Redi, J. (1998). Mobile Computing: Challenges and Opportunities. In Hemmendinger, A. R. D., & Reilly, E. (Eds.), *Encyclopedia of Computer Science* (4th ed.). International Thomson Publishing.

Clarke, I., Sandberg, O., Wiley, B., & Hong, T. W. (2000). *Freenet: A Distributed Anonymous Information Storage and Retrieval System*. Paper presented at the ICSI Workshop on Design Issues in Anonymity and Unobservability.

Clausen, T., & Jacquet, P. (2003). Optimized Link-State Routing Protocol. *IETF*, *RFC-3626*, 2003.

Conti, M., Gregori, E., & Turi, G. (2004). *Towards Scalable P2P Computing for Mobile Ad Hoc Networks*. Paper presented at the Second IEEE Annual Conference on Pervasive Computing and Communications Workshops (PERCOMW '04), Orlando, FL, USA.

Conti, M., Gregori, E., & Turi, G. (2005). *A Cross-Layer Optimization of Gnutella for Mobile Ad hoc Networks*. Paper presented at the 6th ACM international symposium on Mobile ad hoc networking and computing, Urbana-Champaign, IL, USA

Corson, M. S., Macker, J. P., & Cirnicione, G. H. (1999). Internet-Based Mobile Ad Hoc Networking. *IEEE Internet Computing*, *3*, 63–70. doi:10.1109/4236.780962

Cramer, C., & Fuhrmann, T. (2005a). *ISPRP: A Message-Efficient Protocol for Initializing P2P Networks*. Paper presented at the 24th IEEE International, Performance, Computing, and Communications Conference (IPCCC), Phoenix, AZ, USA.

Cramer, C., & Fuhrmann, T. (2005b). *Proximity Neighbor Selection for a DHT in Wireless Multi-Hop Networks*. Paper presented at the Fifth IEEE International Conference on Peer-to-Peer Computing (P2P'05)

Cramer, C., & Fuhrmann, T. (2005c). *Self-Stabilizing Ring Networks on Connected Graphs*. University of Karlsruhe, Germanyo.

Cramer, C., & Fuhrmann, T. (2006). *Performance evaluation of chord in mobile ad hoc networks*. Paper presented at the 1st international workshop on Decentralized resource sharing in mobile computing and networking, Los Angeles, California.

Das, S. R., Perkins, C. E., & Royer, E. M. (2000). *Performance comparison of two on-demand routing protocols for ad hoc networks*. Paper presented at IEEE INFOCOM, Nineteenth Annual Joint Conference of the IEEE Computer and Communications Societies.

Datta, A. (2003). *MobiGrid: Peer-to-Peer Overlay and Mobile Ad-Hoc Network Rendezvous - a Data Management Perspective.* Paper presented at CAiSE 2003 Doctoral Symposium, Klagenfurt, Austria.

Delmastro, F. (2005). *From Pastry to CrossROAD: CROSS-layer Ring Overlay for AD hoc networks.* Paper presented at the Workshop on Mobile Peer-to-Peer Computing - MP2P '05, Kauai, Hawaii.

Ding, G., & Bhargava, B. (2004). *Peer-to-peer File-sharing over Mobile Ad hoc Networks.* Paper presented at the Second IEEE Annual Conference on Pervasive Computing and Communications Workshops (PERCOMW '04), Orlando, FL, USA.

Du, S., & Khan, A. PalChaudhuri, S., Post, A., Saha, A. K., Druschel, P., et al. (2004). *Self-Organizing Hierarchical Routing for Scalable Ad Hoc Networking.* Department of Computer Science, Rice University, Houston, TX, USA.

Dubnicki, C., Ungureanu, C., & Kilian, W. *FPN: A Distributed Hash Table for Commercial Applications* Paper presented at the 13th IEEE International Symposium on High Performance Distributed Computing (HPDC-13 '04)

Eberspacher, J., Schollmeier, R., Zols, S., & Kunzmann, G. (2004). *Structured P2P Networks in Mobile and Fixed Environments.* Paper presented at the International Working Conference on Performance Modeling and Evaluation of Heterogeneous Networks (HET-NETs '04), Ilkley, West Yorkshire.

Franciscani, F. P., Vasconcelos, M. A., Couto, R. P., & Loureiro, A. A. F. (2003). *Peer-to-Peer over Ad-hoc Networks: (Re)Configuration Algorithms.* Paper presented at the International Parallel and Distributed Processing Symposium (IPDPS'03) Nice, France

Franciscani, F. P., Vasconcelos, M. A., Couto, R. P., & Loureiro, A. A. F. (2005). (Re)configuration algorithms for peer-to-peer over ad hoc networks. *Journal of Parallel and Distributed Computing, 65*(2), 234–245. doi:10.1016/j.jpdc.2004.09.007

FreeNet. (2000). Freenet Homepage. Retrieved from http://freenetproject.org/.

Freepastry. Freepastry Homepage. Retrieved from http://freepastry.rice.edu/FreePastry/

Giordano, S. (2001). Mobile Ad-hoc Networks. In Stojmenovic, I. (Ed.), *Handbook of Wireless Network and Mobile Computing.* John Wiley & Sons.

Gnutella. (2000). The Gnutella Protocol specification. Retrieved from http://dss.clip2.com/GnutellaProtocol04.pdf.

Goldsmith, J., & Wicker, S. (2002). Design challenges for energy-constrained ad hoc wireless networks. *IEEE Wireless Communications Magazine, 9*, 8–27. doi:10.1109/MWC.2002.1028874

Gummadi, K., Gummadi, R., Gribble, S., Ratnasamy, S., Shenker, S., & Stoica, I. (2003). *The Impact of DHT Routing Geometry on Resilience and Proximity.* Paper presented at the ACM SIGCOMM 2003 conference on Applications, technologies, architectures, and protocols for computer communications Karlsruhe, Germany.

Heer, T. S, G., S, R., & K, W. (2006). *Adapting distributed hash tables for mobile ad hoc networks.* Paper presented at the Pervasive Computing and Communications Workshops (PerCom Workshops 2006), Fourth Annual IEEE International Conference.

Heer. T, G. S., Rieche. S., & Wehrle. K. (2006). *Adapting distributed hash tables for mobile ad hoc networks.* Paper presented at the Pervasive Computing and Communications Workshops (PerCom Workshops 2006), Fourth Annual IEEE International Conference.

Hu, Y. C., Das, S. M., & Pucha, H. (2003). *Exploiting the Synergy between Peer-to- Peer and Mobile Ad Hoc Networks.* Paper presented at the HotOS IX Workshop, Lihue, HI, USA.

IEEE-draft. (1999). *IEEE p802.11/D10* o. IETF-manet. IETF MANET working Group. Retrieved from http://www.ietf.org/html.charters/manet-charter.html

Jiang, S., Liu, Y., Jiang, Y., & Yin, Q. (2004). Provisioning of adaptability to variable topologies for routing schemes in MANETs. *IEEE Journal on Selected Areas in Communications*, *22*(7), 1347–1356. doi:10.1109/JSAC.2004.829352

Johnson, D. B., & Maltz, D. A. (1996). Dynamic Source Routing (DSR) in AdHoc Wireless Networks. In Imielinski, K. (Ed.), *Mobile Computing*. Kluwer Academic Publishers. doi:10.1007/978-0-585-29603-6_5

Jubin, J., & Tornow, J. D. (1987). *The DARPA Packet Radio Network Protocols*. Paper presented at the Proc. of the IEEE.

KaZaA. (2002). KaZaA Homepage. Retrieved from http://www.kazaa.com.

Klemm, A., Lindemann, C., & Waldhorst, O. (2003). *A special-purpose peer-to-peer file sharing system for mobile ad hoc networks.* Paper presented at the IEEE Vehicular Technology Conference (VTC 2003).

Kortuem, G., & Schneider, J. (2001). *An Application Platform for Mobile Ad-hoc Networks.* Paper presented at the Workshop on Application Models and Programming Tools for Ubiquitous Computing (UBICOMP 2001), Atlanta, Georgia.

Lau. G, J. M., & Ravindran. G. (2005). *RAON: A P2P Network for MANET.* Paper presented at the Wireless and Optical Communications Networks (WOCN 2005) Second IFIP International Conference.

Lawrence, J. (2002). *LEAP into Ad-Hoc Networks.* Paper presented at the Workshop on Ubiquitous Agents on Embedded, Wearable, and Mobile Devices, Bologna, Italy.

Lindemann, C., & Waldhorst, O. (2002). *A Distributed Search Service for Peer-to-Peer File Sharing in Mobile Applications.* Paper presented at the 2nd IEEE Conference on Peer-to-Peer Computing (P2P 2002), Linkoping, Sweden.

Lua, E. K., Crowcroft, J., Pias, M., Sharma, R., & Lim, S. (2005). *A Survey and Comparison of Peer-to-Peer Overlay Network Schemes*. IEEE Communications Surveys & Tutorials.

Minar, N. (2001). *Distributed Systems Topologies*. Paper presented at the O'Reilly P2P and Web Services Conference

Murthy, C. S. R., & Manoj, B. S. (2004). *Ad Hoc wireless Networks, Architecture and Protocols*. Prentice Hall.

Napster. (1999). Napster webpage. Retrieved from http:// www.napster.com

Olaf, L., Stefan, G., & Klaus, W. (2006). *Towards Scalable Mobility in Distributed Hash Tables*. Paper presented at the Sixth IEEE International Conference on Peer-to-Peer Computing.

Oliveira, L. B., Siqueira, I. G., & Loureiro, A. A. (2003). *Evaluation of ad-hoc routing protocols under a peer-to-peer application*. Paper presented at IEEE Wireless Communications and Networking (WCNC 2003)

Oliveira, L. B., Siqueira, I. G., & Loureiro, A. A. F. (2005). On the performance of ad hoc routing protocols under a peer-to-peer Application. *Journal of Parallel and Distributed Computing*, *65*(11). doi:10.1016/j.jpdc.2005.05.023

Papadopouli, M., & Schulzrinne, H. (2001). *Effects of Power Conservation, Wireless Coverage and Cooperation on Data Dissemination among Mobile Devices*. Paper presented at the ACM Symposium on Mobile Ad Hoc Networking and Computing (MOBIHOC 2001), Long Beach, CA.

Perkins, C. E., & Bhagwat, P. (1994). *Highly dynamic destination-sequenced distance-vector routing (DSDV) for mobile computers.* Paper presented at the ACM SIGCOMM's conference on Communications architectures, protocols and applications London, UK.

Perkins, C. E., & Royer, E. M. (1999). *Ad hoc on-demand distance vector routing(AODV).* Paper presented at the 2nd IEEE Workshop on Mobile Computing Systems and Applications, New Orleans, LA.

Pucha, H., Das, S. M., & Hu, Y. C. (2004). *Ekta: An Efficient DHT Substrate for Distributed Applications in Mobile Ad Hoc Networks.* Paper presented at the 6th IEEE Workshop on Mobile Computing Systems and Applications (WMCSA 2004), English Lake District, UK.

Rajagopalan, S. S., & Chien-Chung. (2006). *A Cross-layer Decentralized BitTorrent for Mobile Ad hoc Networks.* Paper presented at the Mobile and Ubiquitous Systems - Workshops, 2006. 3rd Annual International Conference.

Ratnasamy, P. Francis, M.Handley, R.Karp, & S.Schenker. (2001). *CAN: A scalable content-addressable network.* Paper presented at ACM SIGCOMM.

Rowstron, A., & Druschel, P. (2001). *Pastry: scalable, distributed object location and routing for large-scale peer-to-peer systems.* Paper presented at Middleware, Heidelberg, Germany.

Royer, E. M., & Toh, C.-K. (1999). A Review of Current Routing Protocols Ad Hoc Mobile Wireless Networks. *IEEE Personal Communications*, *6*, 46–55. doi:10.1109/98.760423

Schollmeier, R., Gruber, I., & Finkenzeller, M. (2002). *Routing in Mobile Ad Hoc and Peer-to-Peer Networks, A Comparison.* Paper presented at the Revised Papers from the NETWORKING 2002 Workshops on Web Engineering and Peer-to-Peer Computing.

Schollmeier, R., Gruber, I., & Niethammer, F. (2003). *Protocol for peer-to-peer networking in mobile environments.* Paper presented at the IEEE 12th International Conference on Computer Communications and Networks (ICCCN).

Stoica, I., Morris, R., Karger, D., Kaashoek, M. F., & Balakrishnan, H. (2001). *Chord: A Scalable peer-to-peer lookup service for internet applications.* Paper presented at the ACM SIGCOM, California, USA.

Vaidya, N. H. (2004). *Mobile Ad Hoc Networks: Routing, MAC and Transport Issues.* Paper presented at the INFOCOM 2004 (IEEE International Conference on Computer Communication).

Wu, J. (Ed.). (2006). *Handbook on Theoritical and Algorithmic Aspects of Sensor, Ad Hoc Wireless, and Peer-to-Peer Networks.* Auerbach Publications.

Yan, L., Sere, K., & Zhou, X. (2004). *Towards an Integrated Architecture for Peer-to- Peer and Ad Hoc Overlay Network Applications.* Paper presented at the 10th IEEE International Workshop on Future Trends of Distributed Computing Systems (FTDCS '04), Suzhou, China.

Zahn, T., Winter, R., & Schiller, J. (2004). *Simple, efficient peer-to-peer overlay clustering in mobile, ad hoc networks.* Paper presented at the 12th IEEE International Conference on Networks (ICON 2004), Singapore.

Zupeng Li, X. Y. Peiyang Yao, & Jinnan Huang. (2006). *Implementation of P2P Computing in Design of MANET Routing Protocol.* Paper presented at the First International Multi-Symposiums on Computer and Computational Sciences (IMSCCS'06), Washington, DC, USA

KEY TERMS AND DEFINITIONS

Hybrid Routing: This type of protocols combines the advantages of proactive and of reactive routing. The routing is initially established with some proactively prospected routes and then serves the demand from additionally activated nodes through reactive flooding.

Mobile Ad Hoc Networks: is a kind of wireless ad-hoc network, and is a self-configuring network of mobile routers (and associated hosts) connected by wireless links.

Overlay Network: is a computer network which is built virtually or logically on top of another network.

Peer-to-Peer (P2P): can be broadly defined as self organizing; decentralized distributed systems that consist of potentially untrusted, unreliable nodes with symmetric roles.

Proactive Routing: This type of protocols maintains fresh lists of destinations and their routes by periodically distributing routing tables throughout the network.

Reactive Routing: This type of protocols finds a route on demand by flooding the network with Route Request packets.

Structured P2P Overlays: Structured P2P network employ a globally consistent protocol to ensure that any node can efficiently route a search to some peer that has the desired file

Unstructured P2P Overlays: is formed when the overlay links are established arbitrarily. Such networks can be easily constructed as a new peer that wants to join the network can copy existing links of another node and then form its own links over time.

Chapter 12
Middleware Technologies for Ubiquitous Computing

Noha Ibrahim
Université de Lyon, France

Frédéric Le Mouël
Université de Lyon, France

Stéphane Frénot
Université de Lyon, France

ABSTRACT

Middleware handles many important functionalities for ubiquitous computing. The authors distinguish various middleware technologies providing key elements for all applications' requirements such as discovery, adaptation/composition, context management, and management of ubiquitous applications. In this chapter, they propose a classification for some of the most employed ubiquitous middleware. The classification was established upon the challenges raised by ubiquitous computing – effective use of smart spaces, invisibility, and localized scalability - and upon how the various ubiquitous middleware respond to them in terms of interoperability, discoverability, location transparency, adaptability, context awareness, scalability, security, and autonomous management. This classification shows that if many middleware are mature enough and offer specific functionalities respecting the properties of ubiquity, a real lack is noticed in having an interoperable, autonomous and scalable middleware for the execution of ubiquitous applications. The development of the service-oriented paradigm, the semantics, the Web middleware, and the ambient intelligence shows the new trend the middleware research field is engaged in.

INTRODUCTION

Middleware are enabling technologies for the development, execution and interaction of applications. These software layers are standing between the operating systems and applications. They have evolved from simple beginnings — hiding network details from applications — into sophisticated systems that handle many important functionalities for distributed applications — providing support for distribution, heterogeneity and mobility. The evolution of middleware has been influenced by numerous developments and standards efforts. Various middleware paradigms were defined. We

DOI: 10.4018/978-1-60566-250-3.ch012

cite the Remote Procedure Call middleware (RPC), the Message Oriented Middleware middleware (MOM), the Object Request Broker middleware (ORB), and the Service-Oriented Architecture middleware (SOA).

The term middleware first appeared in the late 1980s to describe network connection management software, but did not come into widespread use until the mid 1990s, when network technology had achieved sufficient penetration and visibility. By that time, middleware had evolved into a much richer set of paradigms and services, offered to build distributed applications more easily and in a more manageable way. The term was associated mainly with relational databases for many practitioners in the business world through the early 1990s. By the mid-1990s this was no longer the case. Concepts similar to today's middleware previously went under the names of network operating systems, distributed operating systems and distributed computing environments. Systems such as Apollo's Network Computing Architecture (NCA), Sun's RPC standard, and the Open Software Foundation's Distributed Computing Environment (DCE) are all examples of successful RPC-oriented middleware that has been used for significant production applications. At the same time these systems were being produced in the 1980s and early 1990s, significant research was occurring in the area of distributed objects. Distributed objects represented the confluence of two key areas of information technology: distributed systems and object-oriented design and programming. As a result of research on these and other distributed object systems, combined with an evolution of the Transaction Processing monitor (TP monitor) into the concept of an "object monitor," Common Object Request Broker Architecture (CORBA) was created. The CORBA system is built on top of ORB, which encompasses the entire communication infrastructure necessary to identify and locate objects, handle connection management, and deliver data. At the same time, a specific class of middleware, MOM, that operates on the principles of message passing or message queuing appeared. The MOM middleware unlike RPC and object-orientation is an asynchronous form of communication, i.e. the sender does not block waiting for the recipient to participate in the exchange. While the evolution of distributed objects in the 1990s was marked by significant efforts to establish standards such as CORBA, the evolution of messaging oriented middleware was practically devoid of standards efforts.

In the late 1990s, the growing popularity of Java, the explosive growth of the World Wide Web (WWW), and lessons learned from CORBA and messaging were all combined to form the Java 2 Enterprise Edition (J2EE), a comprehensive component middleware platform. J2EE provides support for numerous application types, including distributed objects, components, web-based applications, and messaging systems. Throughout the development of CORBA, J2EE, and proprietary messaging systems, Microsoft was busy developing its Distributed Component Object Model (DCOM).

In 1999 and 2000, the Web's influence on middleware started to become readily apparent with the publication of the initial version of Simple Object Access Protocol (SOAP). Initially dubbed SOAP was the result of trying to create a system-agnostic protocol that could be used over the Web and yet still interface easily to non-SOAP middleware, including CORBA, DCOM, J2EE, and messaging middleware systems. Given the 1990s "middleware wars" between J2EE, CORBA, and DCOM, and between RPC and messaging, the unified support for SOAP was indeed groundbreaking. At the time, it seemed that SOAP, Web Services and Service-Oriented Computing in general might finally provide the basis for broad industry agreement on middleware standards.

BACKGROUND

If middleware were designed to help manage the complexity and heterogeneity inherent in distributed systems, one can imagine the new role middleware has to play in order to respect the evolution from distributed and mobile computing to ubiquitous one. Ubiquitous computing is all about everywhere and anytime computing. It spreads intelligence and connectivity to more or less everything. So conceptually, machines, clothing, tools, appliances, homes and even the human body will be embedded with chips to connect to an infinite network of other devices. This creates an environment where the connectivity of devices is embedded in such a way that it is unobtrusive and always available.

"*The most profound technologies are those that disappear. They weave themselves into the fabric of everyday life until they are indistinguishable from it*" - so began Mark Weiser's (1991) paper (p.66) that described his vision of ubiquitous computing. The essence of that vision was the creation of environments saturated with computing and wireless communications capability, yet gracefully integrated with human users. If many of these features were unavailable in 1991, many key building blocks needed for this vision are now viable commercial technologies. Realizing the above vision of ubiquitous computing requires dealing with a number of issues, mainly due to the environment's heterogeneity, dynamics, and constraints. Middleware technologies homogenize the diversity of technologies in ubiquitous computing while providing a support to the development of ubiquitous computing software applications.

Ubiquitous computing raises many constraints that have an impact on middleware technologies. Mobile elements are more likely to be everywhere at the same time. Their ubiquity makes it difficult to consider their availability. It leads them to many different places and situations, much more complex than in a static position. Mobile elements are characterized by a variable connectivity. Indoors and outdoors may offer both reliable, high-bandwidth wireless connectivity, or low-bandwidth connectivity with gaps in coverage. Ubiquitous environments are populated with heterogeneous devices and providers, making interaction between these entities more difficult. While hardware devices will undoubtedly improve over time, the problem of power consumption will not diminish. These hazardous mobility, variable connectivity, heterogeneity, hardware constraints, and security and privacy problems affect the software systems and so the middleware, the glue for all software systems.

These constraints define challenges to be taken into account. We distinguish three challenges: effective use of smart spaces, invisibility and localized scalability.

By embedding computing infrastructure to almost everything and everywhere, smart spaces brings together the world of humans and the world of computing in a new unexplored way. If a pervasive computing environment continuously meets user expectations and rarely presents him with surprises, it allows him to interact almost at a subconscious level while leaving him the choice to interact with the system whenever he wants. Future pervasive computing environments will likely face a proliferation of users, applications, networked devices, and their interactions on a scale never experienced before. As environmental smartness grows, so will the number of devices connected to the environment and the intensity of human machine interactions. Scalability is thus a critical problem in pervasive computing. The density of interactions has to fall off as one moves away.

These challenges have already been a subject of researches for numerous fields such as networking and wireless communications, but also for the applications level such as pervasive home and domotic applications. We argue that middleware technologies need also to cope with these new environments, in order to support the development done in the networks and applications layers.

Requirements and main characteristics of middleware for ubiquitous computing were widely studied (Niemela, 2004, Mascolo, 2005, Salminen, 2005). These studies try to define basic requirements for ubiquitous middleware. We will focus on the most relevant characteristics of ubiquitous computing middleware. The following eight fundamental requirements can be identified: interoperability, discoverability, location transparency, adaptability, context awareness, scalability, security, and autonomous management.

Ubiquitous computing environments, quoting Mark Weiser's definition, consist of various kinds of computational devices, networks and collaborating software and hardware entities. Due to the large number of heterogeneous and cooperating parties, interoperability is required at all levels of ubiquitous computing. Interoperability is the ability of two or more systems or components to exchange information and to use the information that has been exchanged.

One of the fundamental challenges of distributed and highly dynamic environments is how the applications can discover the surrounding entities and, conversely, how the applications can be discovered by the other entities in the system. Discoverability is thus a major issue for ubiquity.

In a ubiquitous system, the execution environment of applications can be logically considered a single container including all applications, other components, and resources. Moreover, the idea in distributed environments is that the resources can be accessed without any knowledge of where the resources or the user are physically located. Whether a file is located in the mobile terminal's local file system or in a network server, the user accesses it in the same way. This is called location transparency; the locations of the system components are transparent to the other components, the programmer, and the user.

Changes in applications' and users' requirements or changes within the network, may require the presence of adaptation mechanisms within the middleware. Adaptability is the ability of a software entity to adapt to the changing environment. Moreover, adaptation is necessary when a significant mismatch occurs between the supply and demand of a resource. Adaptability is regarded as one of the most important functionalities in ubiquitous computing. For example, some resources or services in the environment may be vital to the application. As the application's execution environment changes due to the user's mobility, the vital resources need to be substituted by corresponding resources in the new environment in order to ensure continuous operation. The requirement for adaptation is present on many different layers of a computing system.

Ubiquitous middleware need to be context aware in terms of devices coming and leaving, functionalities offered and retrieved, quality of service changing, etc. They need to be aware of all these changes, in order to offer the best functionalities to applications regardless the context around. When considering context-aware systems in general, some common functionalities can be identified that are present in almost every system: context sensing and processing, context information representation, and the applications that utilize the context information. In general, the context information can be divided into low- and high-level context information. Low-level context information can be collected using sensors in the system. Low-level context information sources can be combined or processed further to higher level context information.

Scalability is the ability of the system to accommodate a higher load at some time in the future. A system's load can be measured using many different parameters, such as the maximum number of concurrent users, the number of transactions executed in a time unit, the number of services available in the ubiquitous environment, etc. As mentioned above, scalability is an important challenge for ubiquitous systems.

Security mechanisms, such as authentication, authorization, and accounting (AAA) functions may be an important part of the middleware in order to intelligently control access to computer and network resources, enforcing policies, auditing network/user usage, etc. Another important aspect concerns privacy and trust in ubiquitous environments. In presence of unknown devices, middleware need to respect privacy of users, and provide trust mechanisms adapted to the ever changing nature of the environment.

Autonomous Management concerns the ability for a ubiquitous middleware to control and manage its resources, functions, security and performance, in the face of failures and changes, with little or no human intervention. The complexity of future ubiquitous computing environments will be such that it will be impossible for human administrators to perform their traditional functions of configuration management, performability management, and security management. Instead, one must resort to automate most of these management functions, allowing humans to concentrate on the definition and supervision of high-level management policies, while the middleware itself takes care of the translation of these high-level policies into automated control structures. The challenge is therefore to move from classical middleware support for configuration, performability and security management to support for self-configuration, self-tuning, self-healing and self-protecting capabilities.

CLASSIFICATION OF THE UBIQUITOUS MIDDLEWARE

Several ubiquitous middleware architectures and infrastructures have been introduced in the academic and industrial world. The current middleware treat ubiquity from slightly different perspectives. We distinguish various middleware technologies, ranging from partially-integrated middleware to fully-integrated middleware. We mean by fully-integrated middleware, middleware providing key elements for all applications' requirements such as discovery, adaptation/composition, context management, and management of ubiquitous applications. In this category we cite ubiquitous middleware systems such as Aura (Garlan, 2002), Gaia (Chetan, 2005), Oxygen (Rudolph, 2001), Pcom (Becker, 2004), and One.world (Grimm, 2004). Partially-integrated middleware range from platforms that were specially realized to handle one or two ubiquitous requirements, such as the application discovery in Jini (Kumaran, 2002) and UPnP (UpnP, 2006), to platforms that are being extended to ubiquity for the application management such as OSGi (OSGi Alliance, 2005) and.Net Framework (Chappell, 2006, Prengel, 2006).

We survey the current state-of-the-art architectures from the viewpoint of the core requirements identified above. In this survey, we will highlight the most known and used fully and partially-integrated middleware. We will not deal with the platforms that are being extended to ubiquity as these extensions are still in a preliminary state. Later on, a classification will focus on the strength and weakness of each of the ubiquitous middleware, based on the identified requirements.

Aura (Garlan, 2002) provides user with an invisible halo of computing and information services that persists regardless of location. A personal Aura acts as a proxy for the mobile user it represents. Aura aim is to allow users to execute their tasks regardless their location. It allows users to dynamically realize daily tasks modelled as abstract software applications, in a transparent way, without manually dealing with the configuration and reconfiguration issues of these applications. Aura deals more with adaptation, replacement of services, the dynamic configuration and reconfiguration of user tasks. Project Aura provides several pervasive applications adapted to both homes and offices.

Gaia (Chetan, 2005) is a services-based middleware that integrates resources of various devices. It manages several functions such as forming and maintaining device collections, sharing resources among devices and enables seamless service interactions. It also provides an application framework to develop applications for the device collection. The application framework decomposes the application into smaller components that can run on different devices in this collection. The notion of ad-hoc pervasive computing in Gaia is a cluster of personal devices that can communicate and share resources among each other. The cluster is referred to as a personal active space. The user can program this cluster through a common interface. Mobile Gaia role is to provide services that discover devices that form the personal space, maintain the composition of the cluster, share resources among devices in the cluster and facilitate communication. Similarly to Aura, Gaia focuses on the dynamic aspect of ubiquitous environments and provides the support for dynamically mapping applications to available ressources of a specific active space.

Oxygen (Rudolph, 2001) vision is to bring an abundance of computation and communication within easy reach of humans through natural perceptual interfaces of speech and vision. Computation blends into peoples' lives enabling them to easily do tasks they want to do, collaborate, access knowledge, automate routine tasks and their environment. In other words, it enables a pervasive, human centric computing. The approach focuses on four technological areas: embedded computational devices, handheld devices, networks, and also on adaptive software. Perception is a central issue, however the focus is mainly on vision and speech aiming to replace explicit traditional input mechanisms with conversational and gesture input.

One.world (Grimm, 2004) is a system architecture for ubiquitous computing. It provides an integrated, comprehensive framework for building pervasive applications. The One.world architecture builds on four foundation services. First, a virtual machine provides a uniform execution environment across all devices and supports the ad hoc composition between applications and devices. Second, tuples define a common type system for all applications and simplify the sharing of data. Third, events are used for all communications and make change explicit to applications. Applications are composed from components that exchange events through imported and exported event handlers. Events make change explicit to applications, with the goal that applications adapt to change instead of forcing users to manually reconfigure their devices and applications. Finally, environments host applications, store persistent data, and—through nesting—facilitate the composition of applications and services.

Pcom (Becker, 2004), a Component system for ubiquitous computing is a light-weight component system that offers application programmers a high-level programming abstraction which captures the dependencies between components using contracts. Pcom allows the specification of distributed applications that are made up of components with explicit dependencies modeled using contracts. Pcom relies on a communication middleware, Base. Base is a flexible middleware for Pervasive computing environments. It provides adaptation support on the communication level by dynamically selecting or reselecting communication protocol stacks, even for currently running interaction. Base is written in Java using the Java 2 Micro Edition with the Connected Limited Device Configuration. It assists application programmers by providing mechanisms for device discovery and service registration that can be used to locate and access local as well as remote device capabilities and services. It also provides a simple signaling mechanism to determine the availability of these devices and services.

Jini (Kumaran, 2002) is a Java-based architecture for spontaneous networking. Participants in a Jini community require no previously knowledge of each other, and can take full advantages of the dynamic class loading and type-checking of the Java language, which requires a Java virtual machine (JVM) for all participants. A Jini community is established around one or more LookupServices, which organize the services deployed in the community and respond to requests from clients. The

Lookup service is itself a Jini service, acting as a bootstrapping service. References to these Lookup services are obtained either by unicast or multicast discovery protocols defined by Jini. The main idea of Jini for supporting "spontaneous networking" is achieved by a leasing principle, which means that services are leased into the community. When a service provider registers a service in the Lookup service it obtains a lease, which must be renewed before it expires, otherwise the Lookup service automatically de-register the service. Clients can register for changes in the Jini community, such as new, discarded, or changed services, using remote event registrations. By the same principle clients and service providers can register for events of new or discarded Lookup services. Event-registrations are leased in the community, so automatic cleanup can be initiated for non-responding clients. These are the real benefits of Jini, enabling opportunity to create a self maintaining ubiquitous computing.

UPnP (UPnP, 2006) technology defines an architecture for ubiquitous peer-to-peer network connectivity of intelligent appliances, wireless devices, and PCs of all form factors. It is designed to bring easy-to-use, flexible, standards-based connectivity to ad-hoc or unmanaged networks whether in the home, in a small business, public spaces, or attached to the Internet. UPnP technology provides a distributed, open networking architecture that leverages TCP/IP and the Web technologies to enable seamless proximity networking in addition to control and data transfer among networked devices. It is designed to support zero-configuration, "invisible" networking, and automatic discovery for a breadth of device categories from a wide range of vendors. A device can dynamically join a network, obtain an IP address, convey its capabilities, and learn about the presence and capabilities of other devices. A device can leave a network smoothly and automatically without leaving any unwanted state behind.

Figure 1. Classification of ubiquitous middleware

	Interoperability	Discoverability	Location transparency	Adaptability	Context awareness	Scalability	Security	Autonomous management
Gaia		x	x	x	x			
Aura		x		x	x			
Oxygen		x		x			x	
One.world	x	x	x	x	x			
Pcom		x	x	x	x			
Jini		x	x				x	
UPnP	x	x	x					

We propose a classification of the previously mentioned ubiquitous middleware. The classification was established upon the challenges raised by ubiquitous computing and upon how the various ubiquitous middleware respond to them.

Figure 1, classifies the existent ubiquitous middleware defined above using the requirements of ubiquitous middleware. For each middleware technology, we focused on the requirements it respects and the ones it does not fulfill. If some requirements are relatively well fulfilled by nowadays systems, such as discoverability, context awareness and adaptability, others are far from being fulfilled or even dealt with such as interoperability, security, and especially scalability and autonomous management.

FUTURE TRENDS

Ubiquitous middleware offer more and more functionalities to applications evolving in ubiquitous environments and that by respecting some defined properties. Several domains and technologies have emerged as the new trends for ubiquitous middleware. These new trends tend to deal with the requirements left behind by the actual ubiquitous middleware. The semantic web (McIlraith, 2001), the ambient intelligence (Shadbolt, 2003) and the autonomic computing (Crowley, 2007) are the most spread researches domain for ubiquitous computing.

Semantic modeling tries to solve the interoperability problem. The semantic web is an evolving extension of the World Wide Web in which web content can be expressed not only in natural language, but also in a format that can be read and used by automated tools, thus permitting people and machines to find, share and integrate information more easily. Today's Web was designed primarily for human interpretation and use. Nevertheless, we are seeing increased automation of Web service interoperation primarily in B2B and e-commerce applications. A fundamental component of the semantic web will be the markup of web services to make them computer-interoperable, user-apparent, and agent-ready.

Ambient intelligence refers to electronic environments that are sensitive and responsive to the presence of people. The ambient intelligence paradigm builds upon ubiquitous computing and human-centric computer interaction design and is characterized by systems and technologies that are embedded, context aware, personalized, adaptive and anticipatory. Ambient intelligence introduces proactivity to the world of ubiquitous computing. A proactive middleware needs to interfere in humans everyday, while respecting the invisibility imposed by ubiquitous computing. A balance needs to be found between invisibility and proactivity. Ambient intelligence builds on three recent key technologies: ubiquitous computing, ubiquitous communication and intelligent user interfaces.

Autonomic computing is an initiative started by IBM in 2001. Its ultimate aim is to create computer systems capable of self-management, to overcome the rapidly growing complexity of computing systems management, and to reduce the barrier that complexity poses to further growth. In a self-managing autonomic system, the human operator takes a new role, he does not control the system directly. Instead, he defines general policies and rules that serve as an input for the self-management process. For this process, IBM proposed future computer system not only to adapt to changing environments at runtime but further to be self-configuring, self-healing, self-optimizing and self-protecting. Autonomic computing is a promising opening research field for ubiquitous computing as autonomy can be appreciated by users desiring to use functionalities offered by these environments.

CONCLUSION

Ubiquitous middleware are becoming the nowadays trend in the development of ubiquity in computer science fields. Ubiquitous applications rely upon this layer, to profit from the diverse functionalities it has to offer. Ubiquitous environments brought more constraints and challenges to mobile environments. The main constraints come from, the environment's heterogeneity and dynamics, and the variable connectivity of the devices coming and leaving. The main challenges are in maintaining the computing smartness, scalability, invisibility and proactivity for the users in these environments. The functionalities offered by middleware need to cope with these challenging nature of environments. We sorted the middleware in two groups. The fully-integrated ones, provide functionalities such as discovery, adaptation/composition, and context management. The partially-integrated ones, provide one or two of these functionalities, as they were specifically developed for a specific purpose. We classified these middleware, by analysing if they are interoperable, discoverable, adaptable, context aware, scalable, secure and autonomous. If many of these middleware are mature enough and offer specific functionalities respecting the properties of ubiquity, a real lack is noticed in having an interoperable, autonomous and scalable middleware for the execution of ubiquitous applications. The development of the service-oriented paradigm, the semantics and the Web middleware shows the new trend the middleware research field is engaged in. At the other hand the intersection of this research field with artificial intelligence and autonomic computing leads to the development of the ambient intelligence, the future evolution of ubiquitous computing.

REFERENCES

Becker, C., Handte, M., Schiele, G., & Rothermel, K. (2004). PCOM – A Component System for Pervasive Computing. *2nd IEEE Annual Conference on Pervasive Computing and Communications (PERCOM '04)*, Washington, DC, USA.

Chappell, D. (2006). Introducing the.Net Framework 3.0. Whitepaper "Introducing the.Net Framework 3.0".

Chetan, S., Al-Muhtadi, J., Campbell, R., & Mickunas, M. D. (2005). Mobile Gaia: A Middleware for Ad-hoc Pervasive Computing. *IEEE Consumer Communications & Networking Conference (CCNC 2005)*, Las Vegas, USA.

Crowley, J. L., Hall, D., & Emonet, R. (2007). Autonomic Computer Vision Systems. *International Conference on Computer Vision Systems (ICVS '07)*, Bielefeld, Germany.

Garlan, D., Siewiorek, D., Smailagic, A., & Steenkiste, P. (2002). Project Aura: Towards distraction-free pervasive computing. *IEEE Pervasive Computing / IEEE Computer Society [and] IEEE Communications Society, 21*(2), 22–31. doi:10.1109/MPRV.2002.1012334

Grimm, R. (2004). One.World: Experiences with a Pervasive Computing Architecture. *IEEE Pervasive Computing / IEEE Computer Society [and] IEEE Communications Society, 3*(3), 22–30. doi:10.1109/MPRV.2004.1321024

Kumaran, S., I. (2002). JINI Technology an Overview. *Prentice Hall PTR.*

Mascolo, C., Hailes, S., Lymberopoulos, L., Picco, G.P., Costa, P., Blair, G., Okanda, P., Sivaharan, T., Fritsche, W., Karl, M., Rnai, M.A., Fodor, K., & Boulis, A. (2005). *Survey of middleware for networked embedded systems.* (Technical Report D.5.1).

McIlraith, S. A., Son, T. C., & Zeng, H. (2001). Semantic Web Services. *IEEE Intelligent Systems, 16*(1), 46–53. doi:10.1109/5254.920599

Niemela, E., & Latvakoski, J. (2004). Survey of requirements and solutions for ubiquitous software. *3rd international conference on Mobile and ubiquitous multimedia (MUM '04)* (pp. 71-78). New York.

OSGi Alliance. (2007). OSGi Service Platform. *Core Specification Release 4.1.* Retrieved from http://www.osgi.org/Specifications/HomePage

Prengel, F. (2006). *Net Micro Framework: Programming Small Connected Devices with.* Net, Core Specification. [White paper]

Rudolph, L. (2001). Project Oxygen: Pervasive, Human-Centric Computing - An Initial Experience. *Advanced Information Systems Engineering, 13th International Conference (CaiSE2001)* (LNCS 2068, pp. 765-780) Interlaken, Switzerland.

Salminen, T. (2005). *Lightweight middleware architecture for mobile phones.* Diploma Thesis, University of Oulu, Department of Electrical and Information Engineering.

Satyanarayanan, M. (2001). Pervasive Computing: Vision and Challenges. *IEEE Personal Communication, 8*(4), 10-17.

Shadbolt, N. (2003). Ambient Intelligence. *IEEE Intelligent Systems, 18*(1), 2–3. doi:10.1109/MIS.2003.1200718

UPnP forum (2006). *UPnP Technology – the simple, seamless home network* [White paper].

Weiser, M. (1991). The computer for the 21st century. *Scientific American*, *265*(3), 66–75. doi:10.1038/scientificamerican0991-94

KEY TERMS AND DEFINITIONS

B2B: (Business-to-business) is a term commonly used to describe electronic commerce transactions between businesses, as opposed to those between businesses and other groups, such as business and individual consumers (B2C) or business and government (B2G).

CORBA: (Common Object Request Broker Architecture) is an open infrastructure for distributed systems that is standardized by the Object Management Group (OMG). CORBA specifies a system which provides interoperability between objects in a heterogeneous distributed environment. Furthermore, it provides transparency from the network, different operating systems, and different programming languages in a distributed computing environment.

DCOM: (Distributed Component Object Model) is a proprietary Microsoft technology for communication among software components distributed across networked computers. DCOM, extends Microsoft's COM, a platform for software componentry introduced by Microsoft in 1993.

MOM: (Message Oriented Middleware) is a client/server infrastructure that increases the interoperability, portability, and flexibility of an application by allowing the application to be distributed over multiple heterogeneous platforms. It resides in both portions of client/server architecture and typically supports asynchronous calls between the client and server applications. Message queues provide temporary storage when the destination program is busy or not connected.

ORB: (Object Request Broker) is a piece of middleware software that allows programmers to make program calls from one computer to another via a network. ORB's handle the transformation of in-process data structures to and from the byte sequence, which is transmitted over the network. This is called marshalling or serialization.

RPC: (Remote Procedure Call) is a technology that allows a computer program to cause a subroutine or procedure to execute in another address space (commonly on another computer or a shared network) without the programmer explicitly coding the details for this remote interaction.

SOA: (Service-Oriented Architecture) is a computer systems architecture style for creating and using business processes, packages as services, throughout their life cycle. Main characteristics of the service-oriented architecture are its support for the deployment and the interaction of loosely coupled software systems, which evolve in a dynamic and open environment and can be composed with other services.

SOAP: (Simple Object Access Protocol) is a protocol for exchanging XML-based messages over computer networks, normally using HTTP/HTTPS. SOAP forms the foundation layer of the web services protocol stack providing a basic messaging framework upon which abstract layers can be built.

TP Monitor: (Transaction Processing monitor) is a program that monitors a transaction as it passes from one stage in a process to another. Its purpose is to ensure that the transaction processes completely, or, if an error occurs, to take appropriate actions.

Web Service: is defined by the W3C as a software system designed to support interoperable Machine to Machine interaction over a network. Web services are frequently just Web APIs that can be accessed over a network, such as the Internet, and executed on a remote system hosting the requested services.

Chapter 13
Routing in Asymmetric Wireless Ad-Hoc Networks

Pramita Mitra
University of Notre Dame, USA

Christian Poellabauer
University of Notre Dame, USA

ABSTRACT

The presence of asymmetric links is a common and non-negligible phenomenon in many ad-hoc networks, including MANETs and sensor networks. Asymmetry is caused by node mobility, heterogeneous radio technologies, and irregularities in radio ranges and packet loss patterns. Most existing ad-hoc routing protocols either assume fully symmetric networks or simply ignore any asymmetric links. In the first case, route discovery can fail when the symmetry assumption does not hold true, e.g., many reactive routing protocols rely on a two-phase communication process, where the same path is used to communicate between a sender and a receiver. If a single link on this path is asymmetric, the route establishment may fail. In the second case, asymmetric links are identified and explicitly ignored in the route establishment phase. This can lead to route discovery failure if there is no symmetric path between a sender and a receiver or it can lead to less than optimal routes. This document provides an overview of routing protocols that explicitly consider asymmetric links in the route discovery phase and introduces robust mechanisms that bypass asymmetric links to ensure successful route establishment.

INTRODUCTION

Routing in wireless ad-hoc networks typically assumes that wireless links are bidirectional, i.e., wireless devices have identical transmission ranges. The increasing heterogeneity in wireless ad-hoc networks (i.e., networks consisting of different wireless devices such as laptops, PDAs, and cell phones) leads to a substantial number of wireless links that are asymmetric due to the variation in transmission ranges of these devices. A recent empirical study (Ganesan, Exstein, Woo, & Culler, 2002) shows that approximately 5-15% of the links in a low power wireless network are asymmetric, and an increasing distance between nodes increases the likelihood of an asymmetric

DOI: 10.4018/978-1-60566-250-3.ch013

link. Power aware algorithms running at different nodes in low power networks cause the nodes to function at various transmission ranges at different instances of time for efficient usage of power, thereby increasing the number of asymmetric links in the network. Link asymmetry is also caused by node mobility, heterogeneous radio technologies, and irregularities in radio ranges and packet loss patterns. Based on these key observations, it is to expect that link asymmetry will become more dominant in future wireless ad-hoc networks, including mobile ad-hoc networks (MANETs) and sensor networks.

Routing protocols that assume that all links are symmetric can fail in their route discovery when this assumption does not hold true. For example, many reactive routing protocols such as Dynamic Source Routing (DSR) and Ad-hoc On-Demand Distance Vector (AODV) rely on a two-phase communication process, where the same path is used to communicate between the sender and the receiver. If a single link on this path is asymmetric, the route establishment will fail.

Therefore, other routing protocols (Couto, Aguayo, Bicket, & Morris, 2003) try to locate asymmetric links and explicitly ignore them in the route establishment phase. This can lead to route discovery failure when there is no symmetric path between a sender and a receiver or it can lead to less than optimal routes. In these protocols, it is difficult to achieve high connectivity if the network has many asymmetric links. Further, the average length of routes established using reactive routing protocols that ignore asymmetric links is longer than the average length when asymmetric links are taken into consideration (Marina & Das, 2002).

As a consequence, it is necessary that asymmetric links (with good one-way reliability) are considered in route establishment. This chapter presents an overview of routing protocols that explicitly consider asymmetric links in the route discovery phase and introduces robust mechanisms that exploit asymmetric links to ensure successful route establishment.

BACKGROUND

Ad-hoc routing protocols are mainly classified into on-demand reactive routing protocols and table-driven proactive routing protocols. It has been widely accepted that reactive routing protocols are superior to proactive routing protocols in that the control overhead is smaller when the number of connections handled by the network is also small (Lee, Gerla, & Toh, 1999). Significant work has been done in the development of on-demand reactive routing protocols for wireless ad-hoc networks. In such protocols, when a source node has a packet to send, it invokes a route discovery process to establish a route between the destination node and itself. In addition, nodes along a route (source and intermediate nodes) must perform route maintenance tasks whenever a route fails (e.g., a link breakage has occurred). However, all links are assumed to be symmetric; therefore, this section provides a brief overview of some well-known on-demand reactive routing protocols and how they fail in the presence of asymmetric links.

In AODV (Perkins, Royer, & Das, 2000), each node receiving an RREQ (Route Request) packet, rebroadcasts it until the packet reaches the destination node or a node that already has a path to the destination. This node replies with an RREP (Route Reply) packet, which is sent back to the source node across the newly discovered route (i.e. the reverse path). If there are one or more asymmetric links in the route, the RREP packet will fail to reach the source node and therefore a route discovery failure will occur.

In DSR (Broch, Johnson, & Maltz, 1998), similar to AODV, an RREQ packet is flooded into the network to establish a route when a source node has data to send. The destination node selects the shortest path and an RREP packet containing the path information is routed back to the source node. Thus, in the presence of asymmetric links during the RREP propagation, the recorded source route cannot be successfully sent back to the source node.

In the Zone Routing Protocol (ZRP) (Hass, 1997), a node A defines a routing zone around itself consisting of its k-hop neighborhood, and all nodes within k-hop distance from node A belong to the set of routing zone members of node A. These neighbors are assumed to be connected by symmetric paths. Node A notifies its neighboring nodes of its identity. Upon receiving an identity notification from node A, these neighboring nodes include node A in their routing zone member tables. However, this scheme fails in the presence of asymmetric links in the routing zone. Suppose that node A is able to reach another node B, but not vice versa, due to the presence of an asymmetric link from node A to node B. That is, node A will inform node B of its identity, but A will not receive packets emitted by B. Therefore, node B will mistake node A as a member of its zone. This can cause serious problems when routes are discovered in ZRP.

In Associativity Based Routing (ABR) (Toh, 1997), a BQ-REQUEST (Broadcast Query Request) packet is generated when a source node attempts to obtain an initial path between the source and destination nodes. An intermediate node sends an LQ-REQUEST (Local Query Request) packet to discover a partial path from itself to the destination node after detecting a route failure. At the destination node, the most stable route is selected and this route information is propagated back to the source node via the reverse path. Again, if there exists an asymmetric link during the reply propagation towards the source, the discovered route cannot be established.

As a consequence, the presence of asymmetric links causes the following problems (Kim, Toh, and Choi, 2003): (1) Knowledge asymmetry: In case an asymmetric link exists between node A and node B, B cannot assume that A can receive B's packets, even when B is able to receive A's packets. Such situations lead to an asymmetry of knowledge between the nodes at the two ends of the asymmetric link. (2) Routing asymmetry: Given an asymmetric link between nodes A and B, the path traversed to reach B from A is different from the path required to reach A from B, leading to asymmetry in routing.

Many routing protocols treat asymmetric links as an anomaly and tend to eliminate them from the route discovery procedure. A recent empirical study (Marina & Das, 2002) states that the overheads of maintaining asymmetric links in a route discovery procedure tend to outweigh the benefits derived by using asymmetric links. The study also proposes an extension to AODV for asymmetric links. However, this approach is similar to the blacklisting schemes in terms of avoiding asymmetric links in the route discovery. The modified DSDV protocol (Prakash, 1999) proves that $O(n^2)$ size messages must be exchanged between nodes to account for asymmetric links if distance vector routing is employed. This is significantly greater than the $O(n)$ size messages exchanged in existing routing protocols which assume that all links are bidirectional. The Symmetric Geographic Forwarding protocol (Zhou, Krishnamurthy, and Stankovic, 2004) proposes to advertise already discovered neighbors in periodic beacons during neighbor discovery and marking symmetric and asymmetric links separately in the neighbor table. This protocol only selects neighbors connected with symmetric links for data forwarding and employs a learning function to allow a node to remember asymmetric links, preventing them from being used in the future. The Probabilistic Geographic Routing protocol (Roosta, Menzo, & Sastry, 2005) selects the forwarder node among a set of neighbors using a probabilistic method. The forward and backward reliability values of a link in both directions are stored in the neighbor table, and a neighbor with good forward and backward reliability value (i.e., a neighbor connected with a symmetric link) is chosen as the next hop forwarder.

Such schemes scale well when the number of asymmetric links in the network is insignificant. When the number of asymmetric links is large, and the network still remains a connected graph, the routing protocols that do not utilize asymmetric links perceive the network as an unconnected graph. As a result, the set of asymmetric links remains unnoticed by these protocols. Therefore, the scalability and performance of such schemes decrease drastically with the increase in the number of asymmetric links in the network. It is thereby necessary to discover and utilize asymmetric links during route discovery in wireless ad-hoc networks.

REASONS FOR LINK ASYMMETRY IN WIRELESS NETWORKS

In conventional routing protocols, whenever a node receives a beacon packet from another node, it considers that node to be its neighbor as it assumes the link is bi-directional. Such an assumption may not be realistic for practical wireless ad-hoc networks, since wireless links are often asymmetric. Link asymmetry is mainly caused by the differing transmission ranges of wireless devices. In Figure 1, node A discovers its neighbors B, C, and D by receiving beacons from them, which means that node A is within the wireless transmission ranges of all these three nodes. Both node C and node Dare also within the wireless transmission range of node A, so the links A ↔ C and A ↔ D are symmetric. However, node B

Figure 1. Differing transmission ranges resulting in an asymmetric link between node A and node B

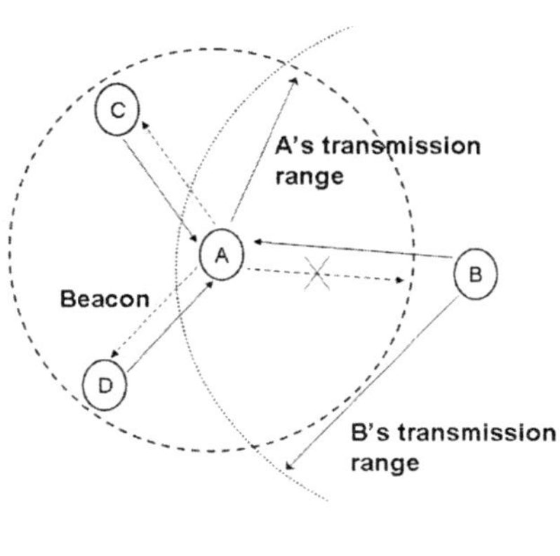

is outside node A's wireless transmission range and therefore, there is an asymmetric link B → A between these two nodes. As a consequence, node A can hear node B but node B cannot hear node A.

Differing transmission ranges of nodes occur due to the following factors:

- The increasing heterogeneity in wireless ad-hoc networks, e.g., networks consisting of different types of wireless devices, where these devices have differing transmission ranges.
- Transmission ranges are also affected by the transmission power used by a wireless device, which can be varied dynamically by resource management protocols in low-power networks.
- Irregularities in radio ranges (caused by interferences and obstacles) and packet loss patterns as well as differing noise floors of wireless devices.

Varying transmit power is one of the main causes of varying transmission ranges leading to link asymmetry. Received Signal Strength Indication (RSSI) is often considered as a good measure of the link quality (Liu, 2008; Srinivasan, 2006). Table 1 shows the effect of varying transmit powers and distances on RSSI and noise floor. In this scenario, five mobile sender devices where placed at distances in the range of 20 to 140m (in steps of 20m) to a receiver device. The device labeled as device 5 transmitted at the highest possible transmit power; while device 1 used the lowest possible transmit power. This table confirms that variations in both distance and transmit power will have visible effects on RSSI and noise floor values.

Figure 2 shows the corresponding *successful packet transmission rate* of those five source devices with varying distance. The figure shows that the successful packet transmission rate drops with increasing distance and decreasing transmit power levels. It also demonstrates that there is a

Table 1. Measured RSSI and noise floor rates with varying distances and transmit powers

Distance	RSSI					Noise Floor				
	Dev 1	Dev 2	Dev 3	Dev 4	Dev 5	Dev 1	Dev 2	Dev 3	Dev 4	Dev 5
20 m	37.78	41.41	44.00	43.66	45.63	28.93	29.86	29.89	30.69	30.8
40 m	35.44	39.04	42.43	43.31	40.87	30.19	29.92	29.92	31.14	31.36
60 m	33.86	37.38	39.44	37.38	40.73	30.79	30.6	30.97	31.3	32.2
80 m	32.67	31.75	34.42	37.11	40.63	31.25	30.67	32.13	31.53	32.23
100 m	32.13	30.20	33.56	34.12	36.32	32.18	30.78	32.4	31.69	33.18
120 m	32.08	29.93	30.17	32.35	34.35	32.2	32.5	34.21	32.27	33.48

high correlation between RSSI and successful packet transmission rate, indicating that RSSI is a good measure of link quality. It can be seen from Table 1 that when the successful packet transmission rate dropped to zero, the measured Noise Floor was more than the measured RSSI value, i.e., no meaningful data reception was possible because the noise was stronger than the received signal. As a consequence, increasing the distance between a sender and receiver or decreasing the transmit power of the sender will affect the link quality.

DISCOVERY AND EXPLOITATION OF ASYMMETRIC LINKS

In Figure 1, node A may choose one of its neighbor nodes for forwarding packets. However, if node A selects node B which is unable to hear node A, node B would never receive the packet forwarded by node A. If node A does not retry its transmission via other neighbors, the transmission of the packet will fail. Therefore, the routing protocol based on such a neighbor discovery technique is subject to failures when the communication is asymmetric. On the contrary, node B can use node A as a forwarder node to send a packet to

Figure 2. Successful packet transmission rate with varying distance and transmit power

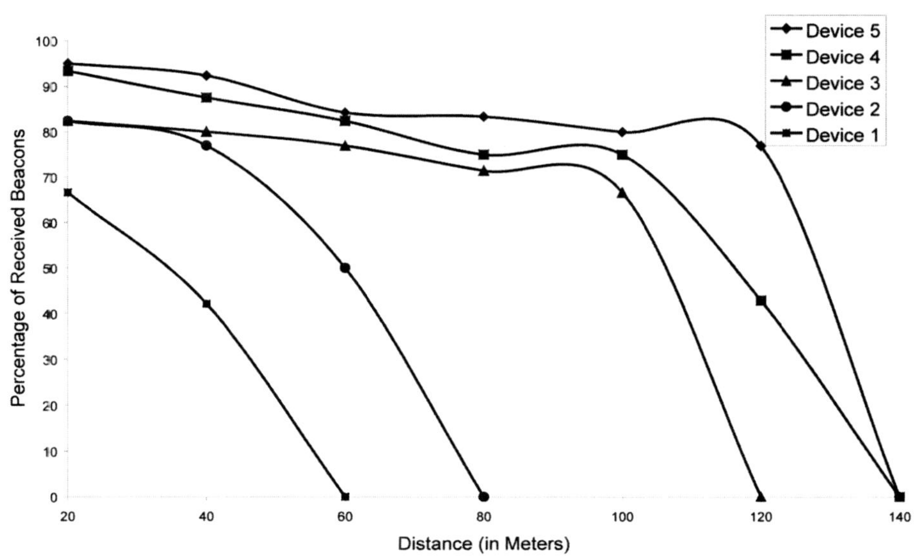

sink node C. If asymmetric links are ignored by the underlying routing protocol, then node B will need to take a detour via a potentially longer route of symmetric links to send a packet to node C. Therefore, it is easy to see the importance to find asymmetric links in the network and exploit them while making routing decisions.

Routing protocols should address link asymmetry in three distinct phases:

- Route discovery: In low power networks, the percentage of asymmetric links is large, and using these asymmetric links for route discovery often results in shorter routes in terms of delay or hop count.
- Route optimization: Asymmetric links also continue to be dynamically formed throughout the lifetime of wireless ad-hoc networks, due to node mobility and variations in packet loss patterns. These newly formed asymmetric paths may yield better routes than the previously discovered ones.
- Route maintenance: Mobility of up-link nodes or down-link nodes can turn an already discovered symmetric path into an asymmetric one leading to the fact that link asymmetry is more of a route maintenance problem than of a route discovery problem.

This section discusses these three phases. Most existing reactive routing protocols use the reverse path of the previously discovered path for destination-to-source communication during route discovery. If that path contains one or more asymmetric links, then the source node will not receive the RREP packet, and thus route discovery fails. An intuitive solution to the problem is to find a different return path for the destination-to-source communication; a method that we describe as reverse route computing. The simplest reverse routing computing involves a two way *flooding* mode for route discovery using asymmetric links. In this mode, the forward route from the source node to the destination node is constructed by flooding RREQ packets. Then, a return route is constructed by flooding RREQ packets from the destination node to the source node (Broch, Johnson, & Maltz, 1998). However, this mode involves a significant amount of control packets, especially in high density networks, in order to achieve high connectivity performance. Further, in this mode, the large number of control packets can collide with data packets of other connections, and thereby degrading the overall communication quality. A similar approach uses two broadcasts for probing independent forward and reverse paths between source and destination nodes (Kim, Toh, & Choi, 2000). This approach requires knowledge of all nodes visited to compute forward and reverse paths. The additional overhead of storing the information about all the nodes visited can prove costly, especially in large networks. Also, the two broadcasts are not required if the shortest path from source to destination is a symmetric path. A possible optimization is to use a *restricted flooding* approach, which limits the search area of the reverse route during the two-way flooding method. Many asymmetric paths are caused by a very few number of asymmetric links along that path, and consequently, it is reasonable to assume that the reverse route will be in close proximity to the newly discovered forward route. Toward that end, only those nodes that received RREQ packets directly from the nodes on the forward route during the forward route discovery will be allowed to relay the RREQ packets for the reverse route discovery (Morino, Miyoshi, & Ogawa, 2007). The third approach for reverse route computing is called *loop based source routing*. It involves finding looped routes including the source and destination nodes to establish the forward and reverse paths. These looped routes can contain asymmetric links (Unoki and Higaki, 2004). This approach uses a single flooding and multiple unicast message transmissions. These reverse route computing protocols have the following drawbacks: (1) they use flooding to find the forward and reverse paths and (2) they maintain predetermined forward and reverse paths, thereby requiring many control packets. Some routing protocols make use of link layer periodic Hello messages to discover asymmetric links; we denote this approach as the *hello approach*. In this approach, each node puts a list of neighbors (i.e., nodes it can hear from) into its periodic Hello messages. During the exchange of Hello messages, each node compares its own neighbor list with the neighbor list of a neighbor node to determine whether an asymmetric link exists. If a node is not listed in a neighbor's neighbor list, it indicates that there is an asymmetric link that uses the neighbor node as the up-link node and this node as the down-link node. Asymmetric paths are often caused by a single asymmetric link along that path, and it is sufficient to find a short multihop *tunneling path* from the down-link node to the up-link node to notify the up-link node of the existence of the outgoing asymmetric link. The *tunneling path* can be formed by using restricted flooding of Link-Info packets by the down-link node. The performance of the network deteriorates when all the neighbor nodes relay the Link-Info packet and broadcast storms may occur when the Link-Info packet is transmitted using multihop paths. A priority-based back-off algorithm can be used to control this problem, and it can also help in minimizing the routing overhead of the *tunneling path* discovery (Suh, Quan, Park, Park, & Kim, 2007). Thus, the link layer based *hello approach* successfully provides a bidirectional abstraction of underlying asymmetric links, and it can be used to improve the efficiency of off-the-shelf routing protocols typically designed for symmetric networks (Ramasubramanian & Mosse, 2008). The *tunneling path* can also be used for transmission of the link layer ACK and network layer routing information data, all encapsulated in a single network layer packet. The compromise made is that the approach is not entirely transparent to the network layer. However, these modifications are still transparent to the routing module, thereby allowing any routing protocol to be used (Nesargi & Prakash, 2000).

Existing routing protocols utilize a discovered route until it is broken, failing to make use of asymmetric paths being newly formed in the system. Routing protocols that use route optimization techniques are able to improve on data transfer latencies, and sometimes reduce the possibility of route failure by replacing the deteriorating routes with newly discovered ones with good reliability. To help with route optimization, mobile nodes should maintain fresh neighborhood information, while keeping the overheads low. Wireless transmission is broadcast by nature and only one pair of communications is possible in a neighborhood at any given time, i.e., all other non-participating nodes in that neighborhood must remain quiet. However, these idle nodes can continue to update their knowledge of the neighborhood by configuring their network interfaces into the promiscuous listening mode (Wu, Tseng, Ni, & Sheu, 2000). Whenever a node overhears an ongoing data transmission in its neighborhood, it checks if it can offer a better partial route in terms of shorter hop count or shorter delay. If there is a better partial route (going through the overhearing node), then the data traffic is *redirected* along the new route (Suh, Quan, Park, Park, & Kim, 2007).

It is not sufficient to only discover optimized paths in wireless ad-hoc networks, i.e., the discovered paths have to be maintained to ensure reliable data transmission. Route maintenance is often realized by supporting hop-by-hop or passive acknowledgment schemes in the link layer. An up-link node discovers route failure through the loss of acknowledgments from the down-link node. In this case, existing route maintenance protocols send a route error packet back to the source node (issued by the up-link node discovering the failure), invoking a *complete* route discovery procedure to obtain a new route between the source node and the destination node. However, a *complete* route discovery is very costly in terms of bandwidth and delay and it should be avoided as much as possible during route reconstruction. It is possible to perform a *partial* route discovery and *reuse* part of the broken route for route reconstruction (Toh, 1997). If an up-link node discovers a link breakage to its down-link node, one of two possible situations will occur.

- Case I: If the up-link node is located in the upper half of the route (i.e., it is closer to the source than the destination), then it directly sends a route error report back to the source node. The source node will invoke a *complete* route discovery to obtain a new route to the destination node.
- Case II: However, if the up-link node is located in the lower half of the route (i.e., it is closer to the destination than the source), the up-link node makes a *partial* route discovery to find a *partial* route from itself to the destination node. If this *partial* route discovery succeeds, the route is repaired and no route error will be reported to the source node. In this case, we *reuse* part of the broken route, i.e., the *partial* route from source node to the up-link node. If this *partial* route discovery fails, the up-link node will report a route error to the preceding node, which will in turn attempt a *partial* route discovery to the destination node again.

This process is recursively repeated until either the broken route is remedied in case II or we find a node in case I. This approach significantly reduces the overheads of a *complete* route discovery. On the other hand, this approach may lead to large bandwidth consumptions or increased delay if the above recursive procedure fails several times.

Route failures typically occur due to the failure of a single link, and it is reasonable to expect that the future route will be built around the broken route (Morino, Miyoshi, & Ogawa, 2005). Thus the flooding of RREQ packets can be constrained to the area around a former broken route, resulting in considerable reduction in control traffic in terms of number of transmitted RREQs. During route reconstruction, neighbor nodes relay the received RREQ packet only if they overheard any data packet or acknowledgments along the previous broken route during a specific wait interval. The relay timing is controlled using a random back-off timer to reduce the number of collisions of rebroadcast RREQ packets by neighboring nodes (Wu, Tseng, Ni, & Sheu, 2000).

However, route maintenance is more than just route reconstruction in wireless ad-hoc networks with asymmetric links. The frequent changes in network topology will have either one of two possible impacts on the discovered paths in a network with asymmetric links:

- Impact I: A symmetric wireless link between node A and node B turns into a broken link if both nodes move out of each other's wireless transmission ranges. In this case, the symmetric path containing the link A-B will fail.
- Impact II: A symmetric link between node A and node B turns into an asymmetric link if node A moves out of node B's wireless transmission range, but node B is still reachable from node A. In this case, the asymmetric path containing the link A-B is still usable in the forward direction.

The previously described route maintenance protocols treat both of these cases as route failures and invoke a new route discovery procedure. However, efficient route maintenance protocols for wireless ad-hoc networks with asymmetric links should be able to differentiate between these two different types of impacts. For impact II, a *local route recovery* procedure will be more efficient than a route reconstruction. This approach performs a local route discovery to find a *tunneling path* in order to patch the asymmetric link, before it is necessary to send a route error packet back to the source node (Wu, Tseng, Ni, &

Figure 3. Successful packet transmission ratio with varying number of nodes

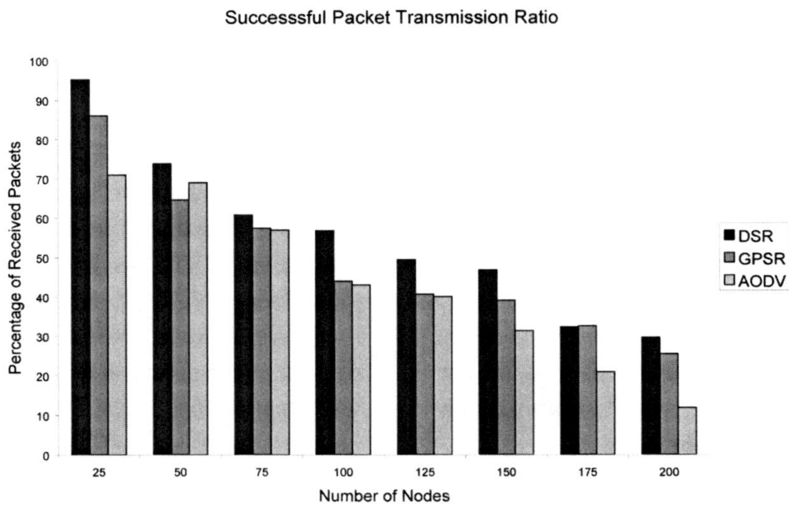

Sheu, 2000). The *local route recovery* approach achieves significantly less network traffic overhead than other route maintenance approaches in that it executes the costly route discovery procedure less frequently. Based on all these key observations, we denote that link asymmetry is more of a route maintenance problem than a route discovery problem.

EXPERIMENTAL ANALYSIS

This section compares the performance of several symmetric routing protocols in a network with asymmetric links. Figure 3 shows the successful packet transmission ratio of three symmetric protocols: DSR, AODV, and Greedy Perimeter Stateless Routing (GPSR) (Karp, 2000). The nodes in the network have varying transmission ranges, as a result of differing transmit powers, radio reception sensitivities, and radio reception thresholds of the nodes. The radio transmission ranges of the nodes are chosen randomly, therefore, it is reasonable to assume that the number of asymmetric links in the network will go up with an increasing number of nodes (in the following experiments, the number of nodes was varied from 25 to 200 in steps of 25).

Figure 3 shows that the successful packet transmission ratio of the studied routing protocols drops with the increase in the number of nodes in the network. The figure also shows that DSR always performs better than GPSR and AODV. Symmetric protocols experience transmission failures on asymmetric links, and usually respond to the problem by triggering route management mechanisms. DSR's route caches have more than one route per sink which enables it to respond quickly to transmission failures along the route without even having to trigger route management, thereby offering better throughput ratio in presence of asymmetric links in the network (Misra & Mandal, 2005). However, with 200 nodes in the network the performance of all the protocols drops below 40%. This graph indicates shows that conventional routing protocols are not suitable in networks with asymmetric links.

In order to show the need for exploiting the asymmetric links in the network, Figure 4-5 compare and evaluate the performance of the performance of Geographic Forwarding (GF) (Karp, 2000) with a modified version of GF that

Figure 4. (a) Throughput ratio with varying asymmetric link percentage. (b) Average route length with varying asymmetric link percentage

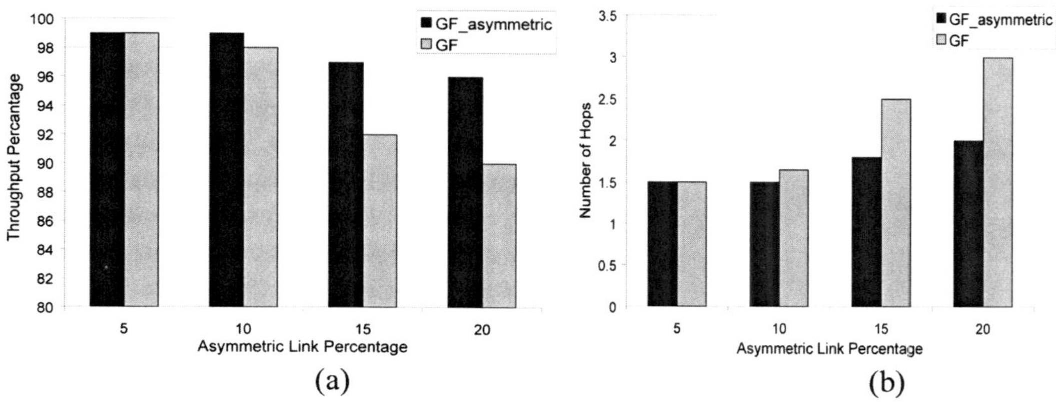

explicitly considers asymmetric links, called GF_asymmetric. GF is a routing protocol that relies on location information, such as obtained through GPS sensors. In this experiment, twenty mobile devices are set up as ad-hoc network, with nodes having different transmission ranges (causing between 5 and 20% of asymmetric links in the network). Figures 4(a) and 4(b) show that GF_asymmetric performs at least as good as the base GF protocol when the network has a small percentage of asymmetric links. However, GF_asymmetric provides (1) better throughput ratios and (2) shorter routes than the base GF protocol when the number of asymmetric links in the network increases. That is, GF_asymmetric discovers and exploits asymmetric links in the network for data transmission, whereas standard GF suffers from low network connectivity as it ignores asymmetric links. However, this performance improvement comes with a small overhead in the form of an increase in network traffic as shown in Figure 5.

FUTURE TRENDS

As stated before, exchanging neighbor lists in link layer Hello packets for the discovery of asymmetric links is a common approach in wireless ad-hoc networks. However, this approach incurs significant overhead due to Hello packets containing potentially large neighbor tables. With the increased deployment of Global Positioning System (GPS) and other localization schemes, geographic routing is becoming an attractive option for scalable routing in wireless networks, and it can be very efficient to exploit the knowledge of geographical locations of nodes for finding and exploiting asymmetric links in wireless ad-hoc networks. Geographic routing protocols do not require a route discovery procedure preceding the data transmission, thereby incurring much less overhead. It is possible to put the geographic location of the up-link node in the data packet. By using the up-link node's location propagated to the receiver node, the down-link node knows whether its transmission range is able to reach the up-link node by simply comparing its transmission range with the Euclidean distance between the up-link node and down-link node. If the radio transmission range of the down-link node is not sufficient to reach the up-link node, then the up-link node has an outgoing asymmetric link to the down-link node (Kim, Toh, & Choi, 2003). A variety of modifications can be built upon this basic *geographic location based neighbor discovery* protocol to

Figure 5. Network traffic with varying asymmetric link percentage

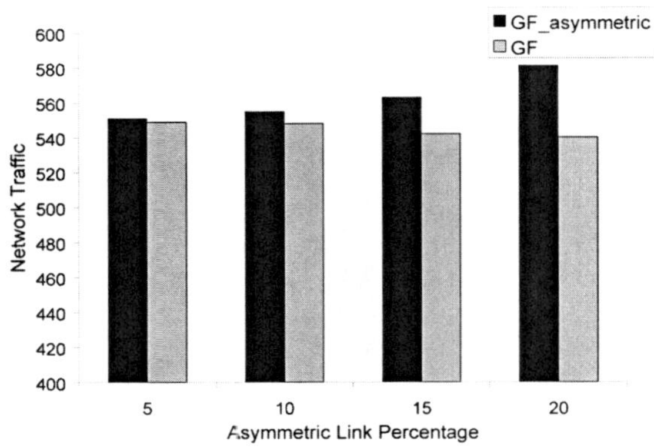

improve the quality of routing in wireless ad-hoc networks in the presence of asymmetric links. It is not sufficient to only find the asymmetric links, i.e., the forward reliability of asymmetric links should be considered in routing decisions in order to improve the reliability of data transmission. A recent study proposes a one-way metric ETF (Expected number of Transmissions over Forward links) to identify highly reliable forward links (Sang, Arora, & Zhang, 2007). Future protocols will also rely on route maintenance schemes for asymmetric paths that exploit the knowledge of the geographic locations of nodes.

Asymmetric path discovery protocols often assume that an asymmetric link always has a *tunneling path*. Therefore, an asymmetric link may remain unnoticed in a low density network where there is no *tunneling path* to notify the up-link node. However, this problem can be addressed if a down-link node is allowed to momentarily increase its radio transmission range (Kim, Toh, & Choi, 2003). Here, the asymmetric links are discovered using the *geographic location based neighbor discovery* protocol. If the radio transmission range of the down-link node is not sufficient to reach the up-link node (in case of an asymmetric link), the down-link node will momentarily increase its radio transmission range to allow the ACK packet to be received by the up-link node. However, this approach is not suitable for power-constrained networks where nodes cannot afford to frequently increase their transmission power.

Most recent research on link asymmetry has focused on the design of protocols for unicast routing. However, link asymmetry has relevance for multicast routing too. If a routing protocol that supports multicast groups selects a path with an asymmetric link, then it will be impossible to use the reverse path for forwarding data packets to the group. Therefore, it is possible to find a detour path in the reverse direction so that the combination of the forward and reverse paths reestablishes bidirectional connectivity for the multicast group (Gerla, Lee, Park, & Yi, 2005).

CONCLUSION

This chapter presents an overview of routing protocols that explicitly consider asymmetric links in route construction in wireless ad-hoc networks. Such networks are found in numerous situations including sensor networks, mobile ad-hoc networks, or networked embedded control systems. Many recent routing protocols achieve high connectivity by utilizing asymmetric links for route construction. Finding a set of forward and reverse routes to create a bidirectional abstraction is the most common approach to utilize asymmetric links in the network. However, most existing protocols use a fixed pair of forward and reverse route to deliver data packets, until one or both of them are broken. Therefore, these protocols are incapable of exploiting the asymmetric links being formed while data packets are being transmitted along the original route. Routing protocols that use route optimization techniques are able to improve on data transfer latencies, and sometimes reduce the possibility of route failure by replacing the deteriorating routes with newly discovered ones with good reliability. Also, a symmetric path may fail when a previously symmetric link becomes asymmetric. Existing protocols will start a new route discovery although the path is still usable for data delivery in the forward direction. In this case, it is sufficient to find a short multihop *tunneling path* from the down-link node to the up-link node using a *local route recovery* method, saving a substantial amount of overhead caused by flooding the entire network with a new route discovery. Therefore, the conclusion is that link asymmetry is more of a route maintenance problem than a route discovery problem. The chapter provides a detailed insight in the design of efficient on-the-fly route optimization and route maintenance mechanisms.

REFERENCES

Broch, J., Johnson, D. B., & Maltz, D. A. (1998). *The Dynamic Source Routing Protocol for Mobile Ad Hoc Networks. Ad Hoc Networking* (pp. 139–172). Addison-Wesley.

De Couto, D. S. J., Aguayo, D., Bicket, J., & Morris, R. (2003, September), A High Throughput Path Metric for Multihop Wireless Routing. In *Proceedings of the Annual International Conference on Mobile Computing and Networking (ACM MOBICOM)*, San Diego, California, USA.

Ganesan, D., Estrin, D., Woo, A., Culler, D., Krishnamachari, B., & Wicker, S. (2002). *Complex Behavior at Scale: An Experimental Study of Low-Power Wireless Sensor Networks* (Technical Report CSD-TR 02-0013). University of California at Los Angeles (UCLA).

Gerla, M., Lee, Y.-Z., Park, J.-S., & Yi, Y. (2005, December). On Demand Multicast Routing with Unidirectional Links. In *Proceedings of the Wireless Communications and Networking Conference (WCNC)*, Volume 4, New Orleans, LA, USA.

Hass, Z. (1997, October). A New Routing Protocol for the Reconfigurable Wireless Networks, *In Proceedings of the 6th IEEE International Conference on Universal Personal Communications (IEEE ICUPC)*, San Diego, CA, USA.

Karp, B., & Kung, H. T. (2000, August), Greedy Perimeter Stateless Routing for Wireless Networks. In *Proceedings of the Sixth Annual ACM/IEEE International Conference on Mobile Computing and Networking (MobiCom)*, Boston, MA.

Kim, D., Toh, C. K., & Choi, Y. (2000, October). RODA: A New Dynamic Routing Protocol using Dual Paths to Support Asymmetric Links in Mobile Ad-hoc Networks. In *Proceedings of the 9th International Conference on Computer Communications and Networks (IEEE ICCCN)*, Las Vegas, NV, USA.

Kim, D., Toh, C. K., & Choi, Y. (2003, April). GAHA and GAPA: Two Link-Level Approaches for Supporting Link Asymmetry in Mobile Ad-hoc Networks. *IEICE Transactions on Communications. E (Norwalk, Conn.)*, *86-B*(4), 1297–1303.

Lee, S., Gerla, M., & Toh, C. (1999, August). A Simulation Study of Table-driven and On-demand Routing Protocols for Mobile Ad-hoc Networks. *IEEE Network*, *13*(4), 48–54. doi:10.1109/65.777441

Liu, R. P., Rosberg, Z., Collings, I. B., Wilson, C., Dong, A. Y., & Jha, S. (2008, September). Overcoming Radio Link Asymmetry in Wireless Sensor Networks, In *Proceedings of the International Symposium on Personal, Indoor and Mobile Radio Communications (PIMRC)*.

Marina, M. K., & Das, S. R. (2002, June). Routing Performance in the Presence of Unidirectional Links in Multihop Wireless Networks. In *Proceedings of the International Symposium on Mobile Ad Hoc Networking and Computing (MobiHoc)*, EFPL Lausanne, Switzerland.

Misra, R., & Mandal, C. R. (2005, January), Performance Comparison of AODV/DSR On-Demand Routing Protocols for Ad-Hoc Networks in Constrained Situation. In *Proceedings of IEEE International Conference on Personal Wireless Communications (ICPWC)*, New Delhi, India.

Morino, H., Miyoshi, T., & Ogawa, M. (2005, April). Ad Hoc Unidirectional Routing Protocol Based on Relay Control of Route Requests. In *Proceedings of the IEEE Autonomous Decentralized Systems (ISADS)*, Chengdu, China.

Morino, H., Miyoshi, T., & Ogawa, M. (2007, March). Unidirectional Ad-Hoc Routing Protocol with Area-Controlled Flooding Using Overheard N Node Information. In *Proceedings of the IEEE Autonomous Decentralized Systems (ISADS)*, Sedona, AZ, USA.

Nesargi, S., & Prakash, R. (2000, April). A Tunneling Approach to Routing with Unidirectional Links in Mobile Ad-hoc Networks. In *Proceeding of the 9th International Conference on Computer Communications and Networks (ICCCN),* Las Vegas, NV, USA.

Perkins, C. E., Royer, E. M., & Das, S. R. (1999, February). Ad Hoc On Demand Distance Vector (AODV) Routing. In *Proceedings of the 2nd IEEE Workshop on Mobile Computing Systems and Applications (WMCSA)*, New Orleans, LA, USA.

Prakash, R. (1999, August), Unidirectional Links Prove Costly in Wireless Ad-Hoc Networks. In *Proceedings of the 3rd International Workshop on Discrete Algorithms and Methods for Mobile computing and Communications (DIAL M)*, Seattle, WA, USA.

Ramasubramanian, V., & Mosse, D. (2008, February). BRA: A Bidirectional Routing Abstraction for Asymmetric Mobile Ad Hoc Networks. *IEEE/ACM Transactions on Networking, 16*(1), 116-129.

Roosta, T., Menzo, M., & Sastry, S. (2005, May). Probabilistic Geographic Routing in Ad Hoc and Sensor Networks. In *Proceedings of the International Workshop on Wireless Ad Hoc Networks (IWWAN)*, London, UK.

Sang, L., Arora, A., & Zhang, H. (2007, September), On Exploiting Asymmetric Wireless Links via One-way Estimation. In *Proceedings of the ACM International Symposium on Mobile Ad Hoc Networking and Computing (MobiHoc)*, Montreal, QC, Canada.

Srinivasan, K., & Levis, P. (2006, May). RSSI is Under Appreciated. In *Proceedings of the Third Workshop on Embedded Networked Sensors (EmNets)*, Boston, MA.

Suh, J.-J., Quan, S. G., Park, J. D., Park, J. H., & Kim, Y. Y. (2007, January). Asymmetric Link Dynamic Routing Algorithm for Ubiquitous Multimedia Services. In *Proceeding of the 13th International Multimedia Modeling Conference (MMM)*, Singapore.

Toh, C.-K. (1997, March). Associativity Based Routing for Ad-hoc Mobile Networks, *Wireless Personal Communications Journal. Special Issue on Mobile Networking and Computing Systems, 4*(2), 103–139.

Unoki, H., & Higaki, H. (2004, March). Loop-Based Source Routing Protocol for Mobile Ad-hoc Networks. In *Proceeding of the IEEE International Parallel & Distributed Processing Symposium (IPDPS)*, Long Beach, CA, USA.

Wang, Z., Lu, J., & Tang, J. (2007, March). Neighbor Monitoring Mechanism to Solve Unidirectional Link Problem in MANET. In *Proceedings of the 3rd International Conference on Wireless and Mobile Communications (ICWMC)*, Guadeloupe, French Caribbean.

Wu, S.-L., Tseng, Y.-C., Ni, S.-Y., & Sheu, J.-P. (2000, January). Route Maintenance in a Wireless Mobile Ad Hoc Network. In *Proceedings of the 33rd Hawaii International Conference on System Sciences (HICSS)*, Maui, Hawaii, USA.

Zhou, G., Krishnamurthy, S., & Stankovic, J. A. (2004, June). Impact of Radio Irregularity on Wireless Sensor Networks. In *Proceedings of the 2nd International Conference on Mobile Systems, Applications, and Services (MobiSys)*, Boston, MA, USA.

Zuniga Zamalloa, M., & Krishnamachari, B. (2007). An Analysis on Unreliability and Asymmetry in Low-Power Wireless Links. *ACM Transactions on Sensor Networks (TOSN), 3*(2).

KEY TERMS AND DEFINITIONS

Down-Link Node: A node having an incoming asymmetric link is called a down-link node. It can only receive packets on this asymmetric link, but cannot use it to send packets in the opposite direction.

Link Asymmetry: This is caused by discrepancy in radio transmission ranges of two neighboring nodes. For two neighboring nodes A and B, if node A is within the transmission range of node B, but not vice versa, then there is a link asymmetry between nodes A and B.

Reactive Routing Protocols: This class of protocols finds a route on demand by flooding the network with Route Request packets.

Reverse Route Computing: This is a method for finding a destination-to-source reverse route, such that the combination of the source-to-destination forward route and destination-to-source reverse route can create a bidirectional abstraction of a symmetric path between the source node and the destination node.

Route Discovery: This is a mechanism using which a source node S wishing to send a data packet to a destination node D obtains a route to the destination. Route discovery is only attempted when S does not already know a route to D.

Route Maintenance: This is a mechanism using which a source node S is able to detect whether the network topology has changed such that the route is no longer usable because one or more links along the route have failed.

Route Optimization: This is a mechanism to refine or optimize routes on-the-fly while they are being used for data transmission. If there is a route R from node 1-to-n: $1 \rightarrow \ldots \rightarrow n$, and there is a shorter partial route through a node j, where $j \notin R$, such that $1 \rightarrow \ldots \rightarrow j \rightarrow \ldots \rightarrow n$, then data traffic will be redirected through the shorter route.

Up-Link Node: A node that has an outgoing asymmetric link is called the up-link node. It can use the asymmetric link to send packets, but cannot receive packets on that link.

Wireless Ad-Hoc Networks: A decentralized wireless network that is ad-hoc because each node is willing to forward data on behalf of other nodes. The determination of which nodes forward data is made dynamically based on the network connectivity.

Chapter 14
Smartphones to Access Hybrid Information Spaces

Spyros Veronikis
Ionian University, Greece

Kyriaki Zoutsou
Ionian University, Greece

Giannis Tsakonas
Ionian University, Greece

Christos Papatheodorou
Ionian University, Greece

ABSTRACT

This chapter presents how mobile phones nowadays can be used as a handy and always around medium to interact with information collections of both physical and electronic form. It presents their evolution into powerful computing devices capable of communicating with other computing infrastructures, like the World Wide Web, and assisting them in finding the information of interest in the most convenient form, right when the need arises. It also provides a description of the current interaction modes between users, devices and information objects as well as some examples of first and second-generation mobile services. Furthermore, the authors hope that understanding the potential introduced by mobile phones in the modern information landscape can bring some insight to new information seeking strategies, that enhance exploration and not just querying. This understanding can be used to create new, innovative and novel services to enhance the seeking experience while interacting with hybrid information spaces.

INTRODUCTION

The total number of mobile phone users in the world is currently estimated at 3.3 billion, accounting for nearly 50% of the population on Earth (Reuters, 2007). Nearly 90% of the world's population is expected to be using mobile phone services by 2010. As broadband wireless communications become available, voice telephony is at the same time converging with electronic

DOI: 10.4018/978-1-60566-250-3.ch014

services. This integration enables mobile phones to access a range of data services. On the other hand, computer evolution and electronic publishing created a wealth of information, available in electronic form. In the mid 90s, the advances in computer networking technology made the Web and the Internet a new medium to distribute electronic content and access content which was not locally held. The success of Internet led to the creation of on-line resources, like magazines and web pages, as well as the wider use of digital libraries and other collections that were universally accessible. The new resources strongly affected the users' seeking behaviour and strategies. As a result, nowadays there is a strong demand for advanced searching and browsing in diverse sources of information, which enhances its completeness, validity and currency.

The need to access electronic sources, anytime available, combined with the potential raised by the wide spread of mobile phones, which can be used as terminals to receive electronic information in almost any place, has been recognized worldwide. Mobile Computing (MC) is a term to describe the ability of using anywhere a portable computing device by means of wireless or wireline communications. This chapter focuses on utilizing MC in hybrid information spaces, i.e. environments that hold information both in physical and electronic form. Mobile interaction with hybrid spaces can be applied in areas such as advertising, tourism and museum guides, peer-to-peer information sharing and remote control of public resources, like printers and displays.

In the rest of the chapter we describe the usage of portable computing devices as a means of interaction with physical and digital spaces. We first provide some background information on smartphones, on the challenges of mobile computing and an overview of a web-based architecture. We then discuss interaction issues with the device and describe their usage in hybrid spaces. Finally we outline the future trends and conclusions.

BACKGROUND: SMARTPHONES AND MOBILE COMPUTING

Apple released the first personal digital assistant (PDA) in 1993 and it was a pocket-sized mobile device used for Personal Information Management (PIM). Typical applications included were calendar, address book, calculator, world-clock and note pad. At the same time, IBM released *Simon*, the first smartphone, which besides PIM functionalities could support mobile telephony and SMS texts over 2G digital cellular telephony networks. In 1995 infrared and wire-line serial communication with nearby devices was implemented as an alternative for data exchange and synchronization. In addition, smartphones could connect to proxies and join other computer networks to retrieve data such as e-mail. The need to transfer quickly and display bigger amounts of data led to faster communication protocols (Bluetooth, EDGE) and 3G telephony networks in early 2000. At that time, cameras and colourful screens were considered standard equipment. By 2005, many smartphones were enabled with open operating systems, rapid transfer protocols (WiFi, USB), touch screens, and solid-state memory cards. Besides telephony, they could support web browsing/email, instant messaging, document viewers and multimedia playback. Nowadays, smartphones are equipped with GPS receivers, multi-touch screens and advanced interaction modalities, such as handwriting recognition and proximity sensing. Smartphones have evolved into powerful communication and computing devices capable of collecting and processing data, anytime/anywhere, and seem to become a universal mobile terminal for communication, control, entertainment and location-based services.

Smartphones' increasing popularity and current computational capabilities meet the mobile application requirements and they compare well against other portable devices, such as PDAs and laptops. However, mobile computing introduces new challenges and constraints, inherent to mobility. Forman and Zahorjan (1994) classify them in three major categories; wireless communication, mobility and portability. Mobile connectivity can be highly variable in performance and reliability due to heterogeneous networking, disconnections, available bandwidth and security risks. Furthermore, the frequently changing location of the device, widely known as roaming, forces it to change its identification number and tracking it down might be considered as privacy violation. Mobility is inherently hazardous since the device might be stolen, damaged or lost. Also, a portable device must depend on a finite energy source, needs to be lightweight, small and durable, and therefore it is resource-poor compared to a desktop device.

However, when using handhelds indoors, in closed and controlled environments, such as exhibitions and libraries, and for short information provision, many of the above mentioned issues are easier to handle. Smartphones can join a single network available, usually a WLAN, which offers full coverage, broadband connections and low traffic due to the limited number of connection requests in the area. Figure 1 shows the two communication modes available in a smartphone, which the user may choose depending on signal strength, cost and speed. Communication through a circuit-switched network is usually much slower and expensive but available almost anywhere and more secure. To reduce security risks, data exchange with the wireless network can be encrypted with a key, and have connection closed after some period of idle time. There is no issue on privacy violation because of user tracking, since users move in a certain public area anyway, exactly as regular visitors. Devices are unlikely to be stolen because of the presence of security or institution's personnel. The small user interface of smartphones makes user interaction with the device cumbersome, but as long as the information exchanged is short, people are willing to tolerate the inconvenience for the benefit of added-value services. To support users in their tasks, nearby resources such as info-kiosks, computer rooms, printers and displays can provide features and services not available by the user's device. Regarding user tracking, we note that precisely extracting the user location inside a building is not an easy task. Most Global Positioning System (GPS) receivers cannot detect satellite signals to inform the user of his location. Hightower and Boriello (2001) developed a taxonomy of several location-sensing techniques for indoor and outdoor applications.

The wide adoption of the Internet and the World Wide Web, the industry trend in the deployment of web services technology and the evolution of

Figure 1. Two alternative communication paths to access data repositories with a smartphone

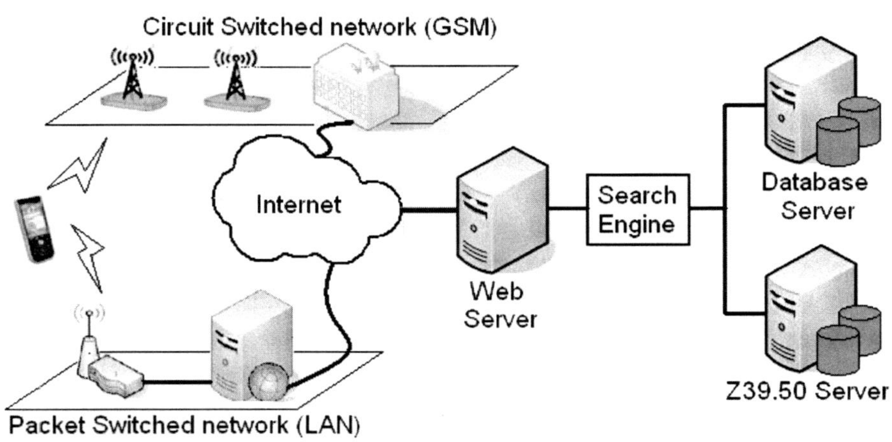

mobile phones to support data exchange, make smartphones ideal for interacting with hybrid information spaces. The CoolTown project (Kindberg et al., 2000) explores the opportunities of using a handheld device to interact with objects, people and places that have a web presence in the WWW. Among several technologies, service oriented architectures (SOAs) are mostly created by using web services. This architecture enhances interoperability since any device capable of running client application software, qualifies. It contains three major components; discovery, description and messaging (Hirsch, Kemp & Ilkka, 2006, p.22). Service providers, such as web servers, must be discovered prior to using the service. Several mechanisms can be used, like sensing or scanning as described above. Alternatively, indirect mechanisms can be used, such as service lists or directories on a web page. Once a service has been detected, the device needs to know which protocols to use for data exchange, how to structure messages and which security or other policies to observe. Web Service Description Language (WSDL) is used to create structured documents that describe the services. These XML-based messages are wirelessly exchanged according to the SOAP protocol, usually over HTTP/HTTPS.

These processes (discovery, description and messaging) are performed in a 3-tier architecture, namely *Presentation*, *Logic* (or application) and *Data*. At the presentation tier the interface (web browser) of the smartphone receives an input from the user or a smart object. The request is processed in the logic tier by the application server which queries the database management system at the data tier and retrieves the necessary information. The extracted data are sent back to the middle tier where they are properly formed by the web server and passed through the presentation tier to the software client in the user's device. Such architectures ensure anywhere/anytime accessibility and therefore are appropriate for the development of mobile information services.

INTERACTING WITH SMARTPHONES

To interact with the user, smartphones provide a user interface similar to Windows, Icons, Menus and a Pointing device (WIMP). Applications run on separate windows, but because of the small screen size, usually only one window is shown at a time, with the rest being minimized. A stylus and a touch screen can be used to make selections from menus, and special purpose commands can be invoked by tapping on corresponding icons. A joystick and special purpose buttons allow for thumb operation of the device. The user can enter some text, by means of a small keyboard that can be numeric, virtual, full or a slide-out QWERTY. Some models also support handwriting recognition for faster and more natural input.

Rukzio (2006) proposed a taxonomy of mobile interaction regarding basic concepts, like sensed property, number of dimensions, direct or indirect interaction, and relative or absolute commands. They refer to either human-computer interaction or interaction with objects that, when prompted, can send electronic information to the device. Hence five interaction categories exist: (a) *Touching* is the interaction technique where the device comes very close or touches the object to trigger a short data exchange session. It requires some sort of contact mechanism or a proximity sensor. (b) *Pointing* is a technique according to which the user points his device to the object in order to retrieve some information, as in the case of infrared connections. (c) *Scanning* is an interaction technique that uses wireless signals and protocols, such as Bluetooth, to search for nearby available resources or services. Once a strong signal is detected, a connection is established and the device exchanges data with the resource. (d) *User-mediated object selection* is an indirect interaction technique between the device and physical objects. The smartphone cannot identify on its own the physical object in the environment. Instead, the user types in some information provided by the object, such as its call number or a URL, to establish a link between them. (e) *Indirect remote control* is an extension of conventional interaction techniques that uses the smartphone's joystick, stylus or keypad, to take control of physical objects, such as public displays. However, there are some interaction barriers such as the small screen size that raises presentation issues, the small keyboard that makes input of long texts cumbersome and the battery life that allows for some hours of operation.

Mobile devices have been used in rich information environments such as large museums, to offer added-value information services, like audio descriptions of artefacts in visitor's native language. Nowadays, smartphones support multimedia playback and wireless connections with nearby computers that serve electronic information, such as info-kiosks, beam stations and web servers. This information provision is particularly important in areas where the objects themselves are not self-explanatory and there is no space available for conventional labels or interpretations. It can be applied in commerce, using advertising posters and interactive display cases, in museums, in libraries and in exhibitions (Rukzio, 2006). An example of such an one-way interaction is the case of a user who photographs a quick response code (QRC), holding information about the bus schedule (time and routes) at a bus station instead of copying it by hand (Liu, Doermann & Li, 2006). The data is transferred to the smartphone and interpreted to something the user can understand, i.e. a short text. In cases that the information quantity is too much to be stored on a visual code, it is usually stored in an HTML document, which can be accessed by a web client running on the smartphone. The visual code only needs to hold the URL of the HTML document.

Moreover a smartphone can assist its user in navigation and provide valuable guiding instructions (Baus, Cheverst & Kray, 2005). GPS-enabled devices and maps that depict floor plans generally support navigation in the physical space. The user can set a destination (point, object or room) or have superimposed on the map items of special interest, as in the case of personal guided tours in museums. Additionally, navigation aids like structured data schemes, such as thematic directories, headers, previous and next page commands and links to other information topics, can help the user browse the intangible space of electronic information. A similar case in a library setting is described in the following scenario; a new visitor accesses the web interface of a library catalogue to search for a book. From its metadata, she ensures it is available and passes its location to a map application running on her smartphone. She follows the instructions of the device to find the book (Aittola, Ryhänen & Ojala, 2003).

SMARTPHONES IN HYBRID INFORMATION SPACES

A hybrid information space is an environment that holds information material in both physical and electronic form. The user can directly interact with books, artefacts, etc in the physical space and also use a computer terminal to access electronic sources of information (texts, multimedia files), held locally or remotely, by means of computer networks such as the Internet. An interface on the terminal allows the user to interact with the electronic portion of the hybrid space. Nowadays, academic libraries are representative examples of hybrid information space, since they hold information in book repositories (physical) and digital libraries. Figure 2 shows the search interface of a prototype developed at Ionian University[1] to support users in searching the library's collections. The search results are displayed on the screen and the user can select a retrieved record to see its metadata (e.g. author, subject area, classification number, availability, location, etc). Floor maps available for reference can help users in locating the books in the library. In addition, several document viewers (e.g. PDF, DOC) are used to display the summaries and full-text of electronic files, retrieved from the digital collections. This prototype was evaluated in regard to the user acceptance and satisfaction (Veronikis, Zoutsou & Papatheodorou, 2007) and the results were encouraging. The ability to access information anytime/anywhere and to transfer it either by saving or by emailing it was strongly acknowledged. New library visitors considered the navigation aids quite useful.

To enhance the user's experience of interaction with the hybrid space, the device operates as a bridge that connects the physical and electronic spaces, and therefore must interact with the items of interest as well. For example, the user can retrieve a detailed description of an object by pointing the smartphone to an infrared beacon, or using the built-in camera to photograph a visual code, which is then interpreted by the device. Alternatively, the device can scan the environment to detect nearby smart objects that broadcast wireless signals containing their call number. Another approach is using the phone to touch an object. It is very convenient when the objects of interest are near the user, such as exhibition items. Want *et al.* (1999) presented a prototype, which incorporates RFID tags into objects and an RFID reader on the device. Dominant companies in the mobile phone arena, such as Nokia, Philips, Benq-Siemens, NTTDoCoMo, and Sony have also built several prototypes to study the use of RFID tags and other proximity sensors for touching, sensing and near-field-communications (NFC). When sensing a nearby object, the mobile phone retrieves its ID number and establishes a connection with a database or other electronic source of information to retrieve related data and display them on the screen.

Thanks to the advanced technological and communication characteristics of mobile devices new services have been developed which enhance the user interaction with hybrid information spaces. This two-way communication involves the iterative data exchange between the user and three components; (a) the user's device, (b) the physical object/person/place and (c) a web-based

Figure 2. A web form interface to search in library's collections

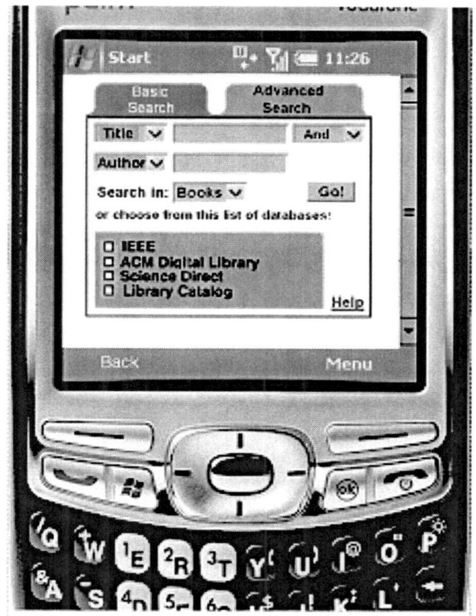

application that retrieves relevant information from the digital space.

Kulfik *et al.* (2007) present an augmented museum visitor's guide, which provides context-aware services to support person-to-person and person-to-system communication, thus enabling users to share the experience and knowledge of other visitors. By sending direct messages to other users or leaving a virtual "post-it" message for others to see, a user can help other visitors and provide recommendations to make the most out of their visit to the hybrid environment. The system uses contextual information, such as user's locations, profiles and history of interactions with exhibits, to interpret the messages and deliver them accordingly.

Yatani, Onuma, Sugimoto and Kusunoki (2004) discuss on the usage of PDAs in museums to support collaborative learning. Children use the handhelds to extract information from the physical space in order to solve puzzles and provide answers to the questions posed on the screen. In addition, an HTML document is used to inform the users of the handhelds about their path and interaction history, during their visit in the hybrid space. The document can be reviewed in a desktop workstation later at home.

Besides ad-hoc interaction, web services can combine a desktop PC at home/office and the portable computing device. Krüger *et al.* (2004) present an approach to a personalized navigation service that interconnects specialized user interfaces for desktop PCs, in-car and on-foot use. The portable navigation system enables the traveller to plan his trip from the PC at home, to choose personalized events of interest at certain locations and to retrieve further data, such as the weather conditions. All this information is stored in a travel itinerary and migrated to his personal handheld, where it can be available whether in car, on-foot or inside other buildings. Krumm and Horvitz (2007) create driver destination models by acquiring GPS data in order to predict the user's destination and notify the user by triggering alerts and recommendations about traffic.

First generation applications of smartphones such as mobile telephony, SMS/MMS, Instant messaging, PIM, e-mail and web browsing, positioning, navigation assistance, data synchronization, and information push-services are mostly extensions of PDA and cell phone functionalities. However, in recent years, second generation applications start to emerge as a result of advancements in mobile computing, wireless networks and proximity sensors.

Smartphones are used for mobile transactions and payments conducted through communication networks that interface with wireless devices (Hu et al, 2004). For example a smartphone can be used to check a bank account and perform transfers regardless of time and location of transaction. It can also be used as an electronic wallet, where an SMS sent to a subscription service will charge the owner's phone bill for the product purchased (e.g. a bus ticket). Smartphones can also be used to buy products from vending machines or grant access to restricted areas by transmitting owner's identification and charging data by means of proximity sensors.

Another promising field for smartphone application services is E-health where apart from patient-profiles and clinical history the devices can keep and access information about prescriptions, medicine administration, and in-home patient monitoring (Chan et al, 2008). For example, eHIT (http://www.ehit.fi) developed a system that uses a smartphone, which communicates with data loggers and databases to deliver round-the-clock vital measurements, such as body temperature, blood pressure and glucose levels.

Raento et al. (2005) present a platform for context-aware mobile applications, such as social networking. Users use their devices to discover and communicate with nearby people having similar interests. In addition, they can submit or read previously posted, geo-referenced comments and share pictures of locations or other content. Data is exchanged via SMS, instant messaging, RSS or a social networking web service like Jaiku[2] and Twitter[3].

FUTURE TRENDS AND CONCLUSION

Next generation mobile networks are expected to converge into a ubiquitous architecture, which includes broadband cellular networks, wireless local networks (WLANs), mobile ad-hoc networks, etc. In addition, personal communication devices are expected to be as common as a wristwatch, connecting to distributed-computing systems, exchanging data with local or remote computers and sensor networks. Next generation interfaces will enable multimodal interaction with the user, i.e. speech and gesture recognition, and with smart objects by means of sensing technologies, such as RFID. Many objects, places and events of particular interest will have an associated electronic description, available anytime/anywhere.

Furthermore, services will be personalized and context-aware (Chen & Kotz, 2000), meeting the users' particular interests and taking into account their preferences, location, nearby resources or people, social situation, network connectivity, available bandwidth, time of day, lighting conditions and noise levels.

As we have seen mobile phones are becoming a ubiquitous computing device used anywhere. Additionally, the WWW interconnects billions of electronic information sources, enabling users with additional information. The devices can be used as information harvesting tools that collect data from the physical space, process them, and subsequently retrieve related electronic data. In turn, this process may start a new interaction cycle. New added-value services can be designed to exploit this interaction, offering opportunities for exploration in hybrid environments, information access and delivery and collaborative activities. However, in order to create usable and effective mobile systems and services for multimedia content delivery, developers have to address new issues and challenges that arise from the user interaction with a hybrid world, such as the selection of proper content and methods to deliver it, its presentation and the most suitable interaction mode.

REFERENCES

Aittola, M., Ryhänen, T., & Ojala, T. (2003, September 8). *SmartLibrary - Location-aware mobile library service*. Paper presented at the 2003 Fifth International Symposium on Human Computer Interaction with Mobile Devices and Services, Udine, Italy.

Baus, J., Cheverst, K., & Kray, C. (2005). A survey of map-based mobile guides. In Meng, L., Reichenbacher, T., & Zipf, A. (Eds.), *Map-based mobile services: Theories, Methods, and Implementations* (pp. 197–216). Springer Berlin Heidelberg. doi:10.1007/3-540-26982-7_13

Chan, V., Ray, P. & Parameswaran, N. (2008). Mobile e-Health Monitoring: an agent-based approach. *IET Communications, 2*(2), 223-230. Retrieved June 20, 2008 from IEEE Xplore.

Chen, G., & Kotz, D. (2000). *A Survey of context-aware, mobile computing research* (Dartmouth Computer Science Technical Report TR2000-381). Hanover, NH.

Forman, G. H., & Zahorjan, J. (1999). The challenges of mobile computing source. In Milojicic, D., Douglis, F., & Wheeler, R. (Eds.), *Mobility: processes, computers, and agents* (pp. 269–285). New York: ACM Press/Addison-Wesley Publishing Co.

Hightower, J., & Borriello, G. (2001). Location Systems for Ubiquitous Computing. *Journal of IEEE Computer, 34*(8), 57–66.

Hirsch, F., Kemp, J., & Ilkka, J. (2006). *Mobile Web Services: Architecture and Implementation*. West Sussex: John Wiley & Sons. doi:10.1002/9780470017982

Hu, W.-C., Lee, C.-W., & Jen, J.-H. (2004). Mobile Commerce Systems. In Nansi, S. (Ed.), *Mobile Commerce Applications* (pp. 1–23). Hershey, PA: Idea Group Publishing.

Krüger, A., Butz, A., Müller, C., Stahl, C., Wasinger, R., Steinberg, K. E., & Dirschl, A. (2004, January 13). *The connected user interface: realizing a personal situated navigation service*. Paper presented at 9th International Conference on Intelligent User Interfaces, Funchal, Madeira, Portugal. Retrieved September 20, 2007 from ACM.

Krumm, J., & Horvitz, E. (2007, July 25). *Driver Destination Models*. Paper presented at 11th International Conference on User Modeling, Corfu, Greece. Retrieved October 27 from SpringerLink.

Kuflik, T., Sheidin, J., Jbara, S., Goren-Bar, D., Soffer, P., Stock, O., & Zancanaro, M. (2007, January 28). *Supporting small groups in the museum by context-aware communication services*. Paper presented at 12th International Conference on Intelligent User Interfaces, Hawaii, USA. Retrieved March 3, from ACM.

Liu, X., Doermann, D., & Li, H. (2006, December 4). *Imaging as an alternative data channel for camera phones*. Paper presented at 5th International Conference on Mobile and Ubiquitous Multimedia, Stanford, California. Retrieved May 15, 2007 from ACM.

Raento, M., Oulasvirta, A., Petit, R., & Toivonen, H., (2005). ContextPhone: A Prototyping Platform for Context-Aware Mobile Applications. *IEEE Pervasive Computing, 4*(2), 51-59. Retrieved December 9, 2006 from IEEE Xplore.

Reuters. (2007). *Global cell phone use at 50 percent.* Retrieved November 30, 2007, from http://news.yahoo.com/s/nm/20071129/tc_nm/cellphones_world_dc

Rukzio, E. (2006). *Physical mobile interactions: mobile devices as pervasive mediators for interactions with the real world.* PhD thesis. Faculty for Mathematics, Computer Science and Statistics. University of Munich.

Veronikis, S., Zoutsou, K., & Papatheodorou, C. (2007, October 1). *An evaluation methodology of ubiquitous access in physical and digital information spaces.* Paper presented at 16th Panhellenic Conference of Academic Libraries, Piraeus, Greece.

Want, R., Fishkin, K. P., Gujar, A., & Harrison, B. L. (1999, May 15). *Bridging physical and virtual worlds with electronic tags*. Paper presented at SIGCHI Conference on Human Factors In Computing Systems: the CHI is the limit, Pennsylvania, USA. Retrieved February 8, 2007 from ACM.

Yatani, K., Onuma, M., Sugimoto, M., & Kusunoki, F. (2004). Musex: A system for supporting children's collaborative learning in a museum with PDAs. *Systems and Computers in Japan 35*(14), 54-63. Retrieved February 20, 2007 from IEEE Xplore.

KEY TERMS AND DEFINITIONS

Global Positioning System (GPS): is the satellite-based positioning technology that allows a GPS receiver to identify its location anywhere on Earth. GPS receivers can be found as handheld devices, installed in automobiles for navigational aids, and as modules that attach to PDAs, such as a PalmPilot.

Human-Computer Interaction: HCI (or CHI) is the scientific area of computer science that studies the design, development and evaluation of interactive systems. HCI uses guidelines and evaluation methods in order to support the design of usable and user adapted computing systems with improved interaction.

Hybrid Space: refers to physical places that support or integrate electronic information delivery. This kind of information delivery may be realized through the provision of physical infrastructure that allow information access, such as computer workstations or info-kiosks, and/or through support of users' mobile devices, usually through some kind of wireless networking. Typical examples include cultural heritage institutions, like museums and galleries, historic monuments, like archaeological sites, and knowledge institutions, like libraries and archives.

Location Dependent Service (LDS): is a context-aware service that adapts itself, depending on user's physical location. LDS is based on the transmission of geographic location information to mobile devices in the same context. Each mobile device can request an LDS and attach geographic location information in order to receive the service.

Near-Field Communication (NFC): is a standards-based, short-range wireless connectivity technology that enables simple and safe two-way interactions among electronic devices, operating in the 13.56 MHz frequency range, over a typical distance of a few centimetres. Interactions are initiated when special purpose tags (RFIDs) are detected by an appropriate scanner.

Recommender Systems: they use information filtering (IF) techniques attempting to present information items (movies, music, books, news, web pages) that are likely of interest to the user. Typically, a recommender system compares the user's profile to some reference characteristics. These characteristics may be from the information item (the content-based approach) or the user's social environment (the collaborative filtering approach).

RFID (Radio Frequency Identification): is an identification method, relying on storing and remotely retrieving short data using devices called "RFID tags" and "transponders". The tag, also known as electronic label, keeps the data (e.g. serial number of an object), which can be retrieved by the transponder (via radio signals) when it comes close to the tag. Unlike barcodes there is no need for line of sight to the scanner and identification is many times faster.

Smartphone: is a mobile phone with advanced capabilities, offering PC-like functionality. It provides digital voice service, as well as text messaging, Web access, still and video camera, audio and video streaming and PIM organizer. Late models integrate GPS receiver, mutli-touch screens, proximity and accelerometer sensors. Compared to PDAs they usually have smaller screens and keypads instead of virtual keyboards and smaller memory capacity. However, late smartphone models such as the iPhone (by Apple) indicate that the distinction between PDAs and smartphones are converging into one device.

WiFi: is the colloquial name for the IEEE802.11 data protocol used to transmit data wirelessly over short distances. WiFi networks have become extremely popular, revolutionising the networking scene over the past five years. Many hotels now offer WiFi access for travelling businesspeople.

ENDNOTES

1. Dept. of Archives and Library Sciences, Digital Libraries and Electronic Publishing Lab (http://dlib.ionio.gr)
2. http://www.jaiku.com
3. http://twitter.com

Chapter 15
Location-Dependent and Context-Aware Computing

Stan Kurkovsky
Central Connecticut State University, USA

ABSTRACT

Systems with context-aware features are typically able to adapt their behavior to the changes of the environment surrounding the system and/or its users without any explicit intervention of the user. Therefore, context-aware systems tend to increase the usability and effectiveness by taking into account the context of their environment. Location-based and context-aware systems are typically associated with mobile devices; they react to reflect their current physical location, time, other surrounding systems, their user(s), as well as other features of their environment, and adapt their operation to the changes of the environment. This information about the many aspects of the system's environment is referred to as context, which can be obtained from a variety of sources, such as RFID and other sensors, wireless network information, physical device status, maintaining information about the user(s) and many other sources.

INTRODUCTION

In his seminal paper, Mark Weiser (1991) presented a vision of pervasive computing, in which the users will be surrounded by a ubiquitous computing environment available to them at anytime and anywhere. In this paradigm, computing devices and sensors are seamlessly embedded in the environment and enable the computing system to possess a certain level of intelligence so that the users would give their undivided attention to specific task at hand instead of focusing on resolving technical issues concerning the operation of and interaction with the computing system and devices. For example, a car equipped with a navigation system integrated with the vehicle audio system and equipped with a voice recognition technology could allow the driver to set a destination location while driving using only voice commands; being able to operate the navigation system without having to touch its controls, such a voice-enabled interface would not distract the

DOI: 10.4013/978-1-60566-250-3.ch015

driver while driving in city traffic if he or she. If this navigation system could sense when the vehicle is low on fuel and it has some information about the driver's preferences, it could alert the driver when the car is in the vicinity of a gas station that sells the preferred brand of gasoline. Such a behavior, when the system can adjust itself to the current conditions without any explicit actions of the user, is typical for context-aware systems. Context awareness enables pervasive computing systems to increase their usability and/or efficiency by reacting to changes in the environmental context.

Driven by rapid progress in development of low-cost power conservative miniature devices that can be embedded in our everyday surroundings, the vision of pervasive computing becomes ever more realistic. Users would interact with such environments not by using a computer but rather by using things they encounter in everyday life that have been enhanced with computing technology. Context awareness allows the devices and systems to react to and anticipate changes in the environment making the process of interaction with such a pervasive environment more natural to the user. This chapter introduces the concept of context and context-awareness, discusses possible architectures of context-aware and location-based systems, and provides examples of existing systems documented in the literature.

BACKGROUND

Active Badge Location System is considered to be one of the earliest context-aware systems (Want et al, 1992). In this system, the user's current location was determined by a series of infrared sensors and small wearable badges; the information about the user's location enabled the system to forward telephone calls to the telephone closest to a particular user. In the following years, a number of other systems leveraging location context information were described in the literature, including systems receiving location information from the global positioning system (GPS) and from the communication infrastructure, e.g. Global System for Mobile communications (GSM) in cellular networks or WiFi access point connectivity in local area networks. In particular, location context was widely used to implement a number of context-aware tour guide applications (Abowd et al, 1997, Cheverst et al 200, Brown 1996, Kreller et al 1998, Fels 1998). Leonhard et al (1996) described a generic framework supporting location awareness based on a formally defined model of location; Indulska and Sutton (2003) and Becker and Durr (2005) provide comprehensive surveys of location management in context-aware systems.

Location information became the first type of context that was widely used in various context-aware systems. However, context is commonly understood to embrace many other characteristics including people and their identity, other systems and objects, date and time, to name a few. The definition of context has been evolving alongside with the research in context-aware computing; since it is an emerging area of research, currently there is no commonly accepted standard definition describing the concept of context (Dourish 2004). Many authors offer their own definitions that are often very specific to a given research project. One of the most precise definitions of context that does not suffer from specificities and is not too broad at the same time is offered by Dey and Abowd (2000): "Context is any information that can be used to characterize the situation of an entity. An entity is a person, place, or object that is considered relevant to the interaction between a user and an application, including the user and applications themselves."

Context information can be derived from many sources; depending on the nature of the specific instance of context, it can be classified along a number of dimensions. A commonly accepted approach is to distinguish between external and internal context (Gwizdka 2000). External context describes the state of the physical environment that can be measured by various sensors or devices; instances of external context may include location, position, date and time, temperature, proximity, light intensity, etc. Internal context refers to the logical state of the system and its user, which is either directly entered by the user or derived by the system based on the previous history of the user's interaction with the system; internal context may include information from the user's calendar (current tasks, events and due dates), current status of the user's projects, personal preferences and past behavior patterns (e.g. preferred brand of gasoline, movie genres, cuisine, etc.), emotional state, etc.

Most external context instances can be classified as places (physical locations of different scope, which may include rooms, buildings, or streets), people (individuals or groups), and things (physical objects or software, e.g. programs, files or documents) (Dey et al, 2001). In such a classification, each entity can be described using a number of attributes broadly assigned to one of four categories: identity (a unique identifier for every entity), location (in addition to the basic positioning information, this also includes data enabling to deduce co-location, proximity or containment relationships among entities), status (intrinsic characteristics of the entity that can be sensed, e.g. temperature, lighting, noise level for a place, vital signs or current activity for a person, or current load and number of processes for a CPU), and time (a timestamp is used to establish temporal relationships and causality between events and entities).

DESIGN AND IMPLEMENTATION OF CONTEXT-AWARE SYSTEMS

Support for context awareness imposes a set of unique requirements on the architecture and functionality of the system. In general, the choice of the system's architectural model depends on whether the sensors providing context data are remote or local, how much computational and storage resources will be available to the system, its support for mobility and roaming, scale of the system, and its support for extensibility. The method used to acquire the data from which context information is derived is one of the most important factors influencing the system architecture. Generally, there are three possible ways of acquiring this data (Chen 2004). Direct sensor access systems use built-in sensors, from which the context information can be gathered directly without the need for any additional facilities to collect and process the raw sensor data. Middleware-based systems are typically larger in scope and have a layered architecture with one of the low-level layers encapsulating all context sensing and processing functionality. Distributed location-dependent and context-aware computing systems often use a dedicated context server, which is responsible for sensing, collecting, processing and maintaining all context information; this approach logically extends the middleware-based architecture and facilitates concurrent access by multiple context-aware applications.

In modern context-aware systems, the bulk of end-user interaction is handled by miniature mobile devices, such as smart phones or specialized miniature devices, all of which typically lack sufficient processing power or storage capacity. Limited capabilities of these devices lead to a wider adoption of system architectures based on the middleware model that hides the details of context data processing or context server model, in which context data processing and context management is offloaded onto a dedicated computational entity. Furthermore, such a separation of collection, processing, and utilization of context naturally allows creating systems with more reusable and extensible architectures.

Sensors used to capture data from which context information is derived can be divided into three types originally proposed to classify location context (Indulska & Sutton, 2003), but equally applicable to other dimensions of context information as well. Physical (hardware) sensors are the most widely used type of sensors used in context-aware and location dependent systems. *Physical sensors* can provide data about location or proximity (GPS, GSM, RFID), optical data (photo, color, IR sensors), audio/video (microphones, still and video cameras), motion/acceleration (motion detectors, accelerometers, etc.), haptic data (touch sensors), ambient conditions (thermometers, hygrometers, anemometers), biological data (heart rate, blood pressure, body temperature), and others. *Virtual sensors* collect context data from software applications and services as well as their respective data; these include application data (e.g. personal calendar may provide information about availability or physical location of a person), operating system events (e.g. high rate of typing on the keyboard indicates that the person is busy), network events (e.g. connection of a WiFi-enabled device to a particular access point can be used to establish proximity of the person using the device to the location of that access point). *Logical sensors* combine data received from physical and virtual sensors with supplementary information received from additional sources in order to provide a higher-order logic inference of context information. For example, current location of a person can be inferred by analyzing and mapping of login information with the database of computer terminal locations.

Context-aware systems use context models to represent, manage and use context information. Construction of an expressive and flexible model for context representation is a very challenging task taking into account the variety of sources and types of context information. A number of surveys (Strang & Linnhoff-Popien 2004, Henricksen & Indulska 2006) outline several context models currently used in context-aware systems. *Key/value pairs* is the simplest type of context model that is widely used in many application (Schilit et al 1994, Samulowitz et al 2001); a matching algorithm can be easily used by a context-aware system to process information represented using this model. *Markup scheme* models employ a hierarchical data structure that contains markup tags with attributes and their content, which may be defined recursively by other tags; XML and related markup languages are most often used by context-aware applications using this model (Brown et al 1997, Capra et al 2001, Indulska et al 2003). *Graphical* models based on Unified Modeling Language (UML) are well suited for conceptual modeling of context and for deriving a relational database schema for the database supporting context storage and management in a context-aware application (Henricksen et al 2002, Sheng & Benatallah 2005). *Object oriented* models offer the benefits of data encapsulation and class reusability to context information representation, which is embedded into software objects and is accessible via well-defined interfaces (Schmidt and van Laerhoven 2001, Cheverst et al 2000). *Logic based* context models use formally specified facts, logical expressions and rules to express the knowledge about context; such a concept allows the system actively manipulate and update context information by adding new and removing old facts, and by creating new facts derived through the inference process (Ghidini & Giunchiglia 2001, Akman & Surav 1997). Finally, *ontology based* context models use the underlying power of ontologies that define concepts and relationships between them; this approach allows applying ontology reasoning methods to the domain of context awareness (Otzturk & Aamodt 1997, Wang et al 2004, Chen et al 2003, Gu et al 2005).

Context-aware systems have been prototyped in a wide range of academic and commercial applications, some of which are discussed below. Many academic projects address broad implications of context awareness and aim at building flexible platforms supporting many dimensions of context that can be further customized for one or more applications of a smaller and narrower scope. These applications are typically designed to prove the concept and demonstrate their usability, acceptance by users, or experiment with many facets of human computer interaction in context aware environments. Commercial applications, on the other hand, take a more pragmatic approach often focusing on implementing context aware features based on one or two sources of contextual data and incorporating it into an application that has a viable revenue model and can generate a substantial user base.

The paradigm of pervasive computing envisioned by Weiser still may not be completely within our reach. However, smart phones available today may be used as a unique tool for implementing some features of pervasive environments including context awareness. Mobile phones are ubiquitous and personal – they offer anytime anywhere network connectivity that enables a wide range of rich applications available to mobile users. Mobile phones are uniquely associated with their individual owners who carry (or wear) them on a nearly permanent basis and use them to store personal information. Such a personal nature of mobile phones makes them very well suited for implementing context awareness. Additionally, many mobile phones are able to collect data from built-in sensors and interfaces used to derive location (GPS sensors) or proximity (WiFi or Bluetooth connectivity) context information. Raento et al (2005) describe ContextPhone, a platform for context-aware applications built for Symbian OS and the Nokia Series 60 Smartphone platform. ContextPhone uses four types of context: location (GPS sensors, GSM network cellular identifier, etc), user interaction (active applications, idle status, alarms, presence of charger, etc), communication behavior (successful and attempted calls, sent and received SMS/MMS and their content), physical environment (nearby Bluetooth devices and network availability and optical marker recognition by built-in camera). ContextPhone middleware provides context awareness functionality supports to a number of applications, which include ContextLogger, ContextContacts and ContextMedia. ContextLogger is an application for collecting data used to derive context information related to the user's mobility – physical location and proximity to known locations. An important feature of many context-aware applications is unobtrusiveness; as such, ContextLogger operates in the background and is completely invisible to the user. Mobility data is accumulated on the smart phone and is periodically uploaded to the server without any user intervention. ContextLogger has been used by the Technical Research Center in Finland to model recurring patterns of interpersonal communication in workgroups and by MIT Media Lab for modeling social network dynamics. ContextContacts enables mobile users to automatically exchange presence information, which provides context information about their activities and availability. Such information supports social awareness of a group and may help others decide whether a person is available for a phone call or which communication channel may be best suited at any given moment (e.g. SMS, email, or phone call). To accommodate the fragmented attention of mobile users (Dourish 2004), presence information provided by ContextContacts is fully integrated with regular calling and messaging features of the phone. ContextMedia facilitates the process of automatic annotation of mobile media. Media captured on the mobile phone (e.g. photos, videos, sound clips) is automatically annotated with the context information describing the situation in which the media was captured (e.g. location, time, surroundings, etc). ContextMedia is used in a number of research projects studying collective publication and syndication of mobile content and media created on mobile devices, including a project at the University of California at Berkeley aimed at exploring automatic annotation and sharing of mobile media.

Context-aware and location-dependent systems have been designed and deployed in a number of specialized application areas. Bardram (2004) describes an application of context aware computing in hospitals. A context-aware hospital bed has a touch-sensitive display and an array of RFID sensors used to identify the patient wearing an RFID-enabled armband, nearby medical staff and equipment. It is connected to a central Electronic Patient Record (EPR) repository and can display patient records tailored to the specialty and tasks of the medical personnel currently attending to the patient. For example, during a checkup by a clinician, the bed can display the patient's chart and allow the clinician to enter a new record into the chart; when a nurse brings the medicine container, the bed can display the medication information or display a warning if the nurse brings a wrong medicine container. The monitor of the bed can also be used by the patient to watch TV or access the Internet. A context-aware pill container has several compartments containing medicine to be taken at different times of day, a fingerprint recognition capability to allow access only by the specified patient, and an RFID sensor to sense the proximity with that patient and provide a visual cue (e.g. flashing a LED or displaying a corresponding message on the monitor of the patient bed). A fully functional prototype of the context-aware hospital framework described by Bardam has been successfully deployed at a Danish hospital.

Krause et al (2006) describe SenSay, a context-aware mobile phone that adjusts its behavior based on the user's current state and the surrounding environment. SenSay acquires context from several wearable sensors measuring the motion of the user, the level of the user's voice and ambient noise, temperature of the body and the ambient environment, user location, and other parameters. This context information is used to predict the settings for the ringer volume or vibrating alarm, and whether incoming calls should be accepted or rejected. SenSay is capable of learning typical actions of the user based on his or her current physiological state, activity and conditions of the ambient environment. For example, high voice activity, ambient noise level of a conversation, and infrequent movements are typical for a meeting when the user would normally set the ringer to vibrate and would rarely answer the phone. To accommodate unique individual preferences of different users, SenSay uses machine learning techniques to provide automatic personalization via classification of possible user states and context-dependent configuration of the mobile phone.

FUTURE TRENDS

Given a tremendous popularity and wide availability of mobile and smart phones, in the foreseeable future they are likely to remain as the best currently existing enabler of pervasive computing environments. Given their highly personalized nature and ever increasing communications, processing and storage capabilities, smart phones are uniquely positioned to become a device of choice for providing their owners with a context aware access to pervasive information services. However, despite the existence of the supporting technology and a wide range of successful research projects supporting context awareness in mobile devices, it is very hard to find any successful implementations of context aware services widely available in the marketplace (Docter et al 2007). This is likely due to the fact that telecommunications industry lacks a clear business model for providing revenue generating context-aware services. Traditionally, telecommunications companies provided network access services, billing and offered content, such as wallpapers, ringtones, etc. A viable business model for providing context-aware services requires fulfillment of a number of additional roles, namely those of context enabler, context provider, and context-aware service provider. Context enablers are responsible for collecting raw data, from which context information is derived. Mobile devices can act as context enablers for such context data as precise location and user activities; other data (such as GSM cell network ID, type of connectivity, or the volume of network traffic) can be supplied by the telecom provider itself, while third party context enablers may be needed for context data that cannot be easily acquired by a mobile phone (road conditions, weather, etc). Context providers are responsible for collecting raw context data from context enablers, processing it and providing access to context information to context-aware service providers. It is very likely that consumer demand for value-added context-aware services will gradually lead to the establishment of a viable business model for such applications.

CONCLUSION

Context-aware and location-dependent services play a vital role in enabling the paradigm of pervasive computing envisioned by Mark Weiser in the early 1990's. Context aware systems adjust their properties and behavior in response to changes in the environment and can anticipate the actions of the user. By doing so, context-aware systems allow their users to focus on the task at hand and eliminate the obstacles of having to interact with the system to adjust its properties – instead, the system adjusts itself and therefore becomes less obtrusive, providing for a more seamless integration into the environment. Context-aware systems have been widely prototyped in many successful research projects, some of which have been sampled in this chapter. Any wide commercial deployment of context-aware services is very likely to come from the telecommunications industry because mobile phones provide a highly capable and personalized platform for providing the interface between their owners and the computing environment.

REFERENCES

Abowd, G., Atkeson, C., Hong, J., Long, S., Kooper, R., & Pinkerton, M. (1997). Cyberguide: A mobile context-aware tour guide. *ACM Wireless Networks*, *3*(5), 421–433. doi:10.1023/A:1019194325861

Akman, V., & Surav, M. (1997). The use of situation theory in context modeling. *Computational Intelligence*, *13*(3), 427–438. doi:10.1111/0824-7935.00046

Bardram, J. (2004). Applications of context-aware computing in hospital work: examples and design principles. *2004 ACM symposium on Applied computing*. Nicosia, Cyprus. (pp. 1574-1579).

Becker, C., & Durr, F. (2005). On location models for ubiquitous computing. *Personal and Ubiquitous Computing*, *9*(1), 20–31. doi:10.1007/s00779-004-0270-2

Brown, P.J. (1996). The stick-e document: a framework for creating context-aware applications, *Electronic publishing, 8*(2-3), 259-272.

Brown, P. J., Bovey, J., & Chen, X. (1997). Context-aware Applications: from the Laboratory to the Marketplace. *IEEE Personal Communications*, *4*(5), 58–64. doi:10.1109/98.626984

Capra, L., Emmerich, W., & Mascolo, C. (2001). Reflective middleware solutions for context-aware applications, *3rd International Conference on Metalevel Architectures and Separation of Crosscutting Concerns* (pp. 126-133).

Chen, H. (2004). An Intelligent Broker Architecture for Pervasive Context-Aware Systems, PhD Thesis, University of Maryland, Baltimore County.

Chen, H., Finin, T., & Joshi, A. (2003). *Semantic Web in a Pervasive Context-Aware Architecture.* Paper presented at Workshop on Artificial Intelligence in Mobile Systems, the 5th Annual Conf. on Ubiquitous Computing, Seattle, Washington.

Cheverst, K., Davies, N., Mitchell, K., Friday, A., & Efstratiou, C. (2000). Developing a context-aware electronic tourist guide: some issues and experiences. *SIGCHI Conference on Human Factors in Computing Systems* (pp. 17–24), New York.

Dey, A., & Abowd, G. (2000). Towards a better understanding of context and context-awareness, *Workshop on the What, Who, Where, When and How of Context-Awareness*. New York: ACM Press.

Dey, A., Salber, D., & Abowd, G. (2001). A conceptual framework and a toolkit supporting the rapid prototyping of context-aware applications. *Human-Computer Interaction*, *16*(2-4), 97–166. doi:10.1207/S15327051HCI16234_02

Docter, J., Licciardi, C., & Marchetti, M. (2007). The Telecom Industry and Context Awareness. *International Conference on the Management of Mobile Business* (pp. 35-42), Toronto, Ontario.

Dourish, P. (2004). What we talk about when we talk about context. *Personal and Ubiquitous Computing*, *8*(1), 19–30. doi:10.1007/s00779-003-0253-8

Fels, S., Sumi, Y., Etani, Y., Simonet, N., Kobayashi, K., & Mase, K. (1998). Progress of C-Map: a Context-aware Mobile Assistant. *AAAI Spring Symposium on Intelligent Environments* (pp. 60-67), Palo Alto, California.

Ghidini, C., & Giunchiglia, F. (2001). Local models semantics, or contextual reasoning = locality +compatibility. *Artificial Intelligence*, *127*(2), 221–259. doi:10.1016/S0004-3702(01)00064-9

Gu, T., Pung, H., & Zhang, D. (2005). A service-oriented middleware for building context-aware services. *Journal of Network and Computer Applications*, *28*(1), 1–18. doi:10.1016/j.jnca.2004.06.002

Gwizdka, J. (2000). What's in the context? Workshop on The What, Who, Where, When, and How of Context-Awareness, Conference on Human Factors in Computing Systems (CHI 2000), April 2000.

Henricksen, K., & Indulska, J. (2006). Developing context-aware pervasive computing applications: Models and approach. *Pervasive and Mobile Computing*, *2*(1), 37–64. doi:10.1016/j.pmcj.2005.07.003

Henricksen, K., Indulska, J., & Rakotonirainy, A. (2002) Modeling context information in pervasive computing systems. *1st International Conference on Pervasive Computing*, Zurich, Switzerland (pp. 167–180).

Indulska, J., Robinson, R., Rakotonirainy, A., & Henricksen, K. (2003). Experiences in using CC/PP in context-aware systems. *4th International Conference on Mobile Data Management* (pp. 247–261).

Indulska, K., & Sutton, P. (2003). Location Management in Pervasive Systems. *Australasian Information Security Workshop Conf. ACSW Frontiers*, Adelaide, Australia (pp. 143–151).

Krause, A., Smailagic, A., & Siewiorek, D. (2006). Context-aware mobile computing: learning context- dependent personal preferences from a wearable sensor array. *IEEE Transactions on Mobile Computing*, *5*(2), 113–127. doi:10.1109/TMC.2006.18

Kreller, B., Carrega, D., Shankar, J., Salmon, P., Bottger, S., & Kassing, T. (1998). *A Mobile-Aware City Guide Application* (pp. 60–65). Rhodos, Greece: ACTS Mobile Communication Summit.

Leonhardt, U., Magee, J., & Dias, P. (1996). Location service in mobile computing environments. *Computers & Graphics*, *20*(5), 627–632. doi:10.1016/S0097-8493(96)00036-2

Otzturk, P., & Aamodt, A. (1997). Towards a model of context for case-based diagnostic problem solving. *Interdisciplinary conference on modeling and using context* (pp. 198–208). Rio de Janeiro, Brazil.

Raento, M., Oulasvirta, A., Petit, R., & Toivonen, H. (2005). ContextPhone: a prototyping platform for context-aware mobile applications. *Pervasive Computing*, *4*(2), 51–59. doi:10.1109/MPRV.2005.29

Samulowitz, M., Michahelles, F., & Linnhoff-Popien, C. (2001). Capeus: An architecture for context-aware selection and execution of services. *New developments in distributed applications and interoperable systems.* Krakow, Poland (pp. 23–39).

Schilit, B., Adams, N., & Want, R. (1994). Context-aware computing applications. *IEEE Workshop on Mobile Computing Systems and Applications*, Santa Cruz, CA.

Schmidt, A., & van Laerhoven, K. (2001). How to Build Smart Appliances. *IEEE Personal Communications*, *8*(4), 66–71. doi:10.1109/98.944006

Sheng, Q. Z., & Benatallah, B. (2005) ContextUML: a UML-based modeling language for model-driven development of context-aware web services, *International Conference on Mobile Business* (pp. 206–212).

Strang, T., & Linnhoff-Popien, C. (2004). A context modeling survey. *Workshop on Advanced Context Modeling, Reasoning and Management, The 6th International Conference on Ubiquitous Computing*, Nottingham, England.

Wang, X., Zhang, D., Gu, T., & Pung, H. (2004). Ontology Based Context Modeling and Reasoning using OWL. *2nd IEEE Conference on Pervasive Computing and Communications* (pp. 18-22), Orlando, Florida.

Want, R., Hopper, A., Falcao, V., & Gibbons, J. (1992). The active badge location system. *ACM Transactions on Information Systems*, *10*(1), 91–102. doi:10.1145/128756.128759

Weiser, M. (1991). The computer for the twenty-first century. *Scientific American*, (Sep): 94–10. doi:10.1038/scientificamerican0991-94

KEY TERMS AND DEFINITIONS

Context Awareness: a property of a computing system that can adapt its behavior to the current context without any explicit user intervention.

Context Data: raw data, from which context information is derived.

Context Model: a formal model describing the representation, management and usage of context information.

Context: any information characterizing the situation of a person, a place, or an object that is relevant to the interaction between the computing system and its user.

External Context: context information referring to the state of the physical environment that can be measured by sensors or devices.

Internal Context: context information referring to the logical state of the system and its user.

Location Dependency: a property of a context-aware computing system, whose context information is limited to the user's physical location or proximity to other objects.

Pervasive Computing: a model of human-computer interaction in which information processing is seamlessly integrated into everyday objects and activities allowing the user to concentrate on the specific task rather than on details of interacting with the system.

Chapter 16
MAC Protocols for Wireless Sensor Networks

Torsten Braun
University of Bern, Switzerland

Markus Anwander
University of Bern, Switzerland

Philipp Hurni
University of Bern, Switzerland

Markus Wälchli
University of Bern, Switzerland

ABSTRACT

The chapter describes related work on medium access control protocols for wireless sensor nodes. We focus on scheduled and contention-based protocols that have been proposed by the research community during the last few years. In particular, we evaluate the potential to save energy of several representative protocols, namely LMAC, TEEM, and WiseMAC. This has been done by measurements of implementations in real sensor networks. The measurement results show that by sophisticated MAC protocol design we can significantly improve the energy-efficiency and increase the lifetime of a sensor node. Real-world measurements are important to determine power consumption parameters of sensor nodes.

INTRODUCTION

Wireless sensor networks (WSNs) use wireless network technologies for exchanging sensor data, signalling messages, management information etc. The wireless medium is shared by the various sensor nodes in a WSN. This requires a Medium Access Control (MAC) layer controlling the access to the wireless medium. Although many MAC layer protocols have been developed for both wired and wireless media, e.g., IEEE 802.11 Wireless Local Area Networks (WLAN), new MAC protocols are required for WSNs, because existing ones do not meet most requirements for sensor networks such as energy efficiency.

DOI: 10.4018/978-1-60566-250-3.ch016

The most important task of a MAC protocol is to avoid collisions that occur when multiple nodes are trying to access the network simultaneously. By avoiding collisions valuable network resources are saved, the utilization of the wireless channel is improved, and the nodes save energy by avoiding useless transmissions. Other sources of energy waste are overhearing and receiving operations. Overhearing occurs when nodes receive packets destined to other nodes and can be dominant in scenarios with heavy load and high node density.

Since overhearing and receiving might not be much cheaper than transmitting, energy-efficient MAC protocols should avoid those operations and switch off the transceiver whenever possible. Other sources of energy waste are over-emitting, i.e. transmissions when receivers are not yet ready to receive as well as excessive control packet overhead.

The complexity of a sensor MAC protocol should be limited, because sensor nodes usually have limited computing and memory resources. Moreover, the MAC protocol should allow fair allocation of the wireless medium among all sensor nodes and aim at high throughput and low delay. In contrast to WLANs, a sensor MAC protocol must support not only small numbers of nodes but scale to hundreds or thousands of nodes depending on the required node density of a WSN by the application. The characteristics of a WSN in terms of size, density, and topology may change quickly requiring a sensor MAC protocol to be automatically adaptable.

The chapter gives an overview of related work in MAC protocols for WSNs including both scheduled and contention-based schemes. After describing the basic principles for MAC protocol operation in WSNs, we describe real-world experiments using implementations on real sensor nodes. The energy-efficiency in terms of sensor node lifetime will be investigated. We show that by proper MAC protocol design, significant energy savings are possible.

BACKGROUND

There are two basic classes of MAC protocols for WSNs: scheduled protocols and contention-based protocols. By combining the basic concepts, hybrid protocols can be designed.

Scheduled protocols are either based on polling or multiplexing. With polling a central controller polls other sensor nodes to detect whether they have pending transmissions. This avoids energy waste caused by collisions but introduces polling overhead and delays. In case of multiplexing, channels are pre-allocated based on time, frequency, or code multiplexing. Scheduling based approaches often form clusters with cluster controllers responsible for the channel allocation. Since only a certain number of channels can be allocated the scalability might be limited then.

Contention-based protocols allow sharing channels and allocating channels on-demand. The main problem is that contention can happen in case of dense networks and highly active sensor nodes. Moreover, collision avoidance is rather difficult to achieve in WSNs due to potentially hidden nodes.

SCHEDULED PROTOCOLS

Cluster-Based Approaches

Scheduled protocols based on cluster formation form clusters with a cluster head acting as base station. The cluster head allocates channels using time, frequency, and / or code multiplexing and assigns the channels to the sensor nodes. Typically, nodes can only communicate with a cluster head. Inter-cluster communication requires special mechanisms and (time, frequency, or code) channels. Cluster heads need to be interconnected to support data forwarding to / from sink, and since the distance among each other is rather high, higher power level for inter-cluster communication is needed.

In case of nodes joining or leaving the cluster, the frame length and the slot assignment need to be adapted. This might generate overhead. If the scheme does not adapt dynamically and uses static channel allocation, the achievable throughput is rather limited. An early example for a cluster-based protocol was Low Energy Adaptive Clustering Hierarchy (LEACH). Nodes organize themselves into clusters with rotating cluster heads as local base stations. Cluster members are only active during their schedule and might then transmit to the cluster head. The cluster head is permanently active, calculates TDMA schedules for other nodes, aggregates data, and forwards it to various nodes or via other cluster heads to the sink. The role of a cluster head is energy consuming and therefore rotated among nodes. A distributed protocol selects nodes as cluster head considering the energy level and the previous times a node served as cluster head.

Scheduled Protocols with Flat Network Organization

Several scheduled protocols avoid the overhead of building clusters and try to form a flat network. Three of those protocols, namely SMACS, TRAMA, and LMAC are described hereafter.

Self-Organizing MAC for Sensor Networks (SMACS) is an infrastructure building protocol that forms a flat topology. Nodes find each other on a fixed frequency and agree on a pair of time slots to send and receive data. Time slots are repeated periodically and nodes wake up for these time slots. The possibility of collisions is reduced by random selection of frequencies. The protocol requires time synchronization among the nodes.

Traffic Adaptive Medium Access Protocol (TRAMA) is a collision-free MAC protocol using a distributed election scheme to determine particular time slots. TRAMA distinguishes between contention-based random access slots used for signaling, and scheduled access periods used for collision-free data exchange.

The TRAMA protocol assumes a time synchronization of sensor nodes. It consists of three protocol phases: Neighbor Protocol, Schedule Exchange Protocol, and Adaptive Election Algorithm. The Neighbor Protocol exchanges short control packets (keep-alive beacons) with 1-hop-neighborhood information during a random access period in order to obtain knowledge about the 2-hop-neighborhood around a node. In the Schedule Exchange Protocol, each node calculates a periodic schedule interval and winning intervals based on priority values: priority (u, t) = hash (u $\oplus$ t), with t = time slot, u = a node in 2-hop-neighborhood. The schedules containing the set of receivers are announced periodically by schedule packets in the scheduled access period. All nodes switch to a sleep state whenever possible, i.e., when there is no traffic to be received or sent. The Adaptive Election Algorithm selects transmitters and receivers according to the information obtained from the two previous protocol phases.

Lightweight MAC (LMAC) is based on time multiplexing: Time is divided into time slots owned by various nodes using a distributed allocation algorithm. It divides a frame into 32 equal slots as shown in Figure 1. Each slot can be allocated by only one node. Furthermore, a node can own only one slot in a frame. Spatial reuse of time slots is possible. LMAC uses control messages to maintain the synchronization among nodes and to address the destination. The slot owner has exclusive transmission rights and can communicate collision-free. Figure 1 shows two slots in more detail. In the first slot, slot owner C has no data to send. It just sends a 12 byte SYNC message at the beginning of the slot and then turns off the radio to save energy. In the second case node A has data to transmit to B: The SYNC message and the data are sent in a single packet. All other nodes not involved in the communication shut down their radios after having received and analyzed the SYNC message.

CONTENTION-BASED PROTOCOLS

Contention-based protocols are characterized by having a common channel shared by all nodes.

Figure 1. LMAC

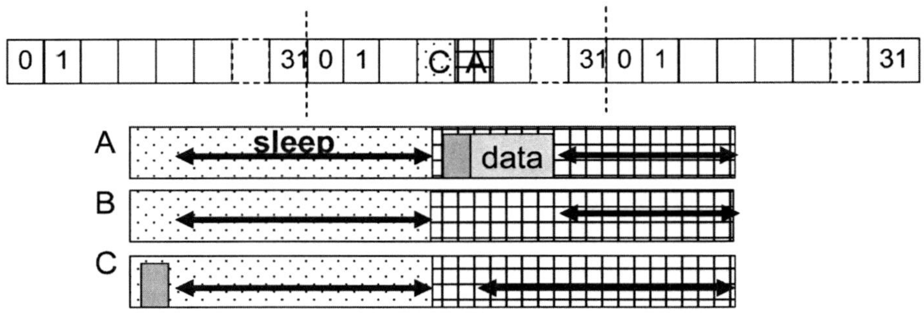

Resources are allocated on demand. This results in high flexibility in case of topology changes. Fine-grained time synchronization is not required. However, there is a high risk of inefficient energy usage due to idle listening and collisions.

Several protocols are based on entering a low duty cycle state and wake up the nodes when they are expected to receive a message. PicoRadio uses a special low-power control channel to indicate an upcoming transmission on a separate data channel. In other protocols the receiving node periodically wakes up to receive packets from sending nodes and falls into sleep again in case no transmissions occur. The transmitting nodes need to know the duty / sleep cycles of their neighbors, which may be achieved by synchronization of the wakeup periods as implemented in S-MAC. A disadvantage of periodic sleep is potentially high delays. An alternative is that the sender might try to wake up the receiver and indicate an upcoming transmission.

In STEM-B (STEM: Sparse Topology and Energy Management), the sending node sends a stream of beacon packets including the MAC addresses of sender and receiver in the paging channel to be answered by the receiving node. The receiver acknowledges the reception of the beacon and activates the data channel. By including the start time of the transmission in the beacon the receiving node can sleep until then. The STEM-T variant uses wakeup tones transmitted by the sender, which are not acknowledged by the receiver. All nodes in the neighborhood remain awake once they have overheard the wakeup tone and await the following transmission.

In WiseMAC, the sending node transmits a long preamble (Figure 2: P) with the length of the preamble larger than the sleep cycle. All nodes sample the medium using signal strength measurements with the same sampling period for activity, but sampling is not synchronized. A node that discovers a busy medium continues to listen,

Figure 2. WiseMAC

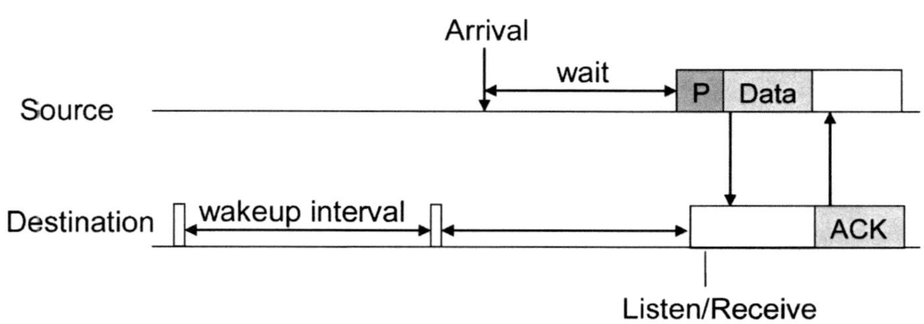

receives the packet, and becomes idle again. Long preambles are avoided by learning the sampling schedules of other nodes: Nodes piggy-back their sampling schedule in acknowledgement packets. The required preamble length Tp is: Tp = min (4·θ·L, Tw), with Tw = sampling period, θ = frequency tolerance, L= time of desired transmission.

Several other protocols based on periodic wakeup periods similar to STEM or WiseMAC have been designed. B-MAC employs an adaptive preamble sampling scheme to reduce duty cycle and minimize idle listening. CMAC uses an aggressive RTS transmission scheme to replace long preambles. Multiple RTS packets are separated by fixed short gaps, which allow receivers to send back CTS packets. X-MAC introduces a series of short preamble packets each containing target address information and thereby avoiding the overhearing problem of low power listening, saving energy on non-target receivers.

In S-MAC, each node has alternating sleep and listen states. In the sleep state, it turns off its radio and sets a timer to awake itself later, while in the listen state, it listens to see whether some other node wants to talk with it. S-MAC implements a so-called coordinated sleeping mechanism: Each node selects its own listen/sleep schedule. If the node receives a schedule from another node it follows that schedule, otherwise it selects its own schedule. The selected schedule is broadcast in a SYNC message. A node might receive a different schedule after announcing its own schedule. If the node does not have other neighbors it discards its schedule and takes over the new one. If the node shares schedules with other neighbors it adopts both (or more) schedules. The collision avoidance scheme of S-MAC is similar to IEEE 802.11 CSMA/CA using RTS/CTS message exchange (upper part of Figure 3).

The Traffic Aware Energy Efficient MAC (TEEM) protocol optimizes S-MAC by combining the SYNC and RTS messages into a single message to be sent when data is available for transmission. The listen period is shorter than in S-MAC and not fixed. If a node has data to send it tries to transmit the $SYNC_{RTS}$ for allocating the medium shortly after slot start (A in lower part of Figure 3). The receiver node addressed in $SYNC_{RTS}$ sends back a CTS message and waits for the incoming data packet. A successful transmission is approved by an ACK packet. If no node has data to send, one node will send a SYNC message in the second interval (B in lower part of Figure 3).

Figure 3. S-MAC and TEEM

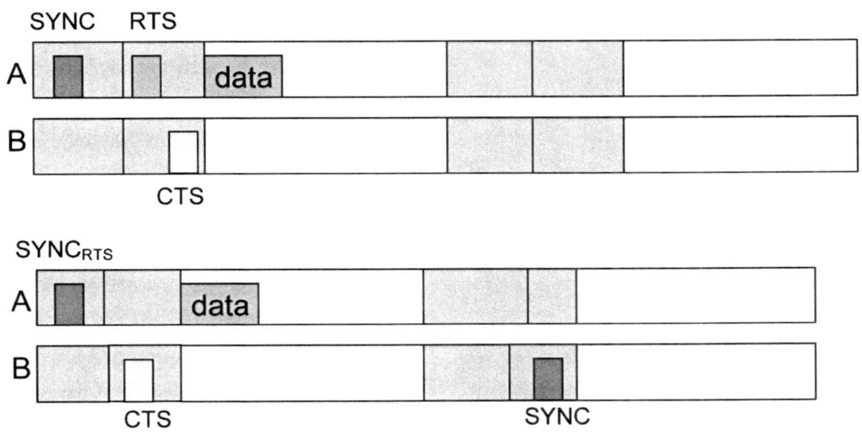

HYBRID MAC PROTOCOLS

Hybrid MAC protocols combine TDMA scheduling and contention mechanisms by allowing contention in TDMA slots. Examples of such protocols are EMACs and Z-MAC. In EMACS each node owns a time slot. EMACS consists of three phases. In the first phase, any node is allowed to send requests for allocating a transmission slot. In the second phase, the time slot owner indicates which communication (his own or that one of the requesting node) will take place in the data phase. Finally, data is transmitted in the third phase.

Z-MAC operation is divided into a setup and a transmission control phase. In the setup phase neighbours discover themselves using ping messages and assign transmission slots using a distributed algorithm ensuring that no two nodes in a 2-hop-neighborhood get assigned to the same slot. This algorithm requires global time synchronization.

In the transmission control phase, each node is in either low or high contention level (LCL / HCL) dependent on received explicit contention notification messages sent based on local estimation of the contention level. If a node is in HCL state, only the slot owner and its 1-hop-neighbors are allowed to send, while in LCL state all nodes are allowed to send in any slot. In general, slot owners have higher priority by using a smaller contention window for back-off.

IEEE 802.15.4 / ZIGBEE

The IEEE 802.15.4 standard describes a physical and MAC layer for sensor networks and is also used for the Zigbee protocol architecture. There are two kinds of devices:

- A full-function device (FFD) is capable of operating as coordinator and implementing the complete IEEE 802.15.4 protocol set.
- A reduced-function device (RFD) is operating with a minimal implementation of the IEEE 802.15.4 protocol set

A coordinator node is a FFD that is configured to provide synchronization services through the transmission of beacons. The coordinator defines an optional super-frame structure with 16 equally sized slots. Beacons synchronize devices and define the start of a super-frame. A super-frame may have an active and inactive period. The coordinator may enter low power mode during inactive period. The active period is further divided into a contention access period and a contention free period. The coordinator may allocate up to 7 guaranteed time slots (GTS) in the contention-free period. A slotted CSMA/CA protocol is used for the contention access period.

EVALUATION OF SELECTED SENSOR MAC PROTOCOLS

This section describes the evaluation of selected MAC protocols for WSNs by measuring the lifetime in real-world test-beds. The implementation in real test-beds is important to determine power consumption parameters accurately. Power consumption parameters from manufacturer data sheets are typically not sufficiently accurate.

In order to investigate how much energy can be saved with sophisticated sensor MAC protocols, we implemented the LMAC and the TEEM protocol on Embedded Sensor Boards (ESBs). In addition to the original protocol versions, we tried to improve the performance of both protocols. For TEEM we allowed more than one packet per slot, and implemented an acknowledgement vector to confirm the reception of multiple packets. In the improved LMAC version, packet markers in the SYNC message helped to identify packet boundaries more easily. Moreover, SYNC messages are protected by CRC checksums. Both

Figure 4. Lifetime Measurements of TEEM and LMAC

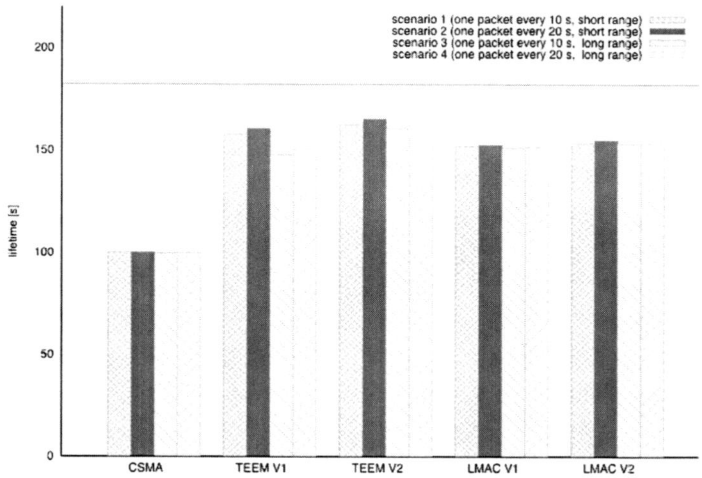

improved protocol versions are denoted as versions 2 hereafter.

Several measurement scenarios consisting of node chains have been used for the evaluation of the two protocols: In a short range scenario, all the five nodes hear each other, but a packet from node 5 is sent via intermediate nodes to the sink. In a long range scenario, the five sensor nodes are distributed in a building and can only communicate with their respective neighbor nodes. Different packet rates (1/10 s and 1/20 s) are used. In all scenarios, the node closest to the sink is equipped with a special capacitor with well-defined, but very limited energy. The lifetime was measured by determining the time when the RS-232 interface of the node has been shutdown when using a 1Farad gold cap capacitor instead of batteries.

The lifetime measurements depicted in Figure 4 show that the different TEEM and LMAC versions significantly increase the lifetime of node 1 compared to the default ScatterWeb CSMA MAC protocol implemented in the ESB nodes. Both improved versions slightly improve the lifetime of the original versions. TEEM performed slightly better than LMAC in terms of both network lifetime, but much less packet errors could be detected for TEEM. LMAC was rather difficult to implement on the available hardware due to difficulties to achieve good time synchronization between the nodes.

The lifetime evaluation in the WiseMAC experiments have been slightly changed compared to the measurements on TEEM and LMAC. We switched off the RS232 interface and determined the lifetime until the supply voltage of the capacitor has fallen below 3 V. We measured the lifetime of WiseMAC in a chain scenario consisting of six nodes with the parameters depicted in Table 1.

Table 1. Measurement Parameters for WiseMAC Evaluation

Basic Interval Duration T	500 ms
Awake Ratio	1%
Retries	3
Minimum Preamble	5 ms
Medium Reservation Preamble	uniform [0,6] ms
Baud Rate	19'200 bps
Bit Rate	9'600 bps
MAC Header	104 bit
Payload	96 bit
Packet Queue Length	20

Figure 5 depicts the lifetime of the fifth ESB node as a function of the traffic rate r when being charged with the initial amount of energy. As the node's energy consumption increases with increasing traffic along the chain, a more or less linear decrease of the node's lifetime can be observed. The lower curve in Figure 9 displays the lifetime of a node using ScatterWeb CSMA. ScatterWeb CSMA keeps the transceiver constantly in the receive state, applying no energy-saving scheme such as periodic switching between sleep and active states. We can see that the WiseMAC implementation achieves a lifetime that is more than twice than for the Scatterweb CSMA implementation. This is a much higher factor than what we achieved for LMAC and TEEM and indicates that WiseMAC is very energy-efficient.

FUTURE TRENDS

Future work will integrate such MAC level protocols with higher-level protocols such as routing and transport protocols. Such integration is called cross-layer design.

Future work might also include protocols for even smaller sensor nodes such as required for body area networks. Nodes must be small and light to be carried conveniently by users. Energy-efficiency might become even more important. However, transmission ranges will be in the order of a few meters only.

CONCLUSION

This chapter described the state of the art in MAC protocols for WSNs. Those protocols can be classified into scheduled and contention-based protocols. The implementation of several representatives of these classes showed that sophisticated protocols significantly increase the lifetime of a sensor node. Due to the lower implementation complexity as well as better lifetime and packet error results, the contention-based protocol has some advantages over the scheduled protocol. Moreover, we discovered that determining power consumption parameters must be determined by real-world implementations, because data sheets from hardware manufacturers are typically not accurate enough and can lead to wrong results when applied in simulation experiments.

Figure 5. Lifetime Measurements of WiseMAC

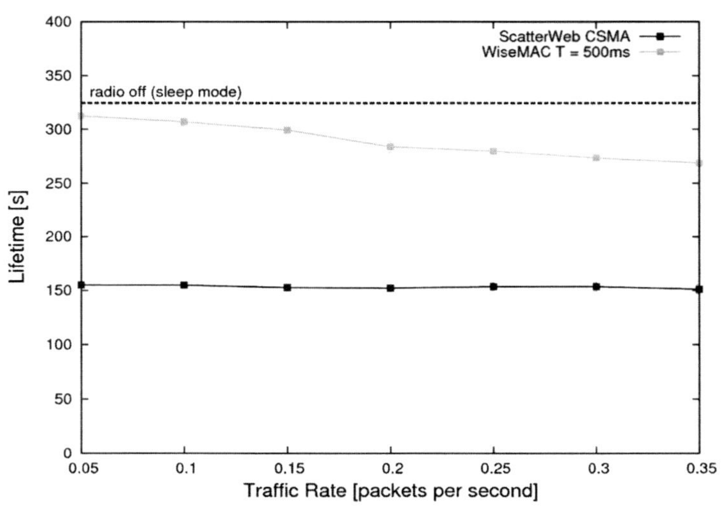

REFERENCES

Braun, T., Voigt, T., & Dunkels, A. (2007). TCP Support for Sensor Networks. *4th IEEE/IFIP Conference on Wireless On demand Network Systems and Services*. Obergurgl, Austria.

Buettner, M., Yee, G., Anderson, E., & Han, R. (2006). X-MAC: a short preamble MAC protocol for duty-cycled wireless sensor networks. *4th international Conference on Embedded Networked Sensor Systems (SenSys)* (pp. 307-320). Boulder: ACM.

Demirkol, A. (2006, April). MAC Protocols for Wireless Sensor Networks. *IEEE Communications Magazine*, 115. doi:10.1109/MCOM.2006.1632658

Enz. (August 2004). WiseNET: An Ultralow-Power Wireless Sensor Network Solution. *IEEE Computer*, 62-70.

Feeney, L. M., & Nilsson, M. (2001). *Investigating the Energy Consumption of a Wireless Network Interface in an Ad Hoc Networking Environment*. IEEE INFOCOM.

Heinzelman, W., Chandrakasan, A., & Balakrishnan, H. (October 2002). An Application Specific Protocol Architecture for Wireless Microsensor Networks. *IEEE Transactions on Wireless Communications*.

Hoesel, V., & Havinga. (2004). A Lightweight Medium Access Protocol (LMAC) for Wireless Sensor Networks: Reducing Preamble Transmissions and Transceiver State Switches. *INSS*.

Hoesel, V., & Havinga. (2004). A TDMA-based MAC Protocol for WSNs. *ACM SenSys*.

Hurni, P., & Braun, T. (2008). Evaluation of WiseMAC on Sensor Nodes. *10th IFIP International Conference on Mobile and Wireless Communications Networks (MWCN2008)*. Toulouse: IFIP.

Hurni, P., Braun, T., & Feeney, L. M. (2006). Simulation and Evaluation of Unsynchronized Power Saving Mechanisms in Wireless Sensor Networks. *4th International Conference on Wired/Wireless Internet Communications* (LNCS 3970, pp. 311-324).

IEEE. (2003). *Specific requirements Part 15.4: Wireless Medium Access Control (MAC) and Physical Layer (PHY) Specifications for Low-Rate Wireless Personal Area Networks (LR-WPANs)*. 802.15.4 IEEE Standard for Information technology - Telecommunications and information exchange between systems — Local and metropolitan area networks.

Ilyas, M., & Mahgoub, I. (2005). *Handbook of Sensor Networks: Compact Wireless and Wired Sensing Systems*. CRC Press.

Karl, H., & Willig, A. (2006). *Protocols and Architectures for Wireless Sensor Networks*. Wiley.

Liu, S., Fan, K.-W., & Sinha, P. (2007). CMAC: An Energy Efficient MAC Layer Protocol Using Convergent Packet Forwarding for Wireless Sensor Networks. *4th Annual IEEE Communications Society Conference on Sensor, Mesh and Ad Hoc Communications and Networks (SECON)* (pp. 11-20). IEEE.

Polastre, J., Hill, J., & Culler, D. (2004). Versatile low power media access for wireless sensor networks. *2nd international Conference on Embedded Networked Sensor Systems (SenSys)* (pp. 95-107). Baltimore.

Raghavendra, C., Sivalingam, K., & Znati, T. (2004). *Wireless Sensor Networks*. Kluwer. doi:10.1007/b117506

Rhee, Warrier, Aia, & Min. (2005). Z-MAC: A Hybrid MAC for Wireless Sensor Networks. *ACM Sensys*.

Schiller, J., Liers, A., Ritter, H., Winter, R., & Voigt, T. (2005). ScatterWeb - Low Power Sensor Nodes and Energy Aware Routing. *Hawaii International Conference On System Sciences (HICSS 2005).*

Schurgers, T. G. (2002, January). Optimizing Sensor Networks in the Energy-Latency-Density Design Space. *IEEE Transactions on Mobile Computing.*

Staub, T., Bernoulli, T., Anwander, M., Wälchli, M., & Braun, T. (2006). Experimental Lifetime Evaluation for MAC Protocols on Real Sensor Hardware. *ACM Workshop on Real-World Wireless Sensor Networks (REALWSN'06)* (pp. 25-29). Uppsala, Sweden: ACM Press.

Suh, C., & Ko, Y.-B. (2005). A traffic aware, energy efficient MAC protocol for wireless sensor networks. *International Symposium on Circuits and Systems (ISCAS'05).*

Ye, W., Heideman, J., & Estrin, D. (June 2004). Medium Access Control With Coordinated Adaptive Sleeping for Wireless Sensor Networks. *IEEE/ACM Transactions on Networking*, 496.

Ye, W., & Heidemann, J. (2003). *Medium Access Control in Wireless Sensor Networks* (USC/ISI Technical Report ISI-TR-580).

Zhong, S. G. (2001). An Ultra-Low Power and Distributed Access Protocol For Broadband Wireless Sensor Networks.

KEY TERMS AND DEFINITIONS

Cluster Formation: Scheduled protocols need often a node that controls channel assignment. In this case, a cluster of nodes is formed and a cluster controller is responsible for channel assignment within the cluster. It also takes over the task of communicating with other clusters.

Contention-Based MAC Protocols: Contention based MAC protocols do typically not reserve channels for individual nodes, but try to resolve collisions by distributed protocol mechanisms, typically based on listening to the wireless medium and transmitting only after silence has been observed on the channel.

Hybrid MAC Protocols: Hybrid MAC protocol apply principles of both contention-based and scheduled MAC protocols.

Lifetime of a Sensor Node: Energy of a sensor node is typically provided by batteries and is a very scarce resource. If the supply voltage falls below a certain threshold, the sensor node is not any more able to work properly and its lifetime can be considered as expired.

Medium Access Control (MAC): The MAC layer is a sub-layer of the data link layer and aims to ensure that nodes access a shared medium in way that they not interfere with each other and avoid collisions.

Scheduled MAC Protocols: Collision avoidance in MAC protocols can be addressed by providing dedicated channels to individual sensor nodes, e.g. by assigning time slots in the case of time multiplexing

Wireless Sensor Network (WSN): a set of sensor nodes that can communicate with each other via wireless communication and possibly using multi-hop communication, i.e., packets can be relayed by sensor nodes

Chapter 17
Mechanisms for Automatic Web Service Composition

Quan Yuan
Florida Atlantic University, USA

Mihai Fonoage
Florida Atlantic University, USA

Ionut Cardei
Florida Atlantic University, USA

ABSTRACT

This chapter introduces the web services composition as a means of studying efficient integration of the existing web services to satisfy users' requirements. It discusses the web services composition definition, combined with the current web services composition methods, and divides those methods into two categories: AI-based methods and Non-AI methods. Also, the authors present the features and the comparison of these two categories, to assist researchers in the understanding of web service composition in a variety of contexts.

INTRODUCTION

Web services are modular elements of application logic that offer functionality to other applications over the Internet. Wide deployment of web services has been made possible by the emergence of XML-based open standards and the ubiquity of internet connectivity. In essence, web services provide request-reply communication semantics similar to remote procedure calls (RPC).

For applications that have complex requirements, more sophisticated services can be composed by orchestrating execution of multiple basic web services. For instance, a web service for travel planning can be implemented by successively invoking web services for airline ticket purchasing, hotel room reservation, and car rental reservation. This composite travel planning web service could become a component of a larger corporate application that handles employee relocation, for instance.

Over a few years many organizations have exposed their products on the internet using web services. For a specific type of service there could be many providers available. For instance, mapping web services are available from Google, Yahoo,

DOI: 10.4018/978-1-60566-250-3.ch017

and Microsoft. Also, the range of service types has exploded, so that sophisticated distributed web applications can now be implemented. Because of the scale of available service options, the task of building a composite web service workflow manually can be difficult and time-consuming. Business process dynamics may also call for web service plans to adapt to changing conditions, such as product cost and availability. These factors have attracted researchers' interest in designing mechanisms for *automated web service composition*. Automated web service composition is very relevant in mobile networking as well. Businesses have to support a growing mobile workforce, and web service provisioning must deal with the inherent variation in connection quality and availability. This issue is also valid for deployments in MANETs. The U.S. DoD Net-Enabled Command Capability (U.S. Department of Defense, 2008) program envisions a service-oriented architecture, with service users and providers executing on a variety of wired and wireless networks, including tactical MANETs.

In this chapter, we review recent developments in web service composition technologies. We review first methods for automatic web service composition that use Artificial Intelligence (AI) techniques, such as planning, theorem proving, expert systems, logic programming. AI techniques for knowledge representation and reasoning assist web service architecture design with specification of message and constraints semantics. XML languages developed for the Semantic Web, such as the Ontology Web Language (OWL) and the Resource Definition Language (RDF) (Klyne, 2004) are used by many of the surveyed techniques.

In the second part of the chapter we review web service composition methods that utilize more mainstream techniques for graph searching and optimization. Finally, a formal method for service composition is described. The chapter ends with an analysis of the advantages and disadvantages of each technique category.

BACKGROUND

In this section, we introduce some web service composition related terms and standards used in this chapter.

WSDL (Web Service Definition Language) is an XML-based language used for describing web services as a collection of endpoints that operate on messages that contain procedure or document-oriented information. When a client program connects to a web service, it can read the WSDL to determine what functions are available to be used. More information can be found in (Chinnici, 2007). WSDL is often used in conjunction with the Simple Object Access Protocol (SOAP).

SOAP is a protocol for exchanging XML-based messages over the internet. It defines an extensible messaging framework containing a message structure that is exchanged over different application-level protocols, such as HTTP or SMTP. SOAP is used in the context of web services to perform remote procedure calls and it has gained a considerable broader user base than CORBA or DCOM due to the ability of HTTP and HTTPS traffic to pass through firewalls. Additional information on SOAP can be found in (Gudgin, 2007).

UDDI (Universal Description, Discovery and Integration) (OASIS, 2008) is platform-independent XML-based registry that provides interfaces for publishing and searching services, defining thus how these services interact over the internet. UDDI is probed by SOAP messages and also provides access to WSDL documents describing the protocol bindings and message formats required to interact with the services listed in its directory.

OWL-S is an ontology built on top of Web Ontology Language (OWL) by the DARPA DAML program, for describing Semantic Web Services. It replaces the former DAML-S ontology. With OWL-S, users and software agents can automatically discover, invoke, compose, and monitor Web resources offering services, under specified constraints. OWL-S consists of three main parts: the service profile for advertising and discovering services, which includes the service name and description, limitations on applicability and quality of service, publisher and contact information; the process model, which gives a detailed description of a service's operation, including the sets of inputs, outputs, pre-conditions and results of services; and the grounding, which provides details on how to inter-operate with a service, via messages, such as communication protocol and message formats. A detailed presentation of OWL-S can be found in (Martin, 2004).

BPEL4WS (Business Process Execution Language for Web Services) provides a language for the formal specification of business processes and business interaction protocols. It extends the Web Services interaction model to support business transactions. Also, BPEL4WS defines an interoperable integration model that should facilitate the expansion of automated process integration in both the intra-corporate and the business-to-business spaces. Additional details can be found in (Microsoft, 2003).

THE ARCHITECTURE OF WEB SERVICE COMPOSITION

Web service composition is the process in which existing (basic) web services are selected and integrated from possibly different organizations on the web at runtime to form a more complex and powerful composite web service. Generally, a basic framework of current web services composition systems consists of four components, shown in Figure 1. They are Ontology Translator, Service Repository, Composition Engine and Execution Engine. Once a user submits his request for a desired service, with a web service description language, the Ontology Translator translates the request into an internal expression, which can be easily recognized by machines, and then transfers it to the Composition Generator. The Composition Generator solves the request from the Ontology Translator by integrating the existing services stored in the Service Repository. Service providers can publish or advertise their services into Service Repository as the basic services in the web service system. If the requested service can be realized, the Composition Generator would build a service process which includes a set of services and the order they are executed for the Execution Engine. When the Execution Engine receives the service process, it executes all the services in the service process based on the defined order.

Most web service composition methods presented here use similar architectures with slight variations.

MECHANISMS FOR AUTOMATED SERVICE COMPOSITION

Planning

Planning methods concentrate on designing or optimizing a web service planner to generate a sequence of services to realize the required composite service. A typical planner takes three inputs: an initial state, a desired goal, and a set of existing actions, all encoded in a service description lan-

Figure 1. The architecture of web services composition

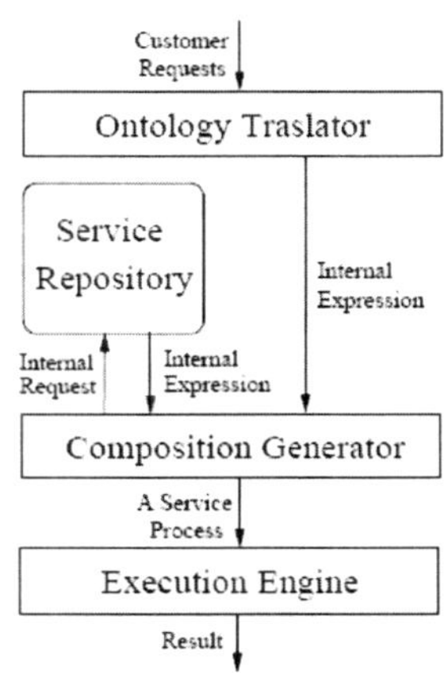

guage. Thus, given the initial state, a web service should return a set of services for achieving the goal state. Two basic methods are forward chaining and backward chaining state-space search. Current techniques are used to enhance these two methods by the use of relationships among conditions or other information.

Shen (2007) develops a framework to compose services through discovery and orchestration for a given goal service. In this method, a goal service is modeled as a stateful web service composed of several activities, which are represented with input/output envelopes for data flow and entry/exit conditions for behavior control. Service composition maps each activity of the goal service to a suitable stateless service described in WSDL in the service registry by matching input/output envelopes and entry/exit conditions. That is, the input envelope and the precondition of the selected service should be guaranteed by the goal service, while the output envelope and the exit condition of the goal service should be covered by the selected service. Two *tightening techniques* are introduced in composition algorithms to achieve "completeness", which means the ability to find a feasible realization if it exists. Tightening Strategy I aims at solving the schema composition problem by pushing down the entry/exit conditions of preceding activities to tighten entry conditions of the following activities, while Tightening Strategy II is used for the instance composition problem, which pushes down both constraints from the goal invocation and conditions from preceding activities through the goal service. The author claims that only with service descriptions, the Tighten Strategy I can be schema complete. However, to guarantee that Tighten Strategy II is instance complete, exhaustive search and the invoking of all the related services are required. Thus, as the number of existing services increases, employing Tighten Strategy II becomes less practical.

Dong (2006) extends the OWL-S ontology framework for dynamic web service composition by raising the level of abstraction of similar services. The authors propose an abstract layer that aggregates all available concrete services which are grouped into different categories according to their functional attributes. A multi-level abstract service hierarchy is built. Each abstract service has an instance pool of all possible concrete service solutions. When detecting that a new concrete service is registered, the service ontology is updated automatically by a backend utility program.

Given a desired composed service, either an existing abstract service can be matched or a new corresponding abstract service is defined with its profile section. For the latter case, the process is implemented in two steps. First, based on the ontology hierarchy of all abstract services for concrete services and the definition of the goal service (functional and non-functional requirements), an abstract service is generated by a composition planner Java Theorem Prover. Then, other functional parameters specified by the goal service are recorded in the Profile section of the generated abstract service, which is used to identify the actual candidate service pool for the abstract service and fill in the new instance section when a satisfied concrete service is found. However, this approach assumes that there are many similar concrete services in the registry and all concrete services of the same abstract service share the same interface information.

The Simple Hierarchical Ordered Planner (SHOP2) (Wu, 2003) is an AI planning system to automatically compose web services with DAML-S web service descriptions. In the DAML-S process ontology, each service is modeled as a process. All processes are divided into three kinds: atomic processes, composite processes and simple processes. SHOP2 is a domain-independent Hierarchical Task Network (HTN) planning system, which creates plans by decomposing tasks. Specially, tasks are decomposed into smaller subtasks until they can be performed directly. SHOP2 can monitor the planning state at each step of the composition process, thus providing users a higher degree of control.

All methods described above use a service description language to define and declare web service forms and functions. Most use WSDL, OWL and DAML-S because of the improved tool support for semantic-based processing. Dong extends the OWL-S ontology framework to improve service reuse, especially for the similar required services, while other methods focus on generating the service plan based on the existing concrete services. Moreover, all these methods do not consider the scalability when the existing service space increases.

Rule-Based Composition

Rule-based methods compile service definition, constraints, and the composed service requirements to rules and facts that are loaded into a knowledge base. Another set of processing rules than build composition plans that satisfy the specifications using a rule-based expert system. Service composition methods in this category usually differ in terms of how services are modeled.

Laliwala (2006) applies the rule-based idea for service composition in an event-driven service-oriented architecture which integrates Semantic Web, Web Services and the computing grids to enable seamless integration of information services from large scale scattered sources. The proposal mainly depends on current Web Service specification: RDF, OWL, OWL-S and SWRL for compliance, and BPEL for composition and execution. The architecture consists of the following components: Ontology Server, Rules Engine, Information Services, Event Manager, Composition Engine, Grid Services and Execution Engine, as shown in Figure 2.

The Information Services component serves as information and service provider. The Event Manager communicates dynamically with the distributed heterogeneous services and other components to furnish precise and event specific composition. Also, it works in coordination with the composition engine to generate the composition schema at runtime. The Composition engine is used to decide the flow of services at run-time and generate the composition schema. The Ontology server stores common vocabularies and concepts to making semantic information more meaningful. The Rules engine defines flow, constraints and controls the behavior of business process. In addition, it models the process, defines the policy, detects and handles the exception, and defines the plan of process. The Execution Engine takes care of executing the schema generated by Composition Engine. And Grid Service supports

Figure 2. The service-oriented architecture

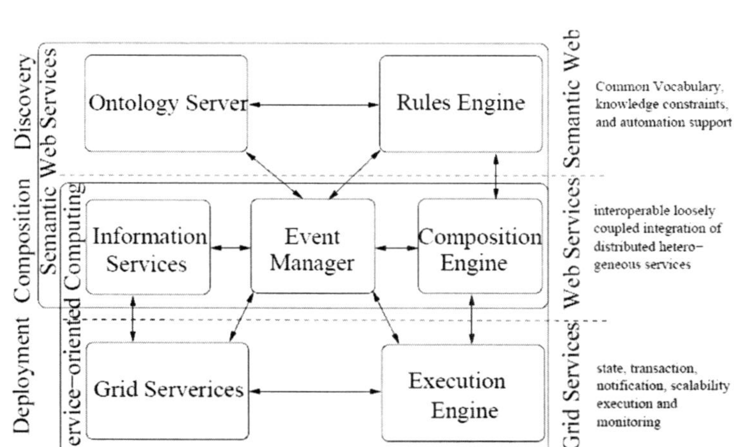

scalability. There are mainly two types of rules used in the method: Event-Condition-Action (ECA) rules and inferences rules. ECA Rules are used to define flow, constraints and control the behavior of a business process by reacting to events. An ECA rule is defined as a tuple (E, C, A), where E is a finite set of events for triggering the rule, A is a finite set of activities and C is a finite set of conditions for judging whether the action A should be executed. Dynamic services composition is realized by using ECA rules composition. An ECA rule-based composition is defined as a six-tuple (R, PRC, IP, OP, D, S), where R is a subset of ECA rules, precondition PRC and input IP define the required parameters for executing WS operation. Output OP defines the output parameters generated after the execution of the WS operation. Data flow D defines the flow of data from one service to another service and S gives the set of services or sequence of services. Inferences rules are if-then type and used to represent facts and required conditions during web service composition. Different from other methods, this approach explicitly solves the scalability issue of web service composition by integrating a Grid component.

SWORD (Ponnekanti, 2002) is another toolset for composition of a class of "information providing" services. In SWORD, each service is represented by its inputs and outputs. Whether a given composite service can be realized with existing services is determined by a rule-based expert system. In detail, every service has two types of inputs and outputs: conditional inputs (for specifying entities and their relationship conditions) and data inputs (for representing the attributes of entities), conditional outputs and data outputs. When modeling a service, it is defined as a rule of the form: precondition => postcondition. The precondition for the service is a conjunction of all the condition inputs and the known facts corresponding to the data inputs of that service. The postcondition is a conjunction of all condition outputs and the known facts corresponding to the data outputs of the service. Given a desired composite service, the initial state is the precondition of that service and the desired final state is the postcondition of the composite service. With the initial and final state, and the provided rules for the existing services, SWORD uses the following steps to determine if a plan exists for the composite service: 1) it asserts the rules corresponding to all the available services in Jess, a Java version rule-based engine; 2) it asserts the initial state facts for the composite service; 3) it runs the engine which causes matching rules to fire; 4) if the final state facts can be derived, then SWORD constructs a plan from the derivation structure used to derive these facts; 5) the constructed plan can be viewed by the developer using the GUI plan-viewer tool.

Another rule-based composition method is proposed by Majithia (2004). This system caches pre-defined and new process templates for abstract service workflows in the form of rules. These abstract workflow are mapped to executable service graphs consisting of concrete services. By distinguishing between different levels of workflow abstraction, this proposed framework allows reuse and provides a high degree of fault tolerance.

Logic Programming Composition

The process of constructing a composite service can be established directly from query proofs using a logic programming language, such as Prolog.

The Semantic Streams architecture (Whitehouse, 2005) uses service composition concepts in sensor network for providing sensor processing services which can combine the data from different sensors and produce a new output. In the proposal, the author presents a semantic service programming model allowing users to program sensor network with a declarative statement such as, "I want to get the speed information of the vehicles near the garage entrance". Two fundamental elements are in the model: event streams and semantic services. Event streams are sequences of asynchronous events (or objects) in time with a set of properties. Semantic services are processes that infer semantic information about the world and incorporate into an event stream. In addition, each service has a first-order logical description of the semantic information of its inputs and outputs. When a user submits his query in first-order logic, an inference engine uses backward-chaining to determine which sensors and services can provide the semantic information the user needs. In other words, the user specifies the global query and the system determines the appropriate local node behaviors. The markup and query language used in the model is built on Prolog and its constraint logic programming extension.

A semi-automatic web service composition approach with semantic descriptions is described by Sirin (2002). Their method keeps a human controller as the decision mechanism. The proposed approach uses a Prolog inference engine for OWL and a composer that handles the communication between human and the engine. Because the system lets users create a DAML-S workflow of services by presenting the possible choices at each step, users have to indicate constraints at each step, which makes the whole process inconvenient and complicated.

Constraint Programming

The constraint programming based approach represents all criteria guiding the selection of the services as constraints, such as QoS requirements, in web service composition, and then chooses an optimal set of services to satisfy customers' requirements. This approach improves efficiency by reducing the complexity and time for web composition.

Channa (2005) formulates dynamic web service composition as a constraint satisfaction problem (CSP). A CSP consists of a finite set of variables and a set of constraints. A constraint is a relation defined on some subset of these variables and denotes valid combinations of their values. A CSP solution finds a consistent assignment of values to the variables, so that all constraints are satisfied. By modeling web service composition to a CSP, an optimal set of services can be discovered. The proposed solution uses service templates. There are three major components of the architecture: OWL-S/UDDI registries, discovery engine and constraint optimizer, as shown in Figure 3. Service providers publish their service description on the OWL-S/UDDI registry. The discovery engine offers interfaces for semantic publication and discovery. Given a query, the discovery engine uses the classes subsumption-relation to compare the ontological concepts specified in the query to those advertised in the registry. The results are returned with rank according to the degree of match. In the discovery engine, the discovery method primarily depends on the semantic descriptions and constraints advertised by the service provider. The constraint optimizer is responsible of producing optimal service sets from the candidate service set returned by the discovery engine, based on business technology and client constraints, which consists of three sub-modules: constraint representation, cost estimation and constraint optimization. Constraint representation supports that users can represent the business constraints in ontologies by providing a set of classes, attributes, instances and relationships. The cost estimation affects the selection of the services for the dynamic composition processes by querying the information stored in the constraint representation for estimating costs according to various factors. The constraint optimization is used to retrieve constraints for the services matching the service template from either UDDI or by invoking an operation of the service specified by the service provider. Then the system checks whether the discovered services satisfy the client constraints and optimizes the service sets according to the objective function which is a linear combination of the process QoS parameters.

Figure 3. The constraint satisfaction-based web service composition architecture

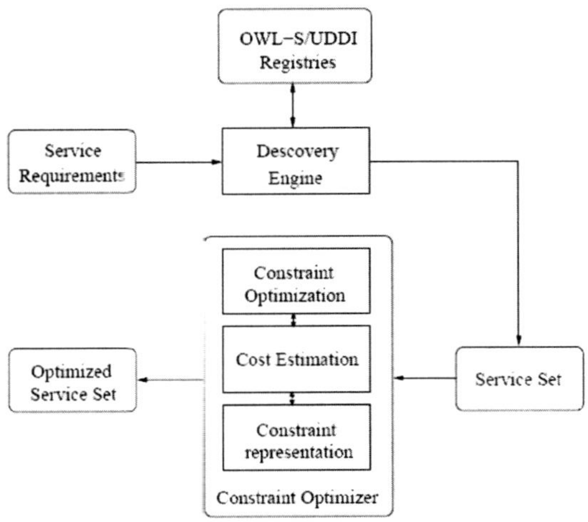

Graph Searching Methodologies

In the graph-based web services composition techniques, services modeled as nodes, with input/output and other execution constraints describing edges. A composite web service workflow is defined by a connected graph of services and an order of execution for the services. Figure 4 describes the general idea behind the graph-based technique. The initial composite service requirements are tailored into inputs to existing services and the output of one service is the input to another. As there may be multiple candidates for a service, an algorithm needs to determine which services should execute and in what order.

Casati (2000) has developed a system called eFlow that is used to specify, establish and monitor complex services that are built on top of other services. These services are modeled as business processes which are managed by the eFlow engine, whose role is to process events and notification of service completions. In eFlow, a composite service is modeled as a graph representing the flow structure which defines the order of execution among the nodes in the process. There are three types of nodes: service nodes, which correspond to the invocation of basic or composite services; decision nodes, which specify the rules controlling the execution flow; and event nodes that enable service processes to send and receive different types of events. Links between nodes can be labeled with transition predicates which, if evaluated to true, imply that nodes connected to that outgoing arc can be executed. Transactional regions define a subgraph that should be executed atomically. The eFlow engine processes messages notifying completion status of service nodes by updating the value of case packet variables accessed by the service node and by subsequently scheduling the next node to be

Figure 4. Graph-based Web services composition

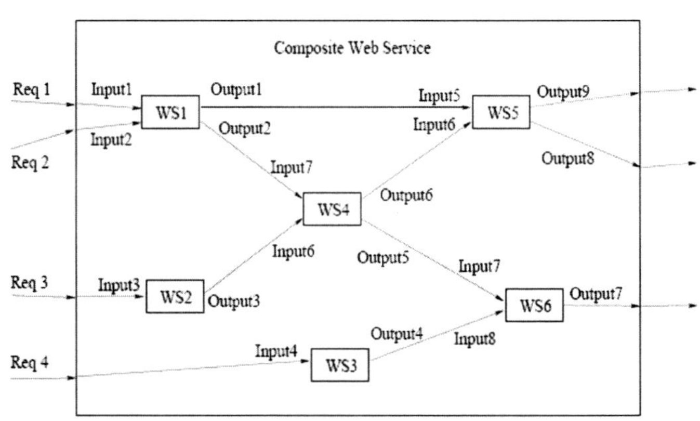

activated in the instance according to the process definition. eFlow will also enforce both authorization rules, to ensure that only authorized users perform modifications, and also consistency rules, to avoid run-time errors that result from the modifications. Changes to the process schema are not fully automated and the scalability of the system is not addressed by the authors.

The main contribution in Arpinar (2004) is peer-to-peer automatic service composition, where each peer can provide one or more web services. These peers are then grouped into communities based on domains for which they provide services. Ontologies expressed in DAML-S (an earlier version of OWL-S) are used to represent the semantic structure of the communities. Each community has a master peer and backup peers. The master peer maintains a list of the masters and back up peers for every other community, and backup peers also have a copy of this list. In the service graph, the links between nodes are based on matching the output of one node to the input of another node. To find the shortest sequence between the source and destination nodes, the authors use a Bellman-Ford shortest-path dynamic programming algorithm. As future work, the authors mention the possibility of validation of the composition based on preconditions and post-conditions, which can show an overall improvement of the entire system.

Interested readers should also consult the graph-based approach from (Fujii, 2006). Their architecture models functional aspects of web services as a directed graph, where components and operations are described graph nodes, and semantic properties as edges. The authors implement a full-fledged framework for natural language requirements processing, service discovery, and composition.

Other Techniques

This section describes a formal method for service composition proposed by Lecue (2006). The Causal Link Matrix (CLM) classifies web services according to causal links – logical dependencies between the inputs and the outputs of a service. The semantic similarity between web services is captures by the SimT function. Web Services are described in a semantic manner, thus ensuring a high level of automation based on the property that semantic descriptions are machine-readable. AI techniques are also used by the authors, in the sense that the composition process consists of a recursive and regression-based approach (McDermott, 2002), where each causal link is represented in the solution plan as a semantic link between web services. The process is recursive - it repeats until all old and new goals are concepts in the knowledge base. The authors describe the consistency, completeness and correctness of the solution plan that contains the necessary service(s). One of the other important characteristic of the proposed methodology is that it finds an optimal solution based on the quality of the semantic connection between web services. The model proposed is also flexible, but its scalability is not addressed by the authors. As an improvement, composition at the process level needs to be added to ensure the correctness of the composition process.

Formal techniques have a great advantage compared to other methods as they allow automatic verification of correctness. Methodologies, such as timed automata, Petri nets, or -calculus can validate such systems. Unfortunately, non-functional requirements are very hard, if not impossible, to express using formal mathematical methods, while with semantic web or is a straightforward task.

FUTURE TRENDS

An important factor for adoption of automated web service composition is the cost of service specification. Carman (2005) proposes a method to automatically induce the specification of web service operations, eliminating the need for reprogramming composite web services when new services become available. The main idea is to take advantage of the background knowledge and experimentation to automatically formulate a model for a new source. The proposed system automatically composes web services using Inductive Logic Programming and Query Reformulation methodologies to generate and test source definitions. The algorithm for achieving these features classifies the input and output of each new service with a semantic markup, and generates hypotheses from these new type signatures for the Datalog deductive database (Ceri, 1989). Some assumptions are made by the authors, namely that the meaning of an operation must be expressible in Datalog, which is a query and rule language used for query evaluations; another assumption is that there are no recursions or negation between the definitions of new sources. One of the positive aspects of the solution is that it also assumes that the description of the sources is incomplete, which is often the case in web services.

With the growing deployment of web services on mobile networks, we expect more emphasis will shift in automated web service composition towards enabling QoS awareness. Current methodologies do not consider service operating constraints such as link availability or service delays. A step towards this direction is done by the work in (Jin, 2005). Their method integrates semantic web techniques with web services agreement management. Negotiation rules and policies are described using the Semantic Web Rule Language (SWRL) (Horrocks, 2004). An agent-based runtime framework is designed with the scope of managing service agreements (contract negotiation) and uses negotiation protocols and strategies.

CONCLUSION

This chapter presented a review of current automatic web services composition methods. The fast growing number of service providers and service types impose performance bottlenecks on manual composition techniques due to the size of the search space. Most current methods for automated service composition use semantic web XML-based technologies for knowledge representation and processing, as they have extensive tool support and also facilitate use of AI techniques, such as planning, expert systems, and logic programming. Such methods provide a high degree of flexibility accommodating service semantics. Other techniques presented take a more low-level approach, using graph searching algorithms to find a process workflow that satisfies the composite service objectives. However, the semantic-based composition methods lack rigorous formal validation.

REFERENCES

Arpinar, I. B., Aleman-Meza, B., Zhang, R., & Maduko, A. (2004). Ontology-driven web services composition. In *Proceedings of IEEE International Conference on E-Commerce Technology (CEC 2004)*.

Carman, M. J., & Knoblock, C. A. (2005). Inducing source descriptions for automated web service composition. In *Proceedings of the AAAI 2005 Workshop on Exploring Planning and Scheduling for Web Services, Grid, and Autonomic Computing*.

Casati, F., Ilnicki, S., Jin, L., Krishnamoorthy, V., & Shan, M. (2000). Adaptive and dynamic service composition in eFlow. In *Proceedings of the Conference on Advanced Information Systems Engineering*.

Ceri, S., Gottlob, G., & Tanca, L. (1989). What You Always Wanted to Know About Datalog (And Never Dared to Ask). *IEEE Transactions on Knowledge and Data Engineering*, *1*(3), 146–166. doi:10.1109/69.43410

Channa, N., Li, S., Shaikh, A. W., & Fu, X. (2005). Constraint satisfaction in dynamic web service composition. In *Proceedings of the Sixteenth International Workshop on Database and Expert Systems Applications*.

Chinnici, R., Moreau, J.-J., Ryman, A., & Weerawarana, S. (2007). *Web Services Description Language (WSDL) Version 2.0 Part 1: Core Language*. W3C Recommendation 26 June 2007. Retrieved February 13, 2008 from http://www.w3.org/TR/wsdl20/

Dong, J., Sun, Y., & Yang, S. (2006). Owl-s ontology framework extension for dynamic web service composition. In *Proceedings of the International Conference on Software Engineering and Knowledge Engineering (SEKE 2006)*.

Fujii, K., & Suda, T. (2006). Semantics-based dynamic web service composition. *International Journal of Cooperative Information Systems (IJCIS), special issue on Service-Oriented Computing*.

Gudgin, M., Hadley, M., Mendelsohn, N., Moreau, J.-J., Nielsen, H. F., Karmarkar, A., & Lafon, Y. (2007). *SOAP Version 1.2 Part 1: Messaging Framework* (2nd ed.). *W3C Recommendation 27 April 2007*. W3C. Retrieved February 13, 2008 from http://www.w3.org/TR/soap12

Horrocks, I., Patel-Schneider, P., Boley, H., Tabet, S., Grosof, B., & Dean, M. (2004). *SWRL: A Semantic Web Rule Language Combining OWL and RuleML*. W3C. Retrieved February 13, 2008 from http://www.w3.org/Submission/SWRL/

Jin, H. & Wu, H. (2005). Semantic-enabled specification for web services agreement. *The International Journal of Web Services Practices*.

Klyne, G., & Carroll, J. (2004). Resource Description Framework (RDF): Concepts and Abstract Syntax. W3C. Retrieved September 20, 2008 from http://www.w3.org/TR/rdf-concepts/

Laliwala, Z., Khosla, R., Majumdar, P., & Chaudhary, S. (2006). Semantic and rules based event-driven dynamic web services composition for automation of business processes. In *Proceedings of the IEEE Services Computing Workshops (SCW'06)*.

Lecue, F., & Leger, A. (2006). A formal model for semantic web service composition. In *Proceedings of ISWC*.

Majithia, S., Walker, D. W., & Gray, W. A. (2004). Automated web service composition using semantic web technologies. In *Proceedings of the International Conference on Autonomic Computing*.

Martin, D., Burstein, M., Hobbs, J., Lassila, O., McDermott, D., McIlraith, S., et al. (2004). OWL-S: Semantic Markup for Web Services. *W3C Member Submission 22 November 2004*. Retrieved on February 13, 2008 from http://www.w3.org/Submission/OWL-S/

McDermott, D. (2002). Estimated-regression planning for interactions with web services. In *Proceedings of the 6th International Conference on AI Planning and Scheduling*.

Microsoft. (2003), *Business Process Execution Language for Web Services Specification*. Retrieved from http://msdn2.microsoft.com/en-us/library/aa479359.aspx

OASIS. (2008), OASIS UDDI. *Specification TC*. Retrieved February 13, 2008 from http://www.oasis-open.org/committees/tc_home.php?wg_abbrev=uddi-spec

Ponnekanti, S. R., & Fox, A. (2002). Sword: A developer toolkit for web service composition. In *Proceedings of the 11th World Wide Web Conference*.

Shen, Z., & Su, J. (2007). *On automated composition for web services. Poster Paper on World Wide Web (WWW)*. Conference.

Sirin, E., Hendler, J., & Parsia, B. (2002). Semi-automatic composition of web services using semantic descriptions. In *Proceedings of Web Services: Modeling, Architecture and Infrastructure workshop in conjunction with ICEIS2003.*

U.S. Department of Defense DISA. (2008). *Net-Enabled Command Capability (NECC) Overview.* Retrieved September 20, 2008 from http://www.disa.mil/necc/

Whitehouse, K. Zhao, F. & Liu, J. (2005). *Semantic Streams: a framework for declarative queries and automatic data interpretation* (Technical Report MSR-TR-2005-45). Microsoft Research.

Wu, D., Sirin, E., Hendler, J., Nau, D., & Parsia, B. (2003). Automatic web services composition using shop2. In *Proceedings of the International Conference on Automated Planning and Scheduling.*

KEY TERMS AND DEFINITIONS

Composed Web Service: a web service that in the course of its execution uses other web services to achieve its objectives.

Planner: a program that generates a sequence of services to realize a composite service. A typical planner takes three inputs: an initial state, a desired goal, and a set of existing actions, all encoded in a service description language.

Provider: organization or system that offers a set of application functions through web services.

Semantic Service Description: machine-readable specification of a service in a knowledge representation language that includes inputs/outputs and execution constraints.

Service Workflow (or Plan): describes the composite web services in terms of the temporal/logical sequence of execution of services part of a composite web service. The workflow can be expressed as a graph where nodes represent services. Two nodes are connected by a directed edge labeled by execution constraints.

Web Service Composition: a method of building a composite web service from the invocation of other web services.

Web Service: an application component that offers functionality to other applications over the internet.

Chapter 18
A Robust Clustering Algorithm for Mobile Ad-Hoc Networks

Zhaowen Xing
University of Oklahoma, USA

Le Gruenwald
University of Oklahoma, USA

K.K. Phang
University of Malaya, Malaysia

ABSTRACT

To mimic the operations in fixed infrastructures and to solve the routing scalability problem in large Mobile Ad Hoc Networks (MANET), forming clusters of nodes has been proven to be a promising approach. However, when existing weighted clustering algorithms calculate each node's weight, they either consider only one metric or rely on some metrics collected from extra devices. This often leads to a higher rate of re-clustering. This chapter presents a robust weighted clustering algorithm, called PMW (Power, Mobility and Workload), to form and maintain more stable clusters. In PMW, the weight of each node is calculated by its power, mobility and workload, which can be easily collected and computed locally and cover the major factors that cause re-clustering. Clustering overhead of PMW is analyzed. The simulation results confirm that PMW prolongs lifetime of MANETs and has a lower cluster head change rate and re-affiliation rate than other existing algorithms.

INTRODUCTION

A mobile ad hoc network (MANET) is a collection of battery-powered mobile nodes (or hosts) connected by relatively lower bandwidth wireless links. Each node has an area of influence called cell, only within which others can receive its transmissions. Due to no fixed infrastructures, all nodes can move freely, the network topology may change rapidly and unpredictably over time, and nodes have to form their own cooperative infrastructures. Thus, each node operates as an autonomous end system and a router for others in the network.

A MANET is of interest because there is no prior investment for fixed infrastructures, it can be easily deployed in a short time, and end users can access and manipulate data anytime and

DOI: 10.4018/978-1-60566-250-3.ch018

anywhere. Examples of MANET applications (Chlamtac et al., 2003) include law enforcement operations, automated battlefield applications, natural disaster recovery situations where the communication infrastructures have been destroyed, self-organizing sensor networks for data collecting, interactive lectures or conferences for data exchanging without pre-installed infrastructures. However, these MANET applications cannot be realized without efficient routing protocols. Current routing protocols over the flat network structure, in which each node participates as a peer, acts as a router and maintains individual routes to other nodes (Shiflet et al., 2004), are either proactive or reactive. Proactive routing protocols store route information in tables and update them periodically, while reactive routing ones search the routes on demand by flooding neighbors with the route requests. However, when the network size is large or the mobility of nodes is high, both the routing protocols over the flat network structure cannot work efficiently (Hong et al., 2002).

One promising way that can solve the routing scalability problem is to divide a MANET into clusters first, and then develop a routing protocol on top of the clustered MANET. A clustered MANET consists of cluster heads and cluster members, where a cluster head (like a mobile support station in a cellular mobile network) manages its clusters, coordinates intra/inter-cluster communication and so on. A cluster member is a node that belongs to a cluster and is not a cluster head.

Many weighted clustering algorithms have been proposed to elect cluster heads, form clusters and maintain clusters (Basagni, 1999; Basu et al., 2001; Chatterjee et al., 2002; Kim, 2006; Liu et al., 2005; Sheu & Wang, 2006). However, when calculating the weight utilized to determine whether a node is eligible to be a cluster head, these algorithms either consider only one metric (like mobility or power of nodes) (Basu et al., 2001; Kim, 2006; Sheu and Wang, 2006) or rely on some metrics collected from extra devices (such as locations of nodes read from Global Positioning Systems) (Chatterjee et al., 2002). This often leads to a higher possibility of re-clustering and, consequently, quality of service cannot be provided.

This chapter presents a robust weighted clustering algorithm, called PMW (Power, Mobility, Power), to form and maintain more stable clusters in MANETs. In PMW, in order to take into account the major factors that frequently cause re-clustering and in order to avoid being dependent on any extra device, the weight of a node is calculated by three parameters: power, mobility and workload. The results of our simulation show that PMW balances the power usage, prolongs the lifetime of clustered MANETs and has a lower cluster head change rate and re-affiliation rate than MOBIC (Basu et al., 2001). The rest of this chapter is organized as follows. Section 2 reviews the related background. Section 3 presents our approach to build stable clusters based on power, mobility and workload. Section 4 analyzes the message overhead and time complexity of PMW. Section 5 evaluates PMW and MOBIC by using the NS-2 simulator. Section 6 discusses some direction of future research. Finally, Section 7 concludes the chapter.

BACKGROUND

Several weighted clustering algorithms have been proposed and surveyed in (Yu and Chong, 2005). Here we briefly review some of the newly published approaches and others, which are already reviewed in (Yu & Chong, 2005), but are strongly related to our algorithm PMW. By considering the system parameters that are utilized to calculate the weight of each node, these approaches are categorized as mobility-only-based (Basu et al., 2001; Kim, 2006), power-only-based (Sheu & Wang, 2006) and combination-based (Basagni, 1999; Chatterjee et al., 2002; Liu et al., 2005).

Mobility-Only-Based

Mobility of nodes triggers re-clustering and makes networks unstable, thus, it becomes the key attribute in the weight computation in the mobility-only-based clustering algorithms.

In MOBIC (Basu et al., 2001), in order to form stable clusters, the Relative Mobility (RM) metric is introduced and calculated as the logarithm of ratio of received signal strengths (RSS): $10 * \log_{10}\frac{rss_1}{rss_2}$, where rss_1 and rss_2 are read from the RSS indicator when two successive HELLO messages, which are sent by the same neighbor, are received. For each node, the variance of RMs among its neighbors with respect to 0 (not the exact mean) is calculated as the aggregate local mobility metric. The nodes with the lowest aggregate local mobility among their neighbors are elected as cluster heads. Unfortunately, it is possible that some elected cluster heads may almost run out of power, thus, the re-election has to be invoked soon.

In ACT (Average Connection Time) (Kim, 2006), to overcome the negative effects caused by nodes moving fast or moving back and forth, the ACT of each node with its neighbors during a time period is introduced as the major parameter to form and maintain the clusters, and nodes having the largest ACT value become cluster heads. However, the ACT is similar to the cumulative time in WCA (Chatterjee et al., 2002) or elapsed time (Liu et al., 2005), it cannot accurately reflect the current level of the battery power because a node may have connected with its neighbors too long and it may almost run out of power.

Power-Only-Based

A node with a higher remaining power level is, of course, a better candidate for the cluster head; so battery power is the only system parameter applied to calculate the weight of each node in power-only-based clustering algorithms.

Because nodes with higher battery power have a higher priority to become cluster heads, it is possible that nodes with the least battery power being left out and claiming themselves as cluster heads. In Sheu's Stable Cluster Algorithm (SCA) (Sheu & Wang, 2006), Sheu et al. set up a battery power level threshold, define nodes whose battery level is below the threshold as bottlenecks, count the number of neighbors that are bottlenecks for each node, and elect nodes with the largest number of bottlenecks as cluster heads. By taking the detour in the election, nodes with the least battery power are kept from becoming cluster heads, thus, the clusters become more stable. Unfortunately, because the mobility of nodes is not considered in the election, the possibility of re-clustering is still high when elected cluster heads have high mobility.

Combination-Based

Each node is assigned with a weight, which is calculated by considering more than one system parameters like node degree, remaining power, roaming speed, and so on (Yu & Chong, 2005). In DCA (Distributed Clustering Algorithm) (Basagni, 1999), each node is assumed to have a different weight, nodes with the biggest weighs are elected as cluster heads, and 1-hop neighbors of elected cluster heads join the cluster as ordinary nodes. However, the calculation of nodes' weights is not discussed (Basu et al., 2001).

In WCA (Weighted Clustering Algorithm) (Chatterjee et al., 2002), to determine whether a node v is suited for being a cluster head, the weight of each node (W_v) is calculated by a formula as shown below that consists of four system parameters: sum of distance with all neighbors (D_v), average running speed (M_v), cumulative time of serving as a cluster head (P_v) and degree difference of nodes (Δ_v), where $\Delta_v = |d_v - \delta|$, in which d_v is the number of neighbors and δ is the ideal number of neighbors that a cluster head can handle. Unfortunately, how to choose δ is not discussed (Yu and Chong, 2005).

$$W_v = f_1 * \Delta_v + f_2 * D_v + f_3 * M_v + f_4 * P_v,$$
$$where,\ f_1, f_2, f_3\ and\ f_4\ are\ weighting\ factors$$
$$and \sum_{i=1}^{4} f_i = 1$$

The values of f_1, f_2, f_3 and f_4 are varied based on different applications. The nodes with the lowest weights are elected as cluster heads. However, how to normalize these parameters is not addressed explicitly. The global positioning system (GPS), the accuracy of which is not ideal for fine computing and the operations of which would drain the limited power of the node quickly, has to be applied to obtain the coordinates of each node for computing the running speed. The cumulative time of a node already serving as a cluster head cannot accurately reflect the current level of battery power because a busy node may almost run out of power and it has never been a cluster head.

In Liu's Group Management (Liu et al., 2005), the resource (*R*) richness and elapsed time of a node being a cluster head (leader) are integrated to evaluate a node's suitability for being a cluster head. The resources are CPU load (*L*), memory (*M*), battery (*B*) and bandwidth (*BW*). The elapsed time (*ET*) is the time between now and the last time a node is a cluster head. The weight of a node *v* is calculated by the following formulas, and nodes with the highest weights are elected as cluster heads.

$$W_v = f_1 * R_v + f_2 * ET_v,\ where\ f_1 + f_2 = 1, and$$
$$f_1 * R_v = f_{11} * L_v + f_{12} * M_v + f_{13} * B_v + f_{14} * BW_v,$$
$$in\ which,\ f_1 = f_{11} + f_{12} + f_{13} + f_{14}$$

However, how to normalize these parameters is not addressed explicitly either. The use of elapsed time in cluster head determination has the same disadvantage as the use of cumulative time in WCA (Chatterjee et al., 2002). In addition, re-clustering may be frequently triggered since the mobility of nodes is not considered in the election.

THE PROPOSED PMW APPROACH

Being inspired by MOBIC (Basu et al., 2001) and WCA (Chatterjeet al., 2002) and considering a new system parameter, called "Power Decreasing Rate (PDR)", we propose a weighted clustering algorithm, called PMW, to build a stable backbone in MANETs. Although our proposed clustering algorithm is also combination-based, it can become mobility-only-based if we tune the weighting factors accordingly. In other words, our approach can build a more stable backbone for MANETs by forming clusters.

The Basis of Our Algorithm

To capture the mobility of nodes, we do not consider the absolute roaming speed, which is actually applied in WCA (Chatterjee et al., 2002). This is because it is easy to calculate the speed's quantity but it is hard to predict the direction of movement. Without the direction, the speed's quantity alone is not appropriate to justify whether a node is a good candidate or not. For instance, the speeds of two nodes are small, but they both move in the opposite directions. As time goes, they will be out of each other's transmission range and get disconnected from each other. Also the utilization of GPS is opted out due to the following reasons:

- When a GPS is utilized, every node must be equipped with one, which incurs high hardware costs. In addition, these GPSs consume the limited battery power of nodes.
- GPS does not work indoor because buildings shield the satellite signals (Basu et al., 2001; Bruning et al., 2007).
- The accuracy of a typical civilian GPS is in the range of 6-12 meters (Zidek et al., 2006), thus, the returned results could be the same when two GPSs are located within 10 meters.

- If data cannot be read from the GPS and no replacement can be found, then the whole system has to wait or fail.

Instead, two mobility metrics, Relative Mobility (*RM*) (Basu et al., 2001) and Mobility Prediction (*MP*), are introduced to monitor the mobility of nodes and applied to determine whether a node is suitable to be a cluster head as follows:

- For each node j ($1 \leq j \leq N$ for N nodes in the network), after receiving two successive HELLO messages from every 1- hop neighbor i ($1 \leq i \leq n$ if there are n neighbors), the RM_{ij} is calculated by the formula (3.1). rss_{ij1} and rss_{ij2} are the received signal strength (RSS) that are read from the RSS indicator when the first and second HELLO message from the same neighbor are received, respectively. Based on the value of RM_{ij}, we can say that if RM_{ij} is equal to 1, then the node j and its neighbor i either do not move at all or move with the same speed in the same direction; if RM_{ij} is less than 1, then they move close to each other; otherwise, they move away from each other.

$$RM_{ij} = \frac{rss_{ij1}}{rss_{ij2}} \quad (3.1)$$

- For each node j, to take into account the mobility of all n 1-hop neighbors, MP_j is calculated as the standard deviation of RM_{1j}, RM_{2j}, ..., RM_{nj} shown in the formula (3.2). However, for the stability of elected clusters, we prefer RM_{ij} to be equal to or less than 1 because we want cluster heads not to move away from their members. Thus, in the MP_j calculation the mean of RM_{ij} ($1 \leq i \leq n$) is 1 instead of the actual mean. A node j with a lower MP_j means that it stays closer to its neighbors, thus, it is a better candidate for the cluster head among its neighbors.

$$MP_j = \sqrt{\frac{\sum_{i=1}^{n} (RM_{ij} - \overline{RM_{ij}})^2}{n}}, \quad \text{where } \overline{RM_{ij}} = 1 \quad (3.2)$$

When dealing with the limited battery power, we consider not only the Remaining Power (RP) of each node but also its Power Decrease Rate (*PDR*) as the workload because nodes with heavier workload consume more energy, so that we can balance the power usage and prevent cluster heads from running out of power quickly. In other words, for each node j, the PDR_j is considered because the RP_j represents only the current state of power level and the power will run out soon if this node usually has a heavy workload (for instance, it provides service as a server and relays packets for many neighbors). The PDR_j at time interval $[t_1, t_2]$ is calculated by using the formula (3.3), where rp_{j1} and rp_{j2} are the remaining power at time t_1 and t_2, respectively.

$$PDR_j = \frac{rp_{j_1} - rp_{j_2}}{t_2 - t_1} \quad (3.3)$$

A node with a lower *PDR* indicates that it was not busy at least during the interval $[t_1, t_2]$. However, when a node had a busy work history, it most likely would be busy in the future as well. Since the larger the time interval is, the more accurate the *PDR* is in indicating a node's workload history, during the initial election, each node saves a copy of its initial remaining power and initial time as rp_{j1} and t_1, such that a more accurate *PDR* can be calculated in the future re-election.

Based on the above analysis about power, mobility and workload, it is obvious that a node j is the best candidate for a cluster head among all its neighbors if its RP_j is the highest, its MP_j is the lowest and its PDR_j is the lowest. In other words, a node with the highest weight is the best candidate for a cluster head when we combine these three metrics together as the weight, which is calculated in formula (3.4). Since these metrics have different units, we apply the inversed exponential function to normalize MP_j and PDR_j and bound their values between 0 and 1. RP_j is left out because it is the remaining power level in percentage and its value is already between 0 and 1.

$$W_j = f_1 * e^{-MP_j} + f_2 * RP_j + f_3 * e^{-PDR_j} \qquad (3.4)$$

In formula (3.4), $RP_j = rp_{j2}$, the weighting factors f_1, f_2 and f_3 are set according to the different scenarios in the applications, and $f_1 + f_2 + f_3 = 1$. When we let $f_2 = f_3 = 0$, that is, we take away the effect of power and workload, our algorithm turns into a mobility-only-based approach just like MOBIC (Basu et al., 2001).

Cluster Formation

Cluster formation involves the following four steps, and messages used in cluster formation and maintenance are summarized in Table 1.

Step 1: Each node j periodically broadcasts (this broadcast interval is predefined (Basu et al., 2001) a HELLO message with the same transmission power. In the mean time, the remaining power level rp_l is recorded at the initial election time t_l; for each HELLO message received from neighbor i, rss_{ij} is recorded. If only one HELLO message from a neighbor is received after j broadcasts three successive HELLO messages, then this neighbor is excluded from the weight calculation (Basu et al., 2001).

Step 2: Immediately after each node j receives two successive HELLO messages from all the 1-hop neighbors, it records the remaining power level rp_{j2} and the time t_2. And then node j calculates the values of RM_j, MP_j, PDR_j and W_j using the formulas 3.1, 3.2, 3.3 and 3.4 defined above, respectively. When the weight is re-calculated, the saved remaining power rp_{jl} at time t_l is applied in the calculation of PDR_j.

Step 3: Each node j broadcasts the value of W_j to all its neighbors in a WEIGHT message, and waits for their WEIGHT messages.

Step 4: Upon receiving the weights from all 1-hop neighbors, the nodes with the highest weight declare themselves as cluster heads among their 1-hop neighbors by broadcasting a CLUSTER-HEAD message, but, no two cluster heads should be 1-hop neighbors. All 1-hop neighbors of elected cluster heads just join them as their members by broadcasting JOIN messages. If two nodes have the same weight, then the node with a smaller ID becomes the cluster head (Basu et al., 2001). If it happens that a node is 1-hop neighbor of two or more cluster heads, then it joins the cluster head with the highest weight and works as a gateway for these cluster heads.

Table 1. Messages Used in PMW

Message	Description
HELLO(my_ID, my_W, CH_ID, my_RP, other_CH)	To notify neighbors about my ID, my weight, my cluster head's ID, my remaining power and any other neighboring cluster heads.
WEIGHT(my_ID, my_W)	To notify neighbors about the value of my weight.
CLUSTERHEAD(my_ID, CH_ID)	To notify neighbors about my role: I am a cluster head, that is, my ID is same as my cluster head's ID.
JOIN(my_ID, CH_ID)	To notify neighbors that I am going to join the cluster whose cluster head's ID is CH_ID. If a cluster head broadcasts a JOIN message, then it informs its members about its resignation and joins a cluster at the same time.

Cluster Maintenance

Because every node can roam and has limited battery power in a MANET, the links between members and cluster heads can be broken, and the links between two cluster heads can be generated (Xue et al., 2006). Consequently, clusters need be re-clustered. In other words, leaving clusters, joining clusters, merging clusters, and re-electing cluster heads are normal re-clustering operations in a clustered MANET. However, these operations should be performed only on demand to reduce the overhead of computation and communication, and to provide consistent quality of service.

In order to detect the link breaks and new link establishments, each node periodically broadcasts a HELLO message, which contains the ID and weight of itself, its cluster head's ID, its remaining power and any other neighbor which is a cluster head (a Boolean variable). Being a cluster head, it has to periodically monitor its remaining power level so that it will resign when the remaining power drops below a predefined threshold. Also each node keeps recording the values of RSS from the last two HELLO messages and re-calculating its weight in case of future re-elections.

Relying on these two periodical operations, cluster maintenance can be done by the following recovery:

- from the link break between a member and its cluster head: after three successive broadcast intervals (*BI*) (Wang and Kim, 2007), if no HELLO message is received from a member, the cluster head will just remove this member from its neighbor and member lists. On the other hand, if a member does not receive a HELLO message from the cluster head after three successive *BI*s, it removes the cluster head from its neighbor list, and joins another cluster head with the highest weight if any. If no other cluster head is available from its neighbor list, this member declares itself as a cluster head.
- from the link establishment because two cluster heads become 1-hop neighbors: if a cluster head has become a 1-hop neighbor of another cluster head for a predefined Cluster Contention Interval (*CCI*) (Basu et al., 2001), then the one with the smaller weight resigns and joins the other. The members of the resigned cluster head cannot join the new cluster head because of non-1-hop neighbors. They have to join other cluster heads with the highest weight or declare themselves as cluster heads instead.
- from the link break because a cluster head resigns: if the current remaining power of a cluster head is less than the Low Power Threshold (*LPT*), and if there exists a member whose power is higher than *LPT*, and this member has no other neighbor that is a cluster head, then the cluster head resigns and triggers a cluster head re-election. But, this re-election is limited to inside the old cluster. That is, the resigned cluster head goes through each member's profile, which is periodically updated after receiving a HELLO message, and finds a replacement that has the highest weight. After the new cluster head is elected, the resigned cluster head joins its cluster. If a member cannot join the new cluster head because they are not 1-hop neighbors, it has to join others or declare itself as a cluster head.

ANALYSIS OF PMW

In this section, we analyze PMW with respect to the message overhead per time step per node and time complexity per network topology change. These terms are defined below. The approach used is inspired by the theoretical analysis in (ER and Seah, 2005).

The price of clustering is that extra time is consumed and additional messages are incurred to form and maintain clusters. The consequence of these additional messages is called message overhead (the more messages are transmitted, the more traffic is in the network and the more battery power of nodes is consumed). Since bandwidth and battery power of each node are limited in MANETs, message overhead is an important metric for evaluating the performance of a clustering algorithm. We analyze the message overhead by analyzing the overhead due to the HELLO protocol and the overheads due to cluster formation and maintenance. In the mean time, the time complexity per network topology change is also computed.

To simplify the analysis, the continuous runtime is divided into discrete time steps, which are the duration between the time when a message is sent and the time when the message is received and processed by a receiver (Bettstetter and Konig, 2002). The random waypoint mobility model with zero pause time is assumed. The following definitions are used in the analysis (ER and Seah, 2005):

- N: the number of nodes in the MANET;
- m: the average number of cluster members in a cluster; $m = \Theta(1)$ because all clusters should have a maximum size constrain to avoid overburdening cluster heads (Banerjee and Khuller, 2001).
- f_{hello}: the number of HELLO messages broadcast by a node per time step; $f_{hello} = \Theta(1)$ because f_{hello} is proportional to average node speed s and inversely proportional to the transmission radius R, and both s and R are less than or equal to some constants (Sucec and Marsic, 2004).
- f_{link}: the average frequency of network topology changes occurred per time step; $f_{link} = \Theta(N)$ (Sucec and Marsic, 2004).
- T: the number of time steps taken by the algorithm after a network topology change to re-establish a valid cluster structure (or called re-clustering);
- M: the number of messages (or called packets) exchanged between nodes after a network topology change to re-establish a valid cluster structure;
- L: the total evaluation time in terms of time steps.

In terms of the average number of messages transmitted by PMW per time step per node, and the number of time steps needed to re-establish a valid cluster structure after a topology change, the following claims are made.

Claim 1: the *message overhead of PMW is O(1) packet transmissions per time step per node.*
Claim 2: *the time complexity of PMW is $T \leq 2$ per network topology change.*

Both claims are proved in the following subsections.

Hello Protocol Overhead

In order to discover its neighborhood and compute its weight, each node broadcasts HELLO messages periodically. Thus, the HELLO protocol introduces an overhead of $f_{hello} * N$ packets per time step for all nodes.

Cluster Formation Overhead

Immediately after each node calculates its weight, it broadcasts a WEIGHT message in one time step. After receiving WEIGHT messages from all its 1-hop neighbors, each node either becomes a cluster head by broadcasting a CLUSTERHEAD message or joins some cluster by broadcasting a JOIN message in one time step. Thus, for all nodes, they broadcast $2N$ messages in 2 time steps, that is, cluster formation overhead is N messages per time step.

Cluster Maintenance Overhead

From the discussion of cluster maintenance, it is obvious that every network topology change is detected by relying on the periodical HELLO messages, and each cluster head resignation is verified by periodically checking the remaining power level. Once a network topology change or a cluster head resignation occurs, relating nodes have to take respective actions to re-establish a valid cluster structure. Because of these actions, cluster formation and maintenance overheads incur, which are investigated in the following four subsections.

Link Break between a Member and its Cluster Head

Since the cluster structure is still valid when a link break occurs between nodes from different clusters or nodes that are members from the same cluster, there is no action. Only a link break between a member and its cluster head triggers the re-clustering.

The cluster head removes this member from its neighbor and member lists, so no message is necessarily transmitted. On the other hand, this member removes this cluster head from its neighbor list as well, and joins another cluster head with the highest weight if any. This case is done by broadcasting a JOIN message in one time step. If no other cluster head is available from its neighbor list, this member declares itself as a cluster head, and broadcasts a CLUSTERHEAD message in one time step. Thus, we have: $T = 1$ and $M = 1$ for one of this kind of link breaks.

Link Establishment Because Two Cluster Heads Become 1-Hop Neighbors

When a cluster head becomes a 1-hop neighbor of another cluster head for a predefined duration of *CCI* interval, the one with the smaller weight resigns and joins the other. The resigned cluster head has to broadcast a JOIN message to inform all its members in one time step. After receiving the JOIN message from their cluster head, each member of this resigned cluster head has to do re-clustering, which is the same as the case in subsection "Link Break between a Member and Its Cluster Head". To summarize, $T = 2$ and $M = m + 1$ for one of this kind of link establishments.

Link Break Because a Cluster Head Resigns

When the current remaining power of a cluster head is less than the threshold *LPT*, and there exists one of its members that can be a new cluster head, this cluster head resigns and triggers a cluster head re-election.

Once a new cluster head is elected, the resigned cluster head broadcasts a JOIN message to inform all its members in one time step. After receiving the JOIN message from its cluster head, each member of this resigned cluster head has to do re-clustering, which is also the same as the case in subsection "Link Break between a Member and Its Cluster Head". In short, $T = 2$ and $M = m + 1$ for one of this kind of link breaks.

Total Cluster Maintenance Overhead

Since $M = 1$ in the case of "link break between a member and its cluster head" and $M = (m+1)$ in the case of "link establishment because two cluster heads become 1-hop neighbors", the total number of messages transmitted per network topology change due to link state changes is $(m+2)$. And the average network topology changes occurred per time step is f_{link}. Therefore, there are totally $f_{link}*(m+2)$ messages per time step due to link state changes.

Since a node cannot become a cluster head any more once it resigns due to its lower remaining power, there are at most N cluster head resignations in an evaluation. An evaluation period consists of L time steps, thus, the average number of cluster head resignations per time step should be N/L. Therefore, the total number of messages is $N(m+1)/L$ per time step due to the cluster head resignation.

In summary, total cluster maintenance overhead is $f_{link}*(m+2) + N(m+1)/L$ messages per time step.

Total Message Overhead

To summarize, the message overhead of PMW (O_{PMW}) is the sum of the overhead due to HELLO protocol, the overhead due to cluster formation and the overhead due to cluster maintenance, that is,

$$O_{PMW} = f_{hello} * N + N + f_{link} * (m+2) + N(m+1) / L$$

Since $f_{hello} = \Theta(1), f_{link} = \Theta(N), m = \Theta(1)$, L is an integer and $L > 1$, given some constants c_1, c_2 and c_3, we have: $f_{hello} \leq c_1, f_{link} \leq c_2 * N, m \leq c_3$ and $(1/L) < 1$. Therefore, O_{PMW} can be expressed as follows:

$$\begin{aligned}
O_{PMW} &= f_{hello} * N + N + f_{link} * (m+2) + N(m+1) / L \\
\Rightarrow O_{PMW} &\leq f_{hello} * N + N + f_{link} * (m+2) + N(m+1) \\
\Rightarrow O_{PMW} &\leq c_1 * N + N + c_2 * N * (c_3 + 2) + N * (c_3 + 1) \\
\Rightarrow O_{PMW} &\leq (2 + c_1 + 2c_2 + c_3 + c_2 c_3) * N \\
\Rightarrow O_{PMW} &= O(N)
\end{aligned}$$

After dividing $O(N)$ by the number of nodes N, the message overhead of PMW is $O(1)$ per time step per node and Claim 1 is proved.

$T = 1$ for a link break between a member and its cluster head, and $T = 2$ for both a link establishment and a link break due to the resignation of some cluster head, therefore, the convergence time is at most 2 time steps per topology change, as per Claim 2.

PERFORMANCE EVALUATION

The performance of PMW and MOBIC (Basu et al., 2001) is evaluated via NS-2 simulator with clustering framework (Basagni et al., 2006). Most of our simulation parameters are the same as the ones in (Basu et al., 2001) listed in Table 2 along with others. Since mobility is the major cause of re-clustering, the weighting factors $f_1 = 0.8$, $f_2 = 0.15$ and $f_3 = 0.05$ are used. The initial power level of each node is randomly distributed between 20% and 100%.

To measure the stability of a clustered MANET, we consider the following metrics:

- The lifetime of the network: the duration from the beginning until any node runs out of its battery power (Choi and Woo, 2006; Sheu and Wang, 2006).
- The cluster head change rate (per second): the total number of cluster heads is divided by the total simulation time (Basu et al., 2001).
- The re-affiliation (joining a cluster and becoming a member) rate (per second): the total number of cluster members is divided by the total simulation time (Choi and Woo, 2006).

All metrics in the following figures are collected from an average value of 50 simulation runs in 50 different scenarios, which are randomly generated using the random waypoint model (built-in in NS-2). To better mimic a real wireless network, 25 constant bit rate (CBR) connections are randomly generated by the traffic-scenario

Table 2. Simulation Parameters

Parameter	Value
Number of nodes (*N*)	50
Network size	670m x 670m
Maximum speed of node movement	1, 10, 20, 30m/s
Transmission range (*TR*)	10m – 250m
Pause time (*PT*)	0s, 30s
Broadcast interval (*BI*)	1s
Cluster contention interval (*CCI*)	3s
Low power threshold (*LPT*)	30%
Simulation time	200s

generator. Each source sends a 512-byte packet through UDP at a rate of one packet per second.

In Figure 1, the lifetime of the network decreases as node mobility and transmission range increase in both MOBIC and PMW. In Figure 1(a), PMW prolongs the lifetime of the network from 9% to 42% (or 23% on average) better compared to that of MOBIC. In addition, PMW outperforms MOBIC from 2% to 27% (or 15% on average) when the transmission range is beyond 50 meters as shown in Figure 1(b). These promising results show affirmatively the effect of taking into consideration the power and workload in the weight calculation and the forced resignation of a cluster head when its power becomes too low.

Figure 2(a) shows that the cluster head change rate increases as node speed increases. This is because cluster heads with higher speed are more likely to become 1-hop neighbors, and consequently, the one with a lower weight has to resign. PMW produces from 5 to 12 (or 7 on average) fewer cluster heads than MOBIC for *PT* (pause time) = 0 and *PT* = 30, respectively. In Figure 2(b), the number of cluster heads increases when the transmission range is less than 50 meters. However, when the transmission range becomes larger than 50 meters, the cluster head change rate decreases as more nodes stay together for a longer period of time, and PMW produces from 5 to 13 (or 10 on average) fewer cluster heads than MOBIC. PMW produces fewer cluster heads mainly because nodes with higher power are likely to get elected as cluster heads and, hence, can function as cluster heads for a longer time.

Figure 3(a) shows that the re-affiliation rate increases as the node speed increases. This is

Figure 1. Lifetime of the network by varying maximum speed and transmission range, respectively

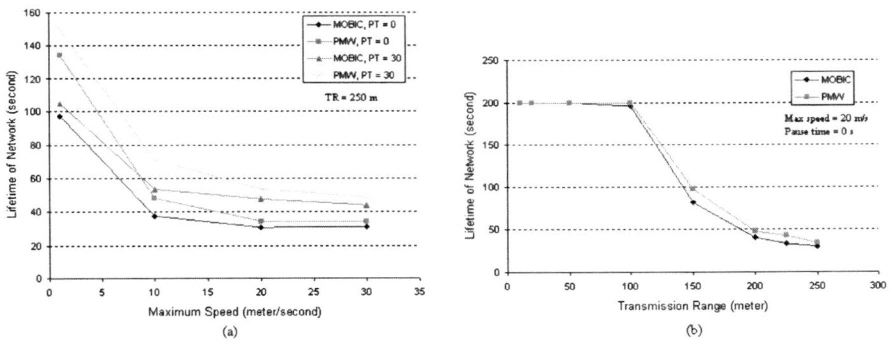

Figure 2. Rate of cluster head changes by varying maximum speed and transmission range, respectively

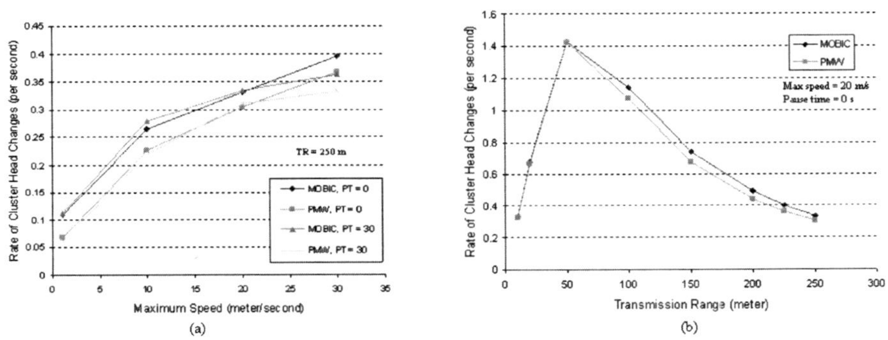

because cluster members with higher speeds are likely to get disconnected from their cluster heads and join the others. PMW produces from 34 to 66 (or 44 on average) fewer cluster members than MOBIC for PT (pause time) = 0 and PT = 30, respectively. In Figure 3(b), the re-affiliation rate has the similar trend as in Figure 2(b), and PMW produces from 34 to 43 (or 40 on average) fewer re-affiliations than MOBIC when the transmission range is greater than or equal to 100 meters. The advantage of PMW having a lower re-affiliation rate is mainly attributed to the less likelihood of the resignation of a cluster head due to power exhaustion.

FUTURE RESEARCH

Since performance of routing protocols over the flat structure suffers as number of nodes grows (Hong et al., 2002; Shiflet et al., 2004), there is the need to develop a reliable routing protocol, which is built on top of a hierarchical structure produced by a light-weight clustering algorithm.

Although replication provides a feasible solution for improving data availability in MANET databases, efficient consistency maintenance of replica remains a challenging problem in a large MANET. Cluster heads are the most stable nodes in the network and can form a virtual backbone, so they can decrease message overhead and help to realize efficient replication service.

Since pessimistic concurrency control techniques limit/block users to access data items and resources (bandwidth, power and storage) in MANET databases are limited, optimistic concurrency control algorithms may efficiently guarantee correctness of concurrent transactions execution in MANET databases, but it is challenging to choose which nodes to maintain the history information that is used for future validation. Again, cluster heads are the best candidates because they are more stable than other nodes, so that the history information is available when it is needed.

CONCLUSION

In this chapter, in order to resolve the mobility and scalability issues in routing protocols in MANETs, we proposed a robust weighted clustering algorithm, called PMW (Power, Mobility, Workload), to form and maintain more stable clusters in MANETs. In PMW, the weight of a node is calculated by three parameters: remaining power, mobility prediction (to check if a node moves along with all its 1-hop neighbors) and workload (represented by power decrease rate because nodes with heavier workload consume more energy). These three metrics are computed

Figure 3. Rate of re-affiliation by varying maximum speed and transmission range, respectively

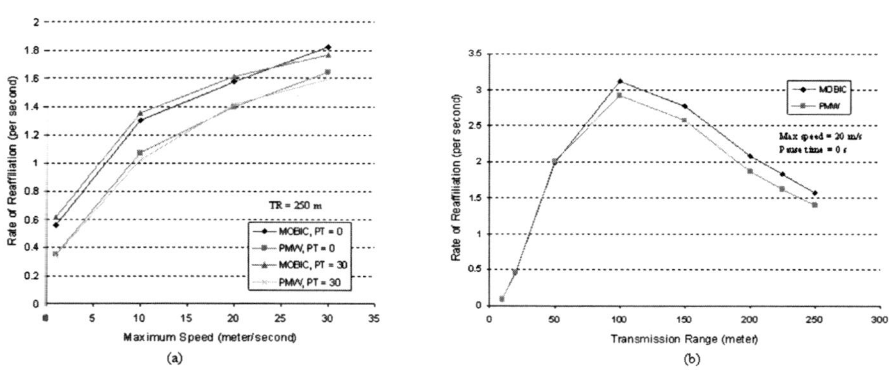

locally, independent of extra devices, and cover the major causes of re-clustering. Thus, in PMW, there is low re-clustering overhead during the cluster maintenance. The message overhead incurred due to PMW was claimed as O(1) per time step per node. The simulation results from the NS-2 simulator did confirm that PMW prolonged the lifetime of a clustered MANET and outperformed MOBIC by as much as 23% on average; PMW elected at least 7 fewer cluster heads on average than MOBIC; and PMW produced at least 40 fewer cluster members on average than MOBIC.

ACKNOWLEDGMENT

This material is based upon work supported by (while serving at) the National Science Foundation (NSF) and the NSF Grant No. IIS-0312746. Any opinion, findings, and conclusions or recommendations expressed in this material are those of the authors and do not necessarily reflect the views of the NSF.

REFERENCES

Banerjee, S., & Khuller, S. (2001). A Clustering Scheme for Hierarchical Control in Mult-hop Wireless Networks. In *Proceeding of 20th IEEE INFOCOM,* (pp.1028-1037). Anchorage, Alaska.

Basagni, S. (1999). Distributed Clustering for Ad Hoc Networks. In *Proceedings of the International Symposium on Parallel Architectures, Algorithms and Networks (ISPAN '99)*, (pp. 310-315). Perth/ Fremantle, WA, Australia.

Basagni, S., Mastrogiovanni, M., Panconesi, A., & Petrioli, C. (2006). Localized Protocols for Ad Hoc Clustering and Backbone Formation: A Performance Comparison. *IEEE Transactions on Parallel and Distributed Systems*, *17*(4), 292–306. doi:10.1109/TPDS.2006.52

Basu, P., Khan, N., & Little, T. D. C. (2001). A Mobility Based Metric for Clustering in Mobile Ad Hoc Networks. In *Proceeding of IEEE ICDC,* (pp. 413-418). Phoenix, Arizona, USA.

Bettstetter, C., & Konig, S. (2002). On the Message and Time Complexity of a Distributed Mobility-Adaptive Clustering Algorithm in Wireless Ad Hoc Networks. *In Proceeding of the 4th European Wireless* (pp.128-134). Florence, Italy.

Bruning, S., Zapotoczky, J., Ibach, P., & Stantchev, V. (2007). Cooperative Positioning with Magic-Map. *Workshop on Positioning, Navigation and Communication 2007 (WPNC'07)*, Hannover, (pp.17-22). Leibniz University of Hannover, Germany.

Chatterjee, M., Das, S. K., & Turgut, D. (2002). WCA: A Weighted Clustering Algorithm for Mobile Ad Hoc Networks. *Cluster Computing*, *5*(2), 193–204. doi:10.1023/A:1013941929408

Chlamtac, I., Conti, M., & Liu, J. J. N. (2003). Mobile Ad Hoc Networking: Imperatives and challenges. *Ad Hoc Networks Publication*, *1*(1), 13–64. doi:10.1016/S1570-8705(03)00013-1

Choi, W., & Woo, M. (2006). A Distributed Weighted Clustering Algorithm for Mobile Ad Hoc Networks. In *Proceedings of the Advanced International Conference on Telecommunications and International Conference on Internet and Web Applications and Services,* (pp. 73-78). Guadeloupe, French Caribbean.

ER. I., & Seah, W. (2005). Clustering Overhead and Convergence Time Analysis of the Mobility-based Multi-hop Clustering Algorithm for Mobile Ad Hoc Networks. In *Proceedings of the 11th International Conference on Parallel and Distributed System,* (pp. 1144-1155). Fuduoka, Japan.

Hong, W., Xu, K., & Gerla, M. (2002). Scalable Routing Protocols for Mobile Ad Hoc Networks. *IEEE Network*, *16*(4), 11–21. doi:10.1109/ MNET.2002.1020231

Kim, K. (2006). A Novel Factor for Robust Clustering in Mobile Ad Hoc Networks. *IEICE Transactions on Communications. E (Norwalk, Conn.), 89-B*(4), 1436–1439.

Liu, J., Sailhan, F., Sacchetti, D., & Issarny, V. (2005). Group Management for Mobile Ad Hoc Networks: Design, Implementation and Experiment. In *Proceedings of the 6th international conference on Mobile Data Management*, (pp.192-199). Ayia Napa, Cyprus.

Sheu, P., & Wang, C. (2006). A Stable Clustering Algorithm Based on Battery Power for Mobile Ad Hoc Networks. *Tamkang Journal of Science and Engineering, 9*(3), 233–242.

Shiflet, C. F., Belding-Royer, E. M., & Perkins, C. E. (2004). Address Aggregation in Mobile Ad-hoc Networks. In *Proceedings of the IEEE International Conference on Communications*, (pp. 3734-3738). Paris, France.

Sucec, J., & Marsic, I. (2004). Hierarchical Routing Overhead in Mobile Ad Hoc Netorks. *IEEE Transactions on Mobile Computing, 3*(1), 46–56. doi:10.1109/TMC.2004.1261816

Wang, Y., & Kim, M. S. (2007). Bandwidth-adaptive Clustering for Mobile Ad Hoc Networks. *International Conference on Computer Communications and Networks,* (pp. 103-108). Honolulu, Hawaii, USA.

Xue, M. ER, I., & Seah, W. K. G. (2006). Analysis of Clustering and Routing Overhead for Clustered Mobile Ad Hoc Networks. In *Proceedings of the 26th IEEE international Conference on Distributed Computing Systems*, (pp. 46-53). Lisboa, Portugal.

Yu, J. Y., & Chong, P. H. J. (2005). A Survey of Clustering Schemes for Mobile Ad Hoc Networks. *IEEE Communications Survey &Tutorials, 7*(1), 32–48. doi:10.1109/COMST.2005.1423333

Zidek, K., Saloky, T., & Polanecka, I. (2006). Usability of GPS Systems for Mobile Robots Navigation. *4th Sloyakian-Hungarian Joint Symposium on Applied Machine Intelligence*. Herlany, Slovakia.

KEY TERMS AND DEFINITIONS

Clustering in a MANET: Nodes are divided into non-overlapping clusters according to certain rules, and are assigned different roles: cluster heads or cluster members. A cluster head manages its clusters, coordinates intra/inter-cluster communication and so on, while a cluster member is a node that belongs to a cluster and is not a cluster head.

Mobile Ad Hoc Network (MANET): A collection of battery-powered nodes connected by relatively lower bandwidth wireless links. Nodes move randomly and organize themselves arbitrarily.

Mobility: Every node has ability to roam in the network.

Mobility Prediction: A quantity is applied to predict whether a node moves along with all its 1-hop neighbors

Multi-Hop Communication: Communication between two nodes is carried out through a number of intermediate nodes that relay packets from one point to another.

Network Lifetime: The duration elapsed from the beginning until the time when a node runs out of its battery power.

Power Decrease Rate: The rate of a node's remaining power decreasing within a time interval.

Relative Mobility: A quantity is used to determine whether a node moves close to or away from another node.

Weight-Based Clustering Algorithm: A clustering algorithm utilizes a node's weight to determine the eligibility of a node's being a cluster head, where the weight can be the value of a local system parameter or a combination of local system parameters.

Chapter 19
Emerging Mobility Applications of Host Identity Protocol

Szabolcs Nováczki
Budapest University of Technology and Economics, Hungary

László Bokor
Budapest University of Technology and Economics, Hungary

Gábor Jeney
Budapest University of Technology and Economics, Hungary

Sándor Imre
Budapest University of Technology and Economics, Hungary

ABSTRACT

This chapter is devoted to give an overview of the Host Identity Protocol (HIP) and to discuss two different mobility management extensions built on it. We introduce the basic ideas and the main paradigms behind the HIP-scheme, and show its wild range usability in next generation mobile and ubiquitous networks. The broad scale of possible advanced mobility applications of HIP will be introduced based on two extensions of the protocol, both developed by the authors. The first one is a micromobility solution based on HIP (μHIP), while the second one provides a scalable and secure network mobility solution (HIP-NEMO).

INTRODUCTION

Mobile telecommunication is facing an enormous evolution. Due to the expansion of wireless technologies, mobile services and applications are becoming more and more popular. The phenomenon of convergence in means of protocols, services and terminals generates increasing demand for more widespread and more sophisticated support of mobility. The Internet is growing into a universal, pervasive and ubiquitous architecture where end-users are expected to consume services through different heterogeneous technologies, enabling accessibility of the converged next generation networks at anytime and anywhere. It is quite obvious that one of the most unique characteristics of future's networks is mobility (J.F. Huber et al., 2004). The growing number of mobile users, the spreading of heterogeneous environments where

DOI: 10.4018/978-1-60566-250-3.ch019

subscribers are roaming among different wired/wireless access networks, and the increasing complexity of mobility scenarios both arises serious challenges for the today's architecture, which is usually the TCP/IP stack. Unfortunately the basics of TCP/IP systems were not designed with any kind of mobility in mind (D. Saha et al., 2004). Many solutions have born to enable mobility support but most of them are relaying on the classic protocol stack thus inheriting its main drawback: hosts and other network level entities are identified by their topological location, which is the IP address assigned to them. In other words IP addresses have dual significance thus being semantically overloaded. On one hand they serve as routing locators, on the other hand they are host identifiers when considering transport level connections, for instance. This property of IP addresses results in a very inconvenient mobility management because every time when a mobile entity changes its actual location, it must tear down all its ongoing upper level connections and must establish them again with the new IP address of the new location. To get rid of the problem and to meet the new requirements, the Host Identity Protocol (HIP) (R. Moskowitz et al., 2007) was designed to separate the dual role of IP addresses described above. The most important characteristic of HIP is that it uses cryptographic keys to identify hosts or other network entities while IP addresses are acting as pure locators. In this architecture transport level connections are not bound to IP addresses, which are dynamically changeable in several cases, but to permanent identifiers, which remain the same for quite a long time. This property provides two main advantages: management of mobility and security becomes very convenient, simple and fast. Additionally, this is inherited by any extension to HIP. This drove us to develop enhanced mobility solutions based on Host Identity Protocol, and by this way to highlight the emerging mobility applications of HIP.

The rest of this chapter is organized as follows. In the next section a general overview of the basic mobility scenarios is presented followed by the introduction of Host Identity Protocol and several existing mobility enhancements and applications of the main standard. After this overview our frameworks for HIP based micromobility (µHIP) and network mobility (HIP-NEMO) are described, and the security strengths of these proposals are also introduced. At the end of the chapter we will discuss the expected future trends along with some concluding remarks.

BACKGROUND

As the background of our work, we first briefly overview the basic mobility scenarios, followed by the basics of the Host Identity Protocol. Finally, existing proposals for making the base protocol more efficient and using the main idea of HIP in several advanced application scenarios will be discussed.

Basic Mobility Scenarios

Wired and wireless telecommunication systems both are converging into an all-IP based architecture where complex and synergistic union of wired and wireless technologies, protocols and terminals will provide integrated services on a universal IP-based infrastructure. As we introduced above, mobility is one of the most unique characteristics of future's convergent architectures but this promising feature also introduces a whole bunch of new challenges to the original Internet. Solutions are needed to manage mobile users and their requirements in the widest range of different scenarios and types of mobility.

When considering mobility management in next generation, IP based wireless networks, at least two main types of mobility should be classified. On one hand single mobile nodes merging from one network point of attachment to another have to be taken into account (i.e. host or terminal mobility). This kind of mobility allows a node to maintain ongoing communication or commence/receive incoming session requests independently from its network point of attachment. On the other hand, mobility should not be limited to single moving nodes or terminals: data communications within entire mobile networks moving between different access points need to be managed as single entities (i.e. network mobility). This kind of mobility requires at least one central entity in a moving network in order to hide the inside operation and to ensure several advantages like saving radio emission power and reducing the handover overhead of the overall architecture.

The emerging wireless technologies and the comprehensive convergence introduced several new types of mobility. We can talk about device-centric, low-level mobility including ad hoc mobility (mobile nodes are reachable and routable in ad hoc structures) and mode mobility (nodes can interchange between ad hoc and infrastructure modes). There is a user-centric, high-level type of mobility consisting of personal mobility (users are generally reachable at different scenarios: one address for many different terminals or many addresses reaching one terminal), session mobility (active communication sessions are interchangeable between terminals) and service mobility (personalized services are maintained during the movements of the terminal).

The appearance of heterogeneous networking architectures introduced another classification: managing mobility between different types of access networks (e.g. UMTS to WLAN) is called "vertical handover", while the usual "horizontal handover" is the case of migration of mobile nodes between homogeneous networks (e.g. when a user leaves the radio cell of a base station and enters a neighboring cell in a UMTS network) (J. Manner et al., 2004.). (Note that the definition of horizontal and vertical handover is little bit vague. E.g. a handover from a 802.11g WLAN link towards an 802.11b link can be considered as either a vertical or a horizontal handover, depending on the point of view.)

However, one thing is surely the same either talking about host or network, vertical or horizontal, or any other kind of mobility: wireless networks cover a definite geographical area consisting of several different domains. Deriving from this, mobile nodes can change their point of attachment to the network basically in two different ways. In the first case users can move inside a single domain, which is usually referred as intra-domain mobility. In the second case users can roam between different domains, which case is called inter-domain mobility. Usually, domains are aggregated and a special protocol is responsible for the local mobility management of this group of domains in order to offer fast and seamless handover control over a limited geographical area. In such cases, we speak about micromobility, and the aggregated group of domains is called micromobility domain. Micromobility protocols are designed for environments where mobile terminals change their network point of attachment so often that the general mobility management protocols (i.e. macromobility solutions) originate excrescent overhead and ineffective operation (P. Reinbold et al., 2003).

All of the above introduced mobility scenarios and types can be supported at different layers of the conventional TCP/IP stack (L. Deguang et al., 2006). Even though network level solutions are the most general (since IP is regarded universal in next generation networks), optimization and extension of the stack to support mobility functions along with more advanced requirements (e.g. security and privacy) are very difficult. This motivated developers to support mobility by introducing new layers. Host Identity Protocol (HIP) is one of the new approaches providing mobility management mechanisms by interposing the Host Identity Layer between the network and the transport layers.

HIP serves quite efficiently and securely as a macromobility protocol, but introduces heavy signaling overhead, significant packet loss and remarkable handover latency in micromobility environments which require fast, seamless and local handover control (T. R. Henderson et al., 2003). Additionally, HIP doesn't comprise any kind of support for network mobility scenarios (R. Moskowitz et al., 2006). This motivated us to develop a HIP based micromobility and network mobility framework in order to make the base protocol capable of effectively managing advanced mobility scenarios as well, and creating the fundamental bedrock for emerging mobile applications and services by this way.

Host Identity Protocol in a Nutshell

There are several solutions providing uninterrupted IP communication for mobile nodes and networks. (D. Johnson, et al., 2004) proposed a general solution for IP macromobility. (V. Devarapalli et al., 2005) extended the previous concept to provide network mobility support while (H. Soliman et al., 2005) and (A. T. Campbell et al., 2002) designed micromobility extensions to the basic protocol. (P. McCann, 2005) introduced a cross-layer method in order to boost up the basic MIPv6 handover procedures. (S. J. Koh et al., 2004) placed their mobility solution into the transport, while (N. Banerjee et al., 2004) into the application layer. All of these mobility solutions are based on the classic TCP/IP stack hierarchy thus sharing the same basic problem: mobile nodes are identified by IP addresses that depend on their actual topological location. That is why IP is semantically overloaded: IP addresses are both identifiers and locators at the same time. This dualism yields a very inconvenient way of mobility management because every single IP address change (i.e. interchange of network point of attachments) indicates the break up of all ongoing communication sessions thus creating the need to re-establishing them again with the new address gained at the new location.

The main concept of Host Identity Protocol (HIP) is to separate the locator role from the identifier function (R. Moskowitz et al., 2006). A very new layer (Host Identity Layer) is inserted between the network and the transport layers (R. Moskowitz et al., 2007). HIP defines a new, cryptographic namespace consisting of the Host Identifiers (HIs). HIs are devoted to hold the identity of nodes while IP addresses are continue acting as pure locators. Based on these cryptographic identifiers special certificates can be generated by the nodes, which method provides the basis of the secure signaling (P. Nikander et al., 2004) or even identity delegation (S. Herborn et al., 2007), offering enormous resource savings, effective session mobility and other promising application possibilities in wireless/mobile environments.

The basic function of HIP is to set up Host Identifier based connections between nodes and to map HIs to IP addresses and vice versa. Using HIP, transport layer connections are no more bound to IP addresses, and connections do not have to be broken if location of hosts change. HIs are usually represented in protocol messages by their 128 bit long hash (called the Host Identity Tag – HIT). In order to establish end-to-end connection, HIP defines a four-packet handshake based procedure called the Base Exchange (BE) (see Figure 1). The BE results in an IPSec Encapsulating Security Payload (ESP) and a Security Association (SA) pair, where SAs are bound to HIs/HITs instead of IP addresses (P. Jokela et al., 2007).

If HIP nodes are moving, the HIP connections may need to be updated. In such cases the macromobility functions of the Host Identity Protocol are used. The basic HIP mobility management is a simple three message sequence, which starts with a HIP UPDATE packet holding the new IP address in the LOCATOR parameter. The other two messages are to securely verify the relayed locator information. The peer host receiving the new LOCATOR notification simply corrects its HI – IP mappings and after successfully completing the UPDATE procedure, communication can continue seamlessly (T. Henderson, 2007). Since in a HIP-based environment upper layer connections are bound to HIs, the changes of the locators (i.e. address changes in the IP layer) are totally

Figure 1. The HIP base exchange

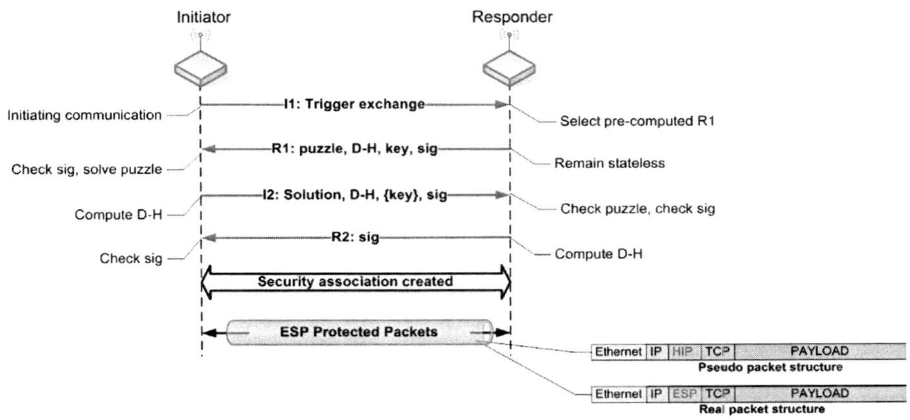

transparent to every application running on a HIP node. Thus, HIP macromobility management and multihoming provisioning are very simple and effective (T. R. Henderson et al., 2003). However, initial accessibility of mobile HIP nodes and improved mobile operation (e.g. simultaneous movement of nodes) might need assistance of a special network entity called the Rendezvous Server (RVS). RVS maps HIs to IP addresses and provides a DNS kind of network directory prepared for frequent updates and queries open for every potential HIP communication partner (J. Laganier et al., 2006). All mobile nodes should register their IP addresses in a known RVS and keep this registration up to date. If we assume that mobile nodes could get all the required information to reach at least one RVS, than in case of address changes the nodes should update their entry at the RVS and report the IP address of the actually used RVS at the DNS. After this action if a peer HIP node wants to reach the HIP mobile, it performs a look up at the DNS (prepared for HIP compatibility according to (P. Nikander et al., 2007)) for the address of the mobile. The DNS answers this query with the address of the mobile's RVS and the peer initiates the HIP connection by sending the I1 packet to the RVS. The RVS – as a relay station – appends some security parameters to the I1 message and forwards the packet according to the actual IP address of the destined mobile node which handles the security measures and finishes the BE in the regular way.

(J. Laganier et al., 2006) proposed a registration extension for HIP in order to allow HIP hosts to register with services such as HIP RVS. This proposal details how HIP enabled hosts can register themselves for an appropriate service provided by another HIP entity in the network. A solution for the problem of discovering the available HIP-based services in the network by discussing two different methods is proposed by.(P. Jokela et al., 2006). In the first method (called On-the-Path Service Discovery) the service provider is located on the packet forwarding path and listens to traffic. It offers its services for nodes communicating on the path by sending Service Announcement Packets (SAP) to them. In the second method (called Regional Service Discovery) the HIP host may initiate the service discovery based on Service Discovery Packets (SDP) packets. The initiator node creates and injects the SDP into the network, fills the destination HIT field with zeros and selects a multicast/broadcast or even anycast IP address to destination. This scheme makes possible to find service providers not located on the packet forwarding path as well.

The above short survey on HIP is devoted to help the reader to get better understanding on the base protocol. This enables one to easily comprehend the new concepts discussed in the upcoming paragraphs.

Overview of Existing HIP Applications

There are already several enhancements and advanced applications those rely on the functions and opportunities of HIP and show its wild range of possible usage scenarios.

(M. Särelä et al., 2004) described the methods and main considerations of applying Host Identity Protocol to tactical networks demanding high security requirements and capabilities for joint combined operations.

(J.Y.H. So et al., 2006) published a HIP based UMTS/WLAN architecture where the Rendezvous Server is integrated into the GGSN (Gateway GPRS Support Node) elements of the 3GPP standards such making the system ready for vertical handovers over a future public-key based communication in such a heterogeneous environment. The main improvement of the scheme over a standard Mobile IP solution is that the vertical handover delay is considerably reduced thanks to the author's handover solution based on HIP multihoming.

(P. Jokela et al., 2005) introduced and analyzed a HIP based decentralized approach for securing distributed computing systems dealing effectively with their organically distributed nature. The authors combine the network and operating system virtualization methods with the Host Identity Protocol and Simple Public Key Infrastructure certificates in order to build virtual domains of trust over insecure interconnection of nodes.

(Zs. Kovacshazi et al. 2007) presented a new multicasting model based on HIP called Host Identity Specific Multicast (HISM). Authors show how this novel multicasting approach handles access control, accounting, and mobility in general, and how it provides multicast support even in environments concurrently employing multiple versions of the IP protocol (IPv4 and IPv6).

(J.Y.H. So et al., 2005) designed new mobility management scheme based on the assembly of Host Identity Protocol and Session Initiation Protocol (SIP) in order to extinguish the mobility management limitations of SIP and create a hybrid macromobility solution for future value added services and applications.

(X. Zhu et al., 2007) introduced a HIP based Web Database Security model. The method proposes a new User Identity exchange scheme to realize authentication of Host Identity and User Identity, and combines this authentication with database systems and encryption methods to ensure security and confidentiality of the data communication. In this way the basics of providing mobile e-commerce, e-banking and e-finance applications in next generation networks were built up.

EMERGING MOBILITY APPLICATIONS OF HIP

In this section we introduce two novel mobility frameworks extending the capabilities and existing application scenarios of HIP. These two original approaches can enable new and fresh applications of the Host Identity Protocol by providing efficient and secure micromobility and network mobility support for HIP-aware nodes and networks.

Micromobility Extension of HIP

This section details an extension to HIP which adds a gateway centric micromobility support to the base protocol. This proposal makes HIP able to effectively serve QoS exigent applications. For example, it is very likely that voice transmission (dynamically extendable with multimedia sessions and/or other value added applications) will be an important service in next generation wireless networks. To efficiently support such services and usage scenarios, the network will have grant low delay and low jitter for packets of such transmissions. The QoS parameters will have to be sustained on the required level even when mobile nodes perform handover operations (i.e. when mobile nodes move from one network point of attachment to another). Therefore mobile nodes applying advanced applications of next generation architectures must be able to perform handover procedures as quickly as possible. Our micromobility extension of HIP is designed to fulfill these demands in the Host Identity layer.

Our proposed HIP micromobility architecture (μHIP) divides the network into micromobility domains maintaining several regular wireless IP points of access (i.e access routers with conventional IP forwarding engine). A newly introduced HIP gateway entity called the Local Rendezvous Server (LRVS) is responsible for managing HIP Mobile Nodes (MNs) in a given domain and for connecting the access network part to the Internet. LRVS gateways provide HIP registration service for MNs in a micromobility area, furthermore LRVSs introduce an IP address mapping function which is used to attach the MNs to the μHIP access network by registering the locally valid IP addresses (IP_L) of MNs. IP_L is valid only in the given domain (i.e. IP_L is not a globally routable address) and the LRVS is responsible for mapping every IP_L to a globally routable address chosen from a private address pool of the LRVS. This IP_G is used to register the MNs at their RVSs and to deliver packets outside the domain during further communication sessions.

The basic operation of this architecture is the following (L. Bokor et al., 2007). If a MN joins a micromobility domain, immediately starts an initialization mechanism (see Figure 2). First the MN physically connects to one of the access routers (AR) of the domain, then gets a serviceable IP address (IP_L) based e.g. on IPv6 stateless autoconfiguration. After this, the MN either may actively initiate a HIP service discovery procedure or passively wait for a service announcement in order to detect the LRVS service provided in the visited micromobility area. Thus the MN will be informed about the HIT and the IP address of the LRVS responsible for its actual domain. Figure 2 shows the case of passive discovery, where the MN sends HIP UPDATE packets to its RVS and current correspondent peer nodes (CNs) (step 1, red arrow). The LRVS – according to (P. Jokela et al., 2006) – intercepts the UDPATE packet, verifies the I1 source HIT and sends back a Service Announcement Packet (SAP) to the MN containing R1 data and information about the LRVS services (HIT and IP address of LRVS). After that the MN continues the service discovery by completing the registration to the LRVS with the final I2-R2

Figure 2. Initiation and connection establishment in μHIP

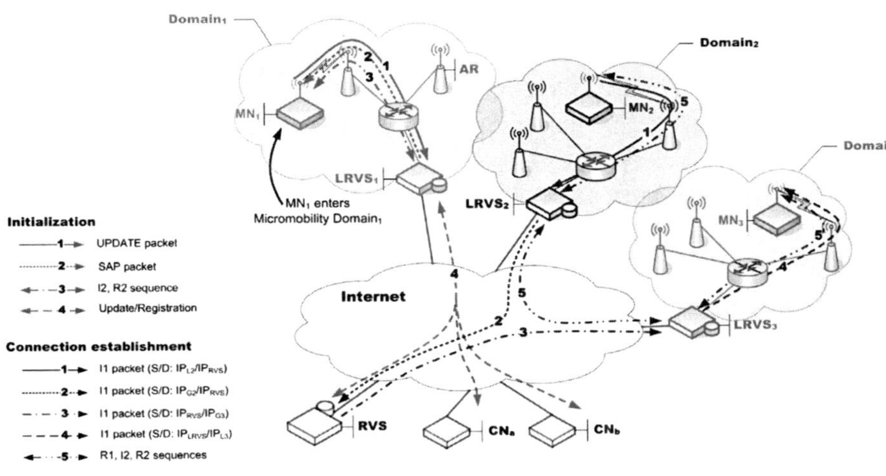

sequence. Till this point everything works almost according to the standards and existing drafts, except that during the service discovery and registration procedure the LRVS not only registers the MN's HIT with the new IP_L, but maps IP_L with the a globally routable address (IP_G) as well. Hiding the inner IP address (IP_L) also provides a simple location privacy solution preventing malicious nodes from learning the users' past or actual location.

After the MN successfully registered at the LRVS, it needs to perform the update and/or registration procedures at its RVS and current CNs as well in order to be reachable for the current and possible future communication partners (step 4, red arrow). Therefore the MN – strongly relying on the self-certifying cryptographic identifiers provided by HIP – delegates its signaling rights to the LRVS at which it is registered. The appropriate certificates are sent during the service discovery of MNs, resulting that the LRVS will own the rights to signal on behalf of all mobile nodes in the micromobility domain under its authority. In possession of these delegated rights the LRVS is able to securely register or update to the RVSs and CNs on behalf of the MNs with the IP_G addresses assigned to them.

The initiation procedure detailed above yields the MN to be registered at the LRVS (with the HIT_{MN}-IP_L-IP_G triplet) and at the RVS (with the HIT_{MN}-IP_G pair), and keeps all the CNs updated with the latter information as well.

The μHIP connection establishment procedures between already initialized MNs (MN_2 and MN_3 in Figure 2) are the following. The initiator (MN_2) sends the first packet (I1) of the Basic Exchange (step 1, blue arrow). The source address of this packet will be set to IP_L of the initiator, the destination address will be the IP address of the responder's (MN_3's) RVS (here we assume that the RVS of MN_2 and MN_3 are identical), and the destination HIT will be set to the responder's HIT. The appropriately constructed and sent I1 packet will be intercepted by the LRVS in the initiator's domain ($LRVS_2$ in our example). This LRVS changes the source IP address of the I1 packet to the globally routable IP address of the initiator (IP_{G2}) and sends the packet to the RVS (step 2, blue arrow). The RVS forwards the packet towards the responder's registered address (IP_{G3}) thus reaching the serving LRVS ($LRVS_3$) which knows the actual location of the responder, so the packet can be forwarded by changing the destination IP address of the I1 packet (IPG_3) to the responder's actual local address (IPL_3) (step 4, blue arrow). The Basic Exchange continues in the regular way, without the inclusion of the RVS, but with the address mapping function of the two LRVSs (step 5, blue arrow). (Note that the CHECKSUM field of IP packets should be recomputed after every address change, similarly as in case of standard RVSs during the rendezvous mechanism described above). This message flow builds up an active HIP association between the initiator and responder, so they can begin sending data packets to each other. Data packets are forwarded by the LRVSs to the actual local IP addresses of the MNs in the same way as they do during a Basic Exchange sequence.

The handover management in the proposed μHIP framework is quite straightforward. In case of intra-domain handovers the MN will receive a new IP_L from the new Access Router belonging to the serving LRVS (see MN_1 and red arrows in Figure 3). In this case the MN – realizing the change of its IP address – simply updates its registration (and if needed its delegation certificate as well) with its new IP_L at serving LRVS. The used update mechanism is detailed in (J. Laganier et al., 2006). It is important to note that neither the CNs of the mobile node nor the RVS has to be informed about the movement as the address changes are locally handled by the proposed micromobility extension. The movements of nodes are completely hidden from the outside world resulting in less signaling overhead, packet loss and handover latency.

Figure 3. Intra-, and inter-domain handover procedures in the proposed framework

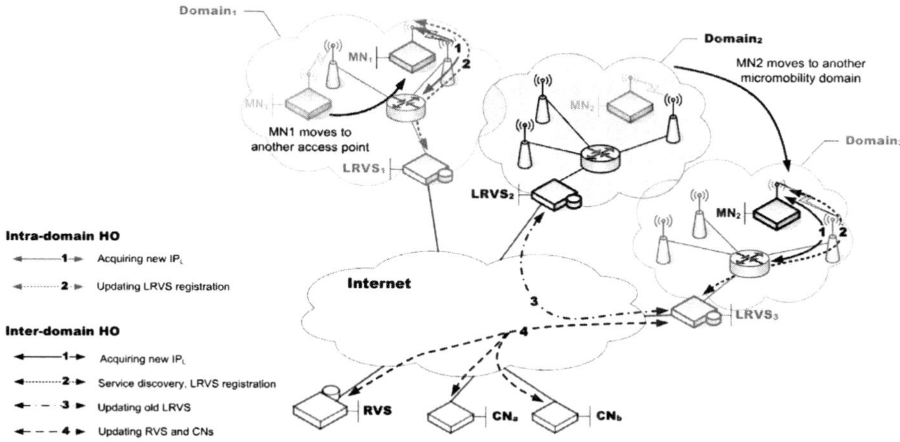

In case of inter-domain handovers the mobile node moves between local administrative domains (e.g. $Domain_2$ and $Domain_3$ in Figure 3) thus invoking the macromobility management procedures of our framework. Arriving at the new domain, the node will receive its new IP_L, and will discover the service parameters of the new LRVS ($LRVS_3$). After the MN realized that it leaved the previously used micromobility domain by entering a new one and learned the HIT and IP address of the new LRVS, it performs the initialization mechanism. This works the same way, as above. Since MN changes its old LRVS, it has to update its RVS and all the correspondent nodes with ongoing communication. But the first thing to do is to update the old LRVS (i.e. $LRVS_2$) to make it able to forward packets sent to the MN's old globally routable IP address as long as the MN has not finished updating the RVS and all of its CNs (step 3, blue arrow). After the old LRVS is updated, the mobile node begins to update its CNs and at last the RVS (step 4, blue arrows). When the MN finished all of the required updates with its CNs and the RVS, it removes the registration association at the old LRVS (this association can overrun the timeout defined for inactive HIP sessions and thus it could be automatically removed) and completes the inter-domain handover procedure.

Network Mobility in the Host Identity Layer

As users realize the benefits of mobile communication they also raise their demand for more sophisticated mobility schemes (E. Perera et al., 2004). Network mobility (NEMO) is one of these scenarios where not only a standalone host but a group of hosts connected together changes its point of attachment to the Internet. NEMO management solutions will enable the dynamic mobility of Personal Area Networks and Vehicular Area Networks among others. With this opportunity users can enjoy emerging applications of Intelligent Transportation System (ITS), which is more or less based on communication between networks applied in moving vehicles.

In this section we introduce how HIP can be utilized to serve as the base protocol in providing Network Mobility (NEMO) support (T. Ernst et al. 2007). First we summarize the main concepts related to mobile networks then we describe the idea of extending HIP, which leads to HIP-NEMO (Sz. Nováczki et al., 2007, 2008), the HIP-based network mobility solution.

In a NEMO scenario there is always a special entity called the Mobile Router (MR) that provides a connection point towards the outside world. There might hosts existing permanently

(Local Fixed Node, LFN) or temporary (Local Mobile Nodes, LMN or Visiting Mobile Node, VMN) connected to the MR. Furthermore a mobile network has the chance to attach to another mobile network forming a nested architecture. Usually a solution providing network mobility do not limit the level of nesting. LFN, LMN, VMN and MR are entities commonly referred as Mobile Network Nodes (MNN). Finally there are scenarios where more than one MR exists in the same mobile network providing alternative ways of connection for MNNs. In this case we speak about multihomed mobile networks. All the scenarios described above can be mixed together in any complex way making network mobility management a very challenging research area.

In order to meet these requirements the IETF (Internet Engineering Task Force) developed the Network Mobility Basic Support Protocol (NEMO-BS) (V. Devarapalli et al., 2005) relying on Mobile IPv6. This and related work in IETF summarizes the main concepts, terms and scenarios related to network mobility and also outlines the most important challenges in this research area.

On the other hand all the complex NEMO scenarios combined with the security requirements of next generation mobile networks are hard to solve with NEMO-BS. The root cause of this is the generic architectural problem of the classical TCP/IP stack. As described above this generic problem indicated the development of the Host Identity Protocol among others. Basically HIP integrates security, mobility and multihoming in a new manner and solutions based on it inherit these nice properties. Thus it is quite straightforward to use HIP as the base for providing network mobility support, HIP-NEMO.

The proposal provides continuous connectivity, secure signaling and data communication for every MNN shall it be even in a nested subnet in a scalable manner. Support for multihomed scenarios is involved too. Furthermore the method uses reduced signaling and packet overhead while providing optimal routes towards any MNN. The cornerstone of the solution is a HIP-capable entity, the Mobile Rendezvous Server (mRVS), which acts as a MR and provides HIP-based services for MNNs in the moving network. The mRVS is the connection point for MNNs to access the Internet and also serves as a rendezvous point for the rest of the network to access MNNs. As the mobile network is expected to dynamically change its point of attachment it can provide only a mobile rendezvous point. Thus the mRVS itself needs to use the standard HIP rendezvous services to ensure a permanent rendezvous point for itself and for all the MNNs inside its territory. Consequently the standard RVS of the mRVS serves as the primary rendezvous point for the rest of the network to access MNNs. This is shown on Figure 4.

The mRVS acts like a signaling proxy for MNNs as they delegate their signaling rights to the mRVS, after which the mRVS is responsible to perform all signaling needs (e.g. connection establishment, handover management) on behalf of the MNNs (step 1 and 2). The mobile rendezvous point is now established. After MNNs delegated their signaling rights to the mRVS, it registers them at the RVS to ensure the permanent rendezvous point for MNNs.

As opposed to standard RVS, which stores single {HIT – IP address} mappings, the mRVS connects this pair with another IP address. This must be a globally routable address pointing to a topologically correct location. This is allocated and assigned by mRVS to the particular MNN. This address is used to provide a global locator for MNNs. As Figure 5 depicts, in our example mRVS assigned IP.LFN.P0 for LFN, while its actual IP address is IP.LFN.P1. One possible way to allocate a proper IP address for LFN is as follows. At registration LFN communicates on IP.LFN.P1. The mRVS simply changes the prefix (P1) of this address to P0, and leaves the remaining part of the address unchanged, as Figure 5 shows.

Finally the mRVS communicates with other mRVSs to efficiently handle all the complex

Figure 4. A simple HIP-NEMO scenario

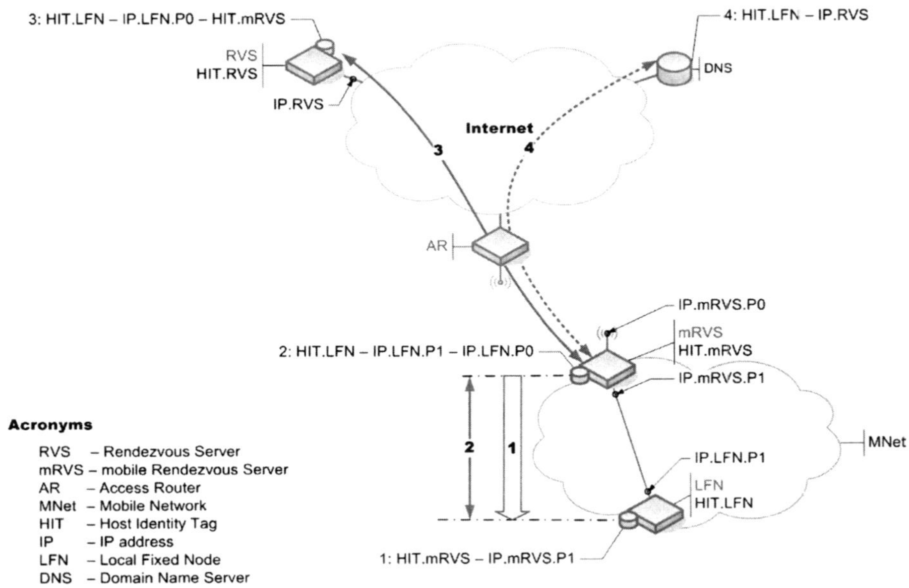

NEMO scenarios, especially nested mobile network situations.

MNNs have to be able to delegate their signaling rights to the mRVS. However, this shall be considered as a HIP-based service, rather than a modification of the base protocol. Thus HIP-NEMO can provide transparent network mobility support for LFNs and LMNs.

Now all necessary information is stored in the network to enable one correspondent node to establish a connection with a MNN. In the DNS system the primary connection point for any MNN can be retrieved. This is the standard RVS of the mRVS. The RVS redirects the packets addressed to the MNNs.

The key issue related to mobile networks is the management of nested mobile subnetworks. Such situations are usually arising serious scalability problems in present NEMO proposals. Thus an efficient NEMO solution has to solve these scenarios in a scalable manner. In HIP-NEMO nested subnetwork management is based on special inter mRVS signaling and a bidirectional tunnel establishment between mRVSs. Inter mRVS communication manages the appearance of new nested subnets and nested subnet handover situations. The bidirectional tunnel is used for data delivery between two mRVSs. The detailed description is out of scope of this summary.

Figure 5. The address allocated by mRVS

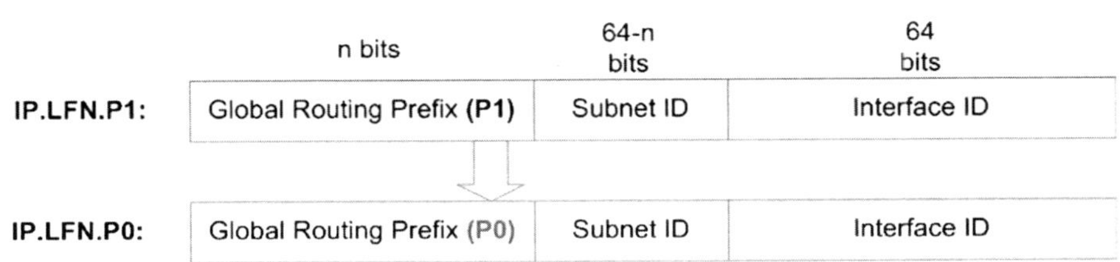

Security Strengths

As HIP-NEMO and μHIP both are built on top of HIP, they inherit the security strengths of the base protocol. In an environment where hosts are identified according to their actual IP addresses (i.e. the current Internet architecture) the most important advantage derived from the usage of HIP is its strong authentication and identification service. This is provided by the cryptographic Host Identities, which are asymmetric key pairs. One can prove its identity by holding the private pair of a public key. Initialization of a HIP association (i.e. the Base Exchange) is designed to protect the responder from most kinds of Denial of Service (DoS) attacks. Confidentiality with HIP is established by encrypting the payload data by the public key. The only way to access the encrypted information is to own the private key. The base protocol specification recommends the usage of ESP, which is used by IPSec to transfer encrypted data. In its current form, the Base Exchange results in an IPSec Security Association pair. Communication partners use this to setup their ESP protected packets and send them to their peer. Furthermore, HIP protects the integrity of payload data as well as integrity of control packets. HIP control packets can also be used to carry cryptographic certificates. Certificates can be used for authentication or authorization purposes by the peer host or intermediate entities. The latter property is a key issue, when considering secure signaling right delegation. Hosts delegate their signaling rights to one or more (i.e. in case of multihomed scenarios) LRVS or mRVS in a secure way by sending registration packets that hold the correspondent certificate. Basic HIP security functions and secure delegation of signaling rights together provide secure location update, which is a key issue in securing mobile communication. Since signaling rights are delegated in a secure way and base HIP signaling messages are signed by the sender, location update signaling is protected.

Service discovery shall be considered as the security bottleneck of the proposals. When a HIP host receives a SAP packet from the network, either as a result of an active service discovery, or passively, it does not know if the service provider can be trusted or not. The SDP packet is unprotected, which yields vulnerability. An attacker can modify the packet, or an attacker can send the packet using the IP address and HIT of someone else. However, there are situations when hosts or moving networks have no other choice but to trust other hosts because there are no other means for them to connect to the Internet. In singlehomed environments this is usually an obvious choice. On the other hand, the decision of to whom shall one delegate its right of signaling becomes a more complex problem in multihomed environments.

FUTURE TRENDS

The next generation of mobile networks is not only envisioned to offer a wild range of high quality services but to provide them on a new technical basis. Novel approaches arise in order to solve the upcoming questions, thus the standards of the present network architectures are in constant evolution. The Host Identity Protocol and its existing and possible emerging applications are the fruits of this evolution too.

But HIP is just one among the many proposals trying to cope with the sophisticated mobility schemes of next generation ubiquitous networks. However, the extensions developed on the top of HIP prove that it might be the best candidate to answer these challenges. Furthermore there are some hot research topics in this field, which aim to further develop and extend the basis concepts introduced by HIP. Here we outline some of these fields and directions of further developments.

Signaling and identity delegation is one of the main potentials of future HIP related development. If this special service can be offered in a simple and secure way it opens the way to radically new applications not clearly provisioned today. On the other hand signaling delegation significantly reduces control overhead which is especially critical on low-bandwidth wireless environments. Using this approach the heavy part of the signaling traffic can be transformed into the wireline network, which leads to enormous resource saving (S. Herborn et al., 2007).

Strongly connected to the previous statement there is another resource related problem in wireless networks: the limited capacity of the batteries applied in mobile devices. Since HIP cryptography requires lots of CPU capacity and thus emits the majority of the battery life there is a strong need to develop a lightweight alternative of the HIP implementation. This branch of HIP-related research is summarized in (T. M. Heer, 2006) and some measurement results on how standard HIP implementation works on lightweight hardware can be found in (A. Khurri et al., 2007).

The Host Identity Indirection Infrastructure (Hi3) (P. Nikander et al., 2004) gives a DHT based structure for the RVS system. The solution combines the positive properties of Secure i3 (D. Adkins et al., 2003) and HIP, resulting an architecture that integrates the effective end-to-end approach of HIP and the flexibility of i3 (i.e. the basis of Secure i3). On the other hand the combined solution seems to be less vulnerable against Denial-of-Service (DoS) attacks than any of the parent proposals.

We should speak about one approach that is not especially related to HIP but, but contains the possibilities to make HIP much more valuable in the future. This approach is the definition of a method which is able to provide routing without structured addresses. (M. Caesar et al., 2006) introduces such a scheme. In case we had an effective solution like this, flat Host Identities used by HIP could be utilized in packet routing as well, without making the same mistake as current IP addresses (i.e. dual significance). However the scaling and efficiency properties of routing methods on flat labels are far from optimal, the results show that the community should take the idea seriously.

As these fields of research work getting more and more attention and the results are turned into practical solutions, the widespread distribution of HIP based networking approach and emerging applications described in this chapter shall be experienced contributing to the evolution of next generation mobile and wireless networks.

CONCLUSION

In this work we presented the Host Identity Protocol approach and especially how it can be utilized to provide advanced applications and to manage complex mobility management problems. The dominant part of the chapter was devoted to introduce two novel HIP-based mobility management schemes both developed by the authors. First a micromobility extension to HIP was discussed, followed by a NEMO solution using HIP as its basis. Our goal was to prove that HIP is a promising concept, on which advanced mobility services and applications can be built. The generic security and architectural properties are both inherited by any extension based on HIP. Developers of these enhancements do not need to pay special attention for fulfilling those basic requirements since HIP gives a ready to use solution. Instead the research efforts can be focused on the concrete application, service or challenge that is meant to be solved. On the other hand it is clearly visible that HIP and all related research work is quite far from being ready or closed. There have to be gained a lot more design and implementation experiences to ensure the widespread usage of HIP and services based on it. Many research groups and organizations are about to reach this common goal.

ACKNOWLEDGMENT

This work is supported by the ANEMONE project (which is partly funded by the Sixth Framework Programme of the European Commission's Information Society Technology) and the Mobile Innovation Center Hungary. The authors would like to thank all participants and contributors who were involved in the work.

REFERENCES

Adkins, D., Lakshminarayanan, K., Perrig, A., & Stoica, I. (2003). *Towards a more functional and secure network infrastructure* (Technical Report UCB/CSD-03-1242). Computer Science Division (EECS), University of California, Berkeley, 2003

Banerjee, N., Acharya, A., & Das, S. K. (2004). SIP-Based Mobility Architecture for Next Generation Wireless Networks. IBM Research Report, September 10, 2004.

Bokor, L. Nováczki, Sz., & Imre, S. (2007). A Complete HIP-based Framework for Secure Micromobility. 5th @WAS International Conference on Advances in Mobile Computing and Multimedia (MoMM2007) (pp. 111-122). Jakarta, Indonesia, 3-5 December 2007

Caesar, M., Condie, T., & Kannan, J. (2006). ROFL: Routing on Flat Labels. In *Proceedings of Special Interest Group on Data Communication (SIGCOMM 2006), Pisa, Italy, September 11-15, 2006.*

Campbell, A. T., Gomez, J., Kim, S., Turányi, Z. R., Valkó, A. G., & Wan, C.-Y. (2002). Internet micromobility. *Journal of High Speed Networks*, *11*, 177–198.

Deguang, L., Xiaoming, F., & Hogrefe, D. (2006). A review of mobility support paradigms for the internet. *IEEE Communications Surveys & Tutorials*, *8*(1), 38–51. doi:10.1109/COMST.2006.323441

Devarapalli, V., Wakikawa, R., Petrescu, A., & Thubert, P. (2005, January). *Network Mobility (NEMO) Basic Support Protocol*. RFC 3963.

Ernst, T., & Lach, H-Y. (2007, July). *Network Mobility Support Terminology*. RFC 4885.

Heer, T. M. (2006, August). *LHIP – Lightweight Authentication for the Host Identity Protocol*. Diploma Thesis.

Henderson, T. (2007, March). *End-Host Mobility and Multihoming with the Host Identity Protocol*. IETF Internet Draft <draft-ietf-hip-mm-05>.

Henderson, T. R., Ahrenholz, J. M., & Kim, J. H. (2003, March). Experience with the Host Identity Protocol for Secure Host Mobility and Multihoming. *IEEE Wireless Communications and Networking*, *3*, 2120–2125.

Herborn, S., Huber, A., Boreli, R., & Seneviratne, A. (2007, January). Secure Host Identity Delegation for Mobility. In *COMSWARE'07* (pp. 1-9).

Herborn, S., Huber, A., Boreli, R., & Seneviratne, A. (2007). *A Scheme for Host Identity Delegation*. Paper presented at the IEEE/Create-Net/ICST International Conference on COMmunication System softWAre and MiddlewaRE (COMSWARE), Bangalore, India.

Huber, J. F. (2004). Mobile next-generation networks. *IEEE MultiMedia*, *11*(1), 72–83. doi:10.1109/MMUL.2004.1261110

Johnson, D., Perkins, C., & Arkko, J. (2004, June). *Mobility Support in IPv6.*, IETF RFC 3775.

Jokela, P., Melen, J., & Ylitalo, J. (2006, June). *HIP Service Discovery*. IETF Internet Draft <draft-jokela-hip-service-discovery-00>.

Jokela, P., Moskowitz, R., & Nikander, P. (2007,. June). *Using ESP transport format with HIP*. IETF Internet Draft <draft-ietf-hip-esp-06>.

Jokela, P., Rinta-aho, T., Jokikyyny, T., Wall, J., Kuparinen, M., Mahkonen, H., & Melen, J. &Kauppinen, T. (2005, November 13-14). *HIP-ernet: a decentralized security infrastructure for large scale grid environments*. Paper presented at 6th IEEE/ACM International Workshop on Grid Computing'05.

Khurri, A., Vorobyeva, E., Gurtov, A. (2007). Performance of Host Identity Protocol on Lightweight Hardware. In *Proceedings of The 2nd ACM International Workshop on Mobility in the Evolving Internet Architecture (MobiArch 2007)* August 27, 2007, Kyoto, Japan

Koh, S. J., Chang, M. J., & Lee, M. (2004, March). mSCTP for soft handover in transport layer. *IEEE Communications Letters*, *8*(3), 189–191. doi:10.1109/LCOMM.2004.823432

Kovacshazi, Z., & Vida, R. (2007). Host Identity Specific Multicast. *International Conference on Networking and Services (ICNS '07)* (p. 1).

Laganier, J., & Eggert, L. (2006). Host Identity Protocol (HIP) Rendezvous Extension. IETF Internet Draft <draft-ietf-hip-rvs-05>

Laganier, J., Koponen, T., &Eggert, L. (2006, June). Host Identity Protocol (HIP) Registration Extension. IETF Internet Draft <draft-ietf-hip-registration-02>

Manner, J., &Kojo, M. (2004, June). *Mobility Related Terminology.* IETF RFC 3753

McCann, P. (2005, November). *Mobile IPv6 Fast Handovers for 802.11 Networks*. IETF RFC 4260.

Moskowitz, R., & Nikander, P. (2006, May). *Host Identity Protocol (HIP) Architecture*. IETF RFC 4423.

Moskowitz, R., Nikander, P., Jokela, P., & Henderson, T. (2007, October 30). *Host Identity Protocol.* IETF Internet Draft (draft-ietf-hip-base-10).

Nikander, P., & Arkko, J. (2004). Delegation of Signaling Rights. In *Security Protocols* (LNCS 2845, pp 203-214).

Nikander, P., Arkko, J., & Ohlman, B. (2004). *Host Identity Indirection Infrastructure (Hi3).* Paper presented at the Second Swedish National Computer Networking Workshop 2004 (SNCNW2004).

Nikander, P., & Laganier, J. (2007, April). *Host Identity Protocol (HIP) Domain Name System (DNS) Extensions*. IETF Internet Draft <draft-ietf-hip-dns-09>.

Nováczki, S., Bokor, L., Jeney, G., & Imre, S. (2007). A HIP based Network Mobility Protocol. In *Proceedings of the 2007 International Symposium on Applications and the Internet Workshops* (SAINTW'07)

Nováczki, S., Bokor, L., Jeney, G., & Imre, S. (2008). Design and Evaluation of a Novel HIP-Based Network Mobility Protocol. *Journal of Networks*, *3*(1). doi:10.4304/jnw.3.1.10-24

Perera, E., Sivaraman, V., & Seneviratne, A. (2004). Survey on network mobility support. *ACM SIGMOBILE Mobile Computing and Communications Review, 8*(2).

Reinbold, P., & Bonaventure, O. (2003). *IP Micro-Mobility Protocols* (pp. 40–57). IEEE Communications Surveys & Tutorials Third Quarter.

Saha, D., Mukherjee, A., Misra, I. S., Chakraborty, M., & Subhash, N. (2004). Mobility support in IP: a survey of related protocols. *IEEE Network*, *18*(6), 34–40. doi:10.1109/MNET.2004.1355033

Särelä, M., & Nikander, P. (2004). Applying host identity protocol to tactical networks. In *Proceedings of IEEE Milcom 2004, Monterey, California, USA, Nov 2004.*

So, J. Y. H., & Jidong, W. (2006). *HIP Based Mobility Management for UMTS/WLAN Integrated Networks*. Paper presented at the Australian Telecommunication Networks and Applications Conference 2006

So, J. Y. H., Jidong, W., & Jones, D. (2005). SHIP mobility management hybrid SIP-HIP scheme. *6th International Conference of Software Engineering, Artificial Intelligence, Networking and Parallel. Distributed Computing, 05*, 226–230.

Soliman, H., Castelluccia, C. El Malki, K. &Bellier, L. (2005, August). *Hierarchical Mobile IPv6 Mobility Management (HMIPv6)*. IETF RFC 4140.

Zhu, X., & Atwood, W. J. (2007). A Web Database Security Model Using the Host Identity Protocol. *11th International Database Engineering and Applications Symposium'07* (pp. 278-284), 6-8 Sept. 2007

KEY TERMS AND DEFINITIONS:

Local Rendezvous Server (LRVS): is a special HIP enabled gateway router which is responsible for managing HIP Mobile Nodes (MNs) in a given micromobility domain of µHIP. LRVS gateways provide HIP registration service for MNs, furthermore LRVSs introduce an IP address mapping function which is used to attach the MNs to the µHIP access network by registering the locally valid IP addresses of MNs.

Micromobility: If wireless networking domains are aggregated and a special protocol is responsible for the local mobility management of this group of domains in order to offer fast and seamless handover control over a limited geographical area, than we speak about micromobility, the aggregated group of domains is called micromobility domain, and the special control protocol is called micromobility protocol.

Mobile Rendezvous Point (mRP): is a RP that is mobile. This term describes the key concept of HIP-NEMO. It is provided by a mobile RSV to enable network mobility support. The mRP must have a permanent rendezvous point, which ensures global reachability of the nodes in the mobile network.

Mobile Rendezvous Server (mRVS): is a special HIP enabled mobile router. On one hand it is the mobile rendezvous point of nodes in the moving network. On the other hand it serves as a signaling proxy for MNNs.

Network Mobility: a special mobility scenario, which arises when a router connecting a network to the Internet dynamically changes its point of attachment to the fixed infrastructure, thereby causing the accessability of the entire network to be changed in, relation to the fixed Internet topology.

Rendezvous Point (RP): is a networking abstraction, which usually refers to a middlebox-like device that enables two nodes in the network to establish some kind of connection in situations where each party have insufficient knowledge about the other to initiate the connection establishment in a direct manner.

Signaling Delegation: a method used by a network entity the delegate to securely offer to another one (usually a middlebox) the ability to send signaling messages on behalf of it.

Signaling Proxy: a special middlebox in the network that can accept and process signaling delegation certificates.

Wireless Heterogeneous Network: in our context wireless communication networks using different access technologies (e.g. when a wireless network providing services through WLAN is able to maintain these services when merging to a cellular network as well) is called a wireless heterogeneous network.

Chapter 20
Service Discovery in Mobile Ad-Hoc Networks

Haidar Safa
American University of Beirut, Lebanon

Zeina Koteiche
American University of Beirut, Lebanon

ABSTRACT

Service discovery plays a relevant role in mobile ad hoc environments. Indeed, upon joining a self-organizing network, mobile nodes should be able to explore the environment to learn about, locate, and share the available services. As a result, many researches were performed in this area. Recently, research is being directed towards integrating the service discovery into the routing protocols. This chapter presents the basic concepts of service discovery and their related issues. It also describes the service discovery challenges that arise due to the properties of mobile ad hoc networks. The chapter is concluded by presenting some observations and discussing the current challenges to provide guidelines for possible improvements.

INTRODUCTION

A mobile ad hoc network (MANET) consists of a set of mobile nodes operating without the aid of established infrastructure of centralized administration. These nodes are free to move randomly, organize themselves arbitrarily, and may have very little or no knowledge about the identities and capabilities of each other. Ad hoc networks have attracted new applications specialized fields such as emergency services, disaster recovery, environment monitoring, law enforcement, search and rescue, plus commercial and educational applications. Service sharing and discovery play a relevant role in mobile ad hoc environments. Upon joining a self-organizing network, mobile nodes should be able to explore the environment to learn about, locate, and share the available services.

Service discovery is a topic that has been researched relatively thoroughly in stable wired networks. Several protocols such as Service Location Protocol (SLP) of IETF (Guttman et al., 1997), Sun's Jini (Arnold et al., 1999), Microsoft's Universal Plug and Play (UPnP) (John, 1999)

DOI: 10.4018/978-1-60566-250-3.ch020

have been developed to help applications discover remote services residing on machines in a wired network. In these protocols, a service advertises and registers itself to a registry that keeps track of networked services. Their architectures are primarily centralized or semi-centralized and have an implicit assumption that the underlying network is stable and capable of providing reliable communication. These protocols do not directly dwell on MANET environments where self-configurability is the key. Clearly, service discovery in a mobile computing environment requires a decentralized design approach that better copes with dynamic environments such as those normally encountered in MANETs (Kozat & Tassiula, 2004).

With the deployment of wireless ad hoc networks, several service discovery protocols have been proposed for MANETs. In this chapter, we present the basic concepts of service discovery and their related issues by dividing the existing solutions into different comparable categories. We then describe the service discovery challenges that arise due to the properties of mobile ad hoc networks. We conclude the chapter by presenting our observations and discussing the current challenges to provide guidelines for possible improvements.

SERVICE DISCOVERY IN MANETS

The traditional service discovery protocols, which assume a stable network condition, cannot be adapted to the resource-limited and dynamic mobile ad hoc environment. There has been significant research on service discovery solutions for MANETs that take into consideration its infrastructureless nature (Doval & O'Mahony, 2002; Preu, 2002; Ratsimor et al., 2002; Safa et al., 2007; Artail et al., 2007; Dhanakoti et al., 2005; Helal, 2003). These solutions can be categorized as broadcast-based, advertisement-based, application-based, network-supported, probability-based, semantic routing-based, directory-based, directory less, or a combination of several categories.

Broadcast Based Approaches vs. Advertisement Based Approaches

In the broadcast-based solution, a service discovery request is broadcast throughout the network. Every node that receives the service request applies the service matching process to find out whether the requested service description matches one of the services that it offers. If it matches, it responds with a service reply that contains the service description that can be used to invoke the service (Toh, 2002). When the requesting node receives a positive service reply, it launches the service invocation process in which it establishes a session with the service provider to invoke the service. The requesting node may receive several replies from different service provider nodes that offer the same service. In such case, it selects the most optimal service provider through a process referred to as service selection as shown in Figure 1. This architecture offers high reliability in discovering services but suffers from several drawbacks. First, global broadcast scales poorly with increasing network diameter and size. Second, it utilizes resources and computation power on all nodes of the network including those that do not have the service and the ones that are not on the route to the desired service. Third, it is heavy on network bandwidth since requests traverse all the nodes on all paths possible.

In the advertisement-based solution the services advertise themselves to all of the nodes and each node interested in discovering services will cache the advertisements. When a node wants to request a service, it first runs the service matching locally and the service selection to identify the service provider, then the service invocation process to invoke the service from the appropriate service provider (Koodli and Perkins, 2002). In this architecture, the cache size increases with the number of services while many of the nodes may have limited memory that does not allow them to store all the service advertisements. Similar to the broadcast method, this approach is also inefficient in terms of bandwidth usage, since the whole

Figure 1. Broadcast-based service discovery architecture

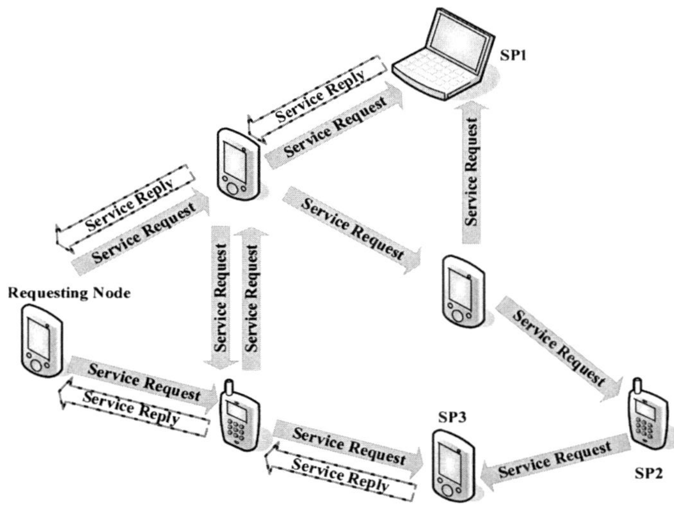

network has to be periodically flooded with advertisements. It should be noted that there are solutions that offer both advertisements and broadcast of requests, but nevertheless do not effectively address the problems of network load, network-wide reachability, and scalability.

Application Layer Based vs. Network Layer Based Approaches

Traditionally, service discovery has been considered as an application layer function. In such approach, the service discovery protocol finds the address of the service provider node and the service description and access mechanisms that should be used to invoke the service; whereas a routing protocol finds the route between requesting client and the service provider. This approach has been adopted by several MANET service discovery protocols like NOM (Doval and O'Mahony, 2002), JESA (Preu, 2002), and Allia (Ratsimor et al., 2002). However, Torres and Garcia-Macias, 2005, argue that routing and service discovery are not independent activities, since once a service request is received by a service provider, it is necessary to establish a route to notify the client about the existence of the service; similarly the client needs a route to the service provider for service invocation reason. Moreover there are several intrinsic characteristics of ad hoc networks that motivate a new approach combining routing and service discovery (Torres & Garcia-Macias, 2005). As nodes move frequently, it is not possible to guarantee that the route to a service provider will remain untouched along the service lifetime. The unpredictable nature of ad hoc networks implies varying conditions in nodes density, link status, and available bandwidth. Frequent mobility in ad hoc networks implies that the network topology may vary; consequently routes in the network can change, network partitioning can occur, and previously available services may become unavailable. Thus, one can say that 1) ad hoc networks cannot rely on registry servers; and 2) service discovery and routing are more tightly coupled than in fixed networks. However, a cross layer service discovery protocol that violates the principles of strict layering found in traditional network results in a low cohesion and a high coupling between the layers which makes maintenance, improvements and updates a very inflexible task.

Several cross layer-based approaches have been proposed for service discovery. These approaches are focusing not only on the actual integration of the service discovery with the network layer but also targeting additional topics that help improving their proposed protocols. For instance, a new protocol for service advertisement and discovery (Ad/D) namely service Ad/D protocol with independent zones (SPIZ) was proposed by Noh et al., 2007. SPIZ is a cross layer approach that combines the push-based with the pull-based advertisement and discovery strategy based on a variable zone radius in order to reduce flooding and system overhead. The push-based model is used by service providers for advertising their service whereas the pull-based model is used by each service requester for querying any desired services. Similar other approaches were proposed such as the hybrid adaptive protocol for integrated discovery (HAID) (Oh et al., 2005) and the modified zone routing protocol (M-ZRP) (Ververidis et al., 2005). Furthermore, Seno et al., 2008, proposed a new hierarchical approach for service advertisement and discovery which is based on cluster based routing protocols (CBRP) that combine the notion of clustering with the cross layer approach. Varshavsky et al., 2005 introduced another cross layer protocol that addresses service selection and reselection.

Due to the advantages and disadvantages of both approaches, new proposals are constantly emerging for both combining and separating routing and service discovery. However, network layer approaches are getting more focus currently in the research community.

Probability-Based Schemes vs. Semantic-Routing-Based Schemes

In probability-based service discovery schemes, when receiving a service request, nodes that fail the service matching process forward the packet with probability P (Gao et al, 2005; Gao et al, 2004). When the value of P is 1, all nodes will be forwarding any request they receive, which may cause flooding and lead to request forwarding redundancy, serious contention, and frequent collisions. When the value of P is less than 1, the redundancy of request forwarding will be reduced. Meanwhile the coverage of service request will also be reduced, which may affect the ability of finding matched services. Additionally, when the value of P is constant and the maximum number of hops that a service request packet can travel is relatively small, the number of succeeded requests will gradually decrease almost in the same speed as that of the value of P.

In semantic routing based schemes, nodes can intelligently select the next-hop nodes for service request packets by inspecting service description semantics as well as current network topology. When receiving a service request, only selected nodes will forward the packet. Chakraborty et al., 2006, classified services into several groups. Each server generates periodically a service advertisement packet that includes the information about: 1) the services provided by the server itself; and 2) the group of services provided by some servers in the sender's vicinity. When forwarding service requests, some neighbors of the current node may have seen some services that belong to the same group as the requested service. Such nodes are referred to as candidate nodes. Service requests will be matched at those nodes with higher probability. Thus, instead of broadcasting the request packet to all neighbors, the request packet is selectively forwarded towards these candidate nodes in unicast mode.

Directory-Based vs. Directory-Less Approaches

A directory-less architecture is a client-server based architecture in which the service requester (i.e. clients) reactively sends out service request messages. Service providers (i.e., servers) listen to such messages at a well-determined network interface and port, and apply the service matching process on the received requests. If the requested service is matched, then a reply message is generated and sent to the clients. Service requesters

Figure 2. Directory-based service discovery architecture

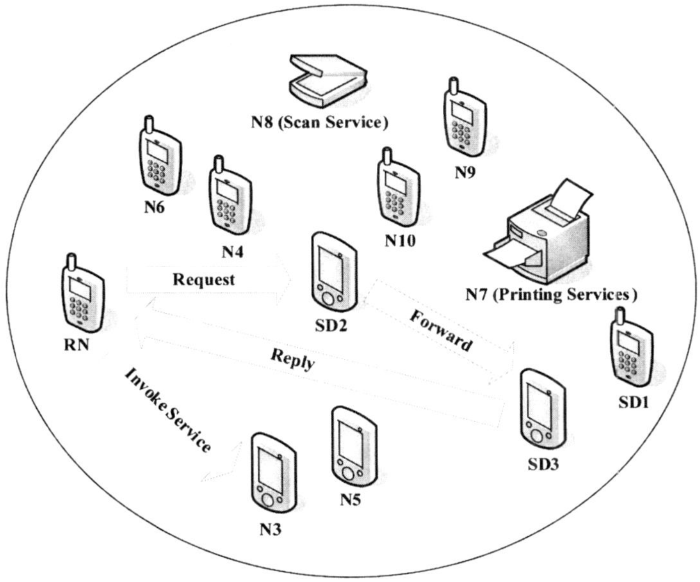

can also passively learn about the available services by listening to the service advertisements that are proactively generated by the servers. A directory-based architecture involves service directories (i.e., directory agents or service brokers) that reside between service requestors and the service provider logical entity (Safa et al., 2007; Artail et al., 2007) as shown in Figure 2. Service requesters direct their requests to well-known service directories with which service providers register their services. In return, service directories send back service reply messages to the clients and registration acknowledgments to the service providers. Upon receiving a reply message, the requestor starts the service selection and service invocation processes.

CHALLENGES FACING SERVICE DISCOVERY IN MANETS

While designing a service discovery and sharing protocol for MANETs, challenges related to their nature should be considered. These challenges can be summarized as follows.

Dynamic Network Topology

Unlike wired networks which consist of more or less stable networks, mobile ad hoc networks involve highly dynamic networks. Therefore, traditional service discovery approaches are not suitable for mobile ad hoc networks since MANETs are infrastructure-less and known by their rapidly changing network topology. Consequently, a suitable service discovery protocol should not be based on any infrastructure or fixed/centralized nodes. In addition, the use of an approach that requires a lot of communication is not suitable for ad hoc networks because of the limited energy resources. If a centralized service directory is used, its data would soon be inconsistent due to frequent disconnections of mobile nodes (i.e., service providers). Moreover, if the directory nodes fail, the whole system would be paralyzed.

Limited Power Sources

The majority of ad hoc networks consist of wireless devices that usually rely on batteries or other limited power sources. This adds further restrictions on the service discovery protocol since a good design should consider prolonging the lifetime of such networks. Since mobile nodes spend reasonable amount of energy in their communications, they switch periodically to sleep mode in order to save their battery resources. This leads to more challenges since when a node switches to the sleep mode, the services that it offers become unavailable. A good approach should enable these resource-constrained devices to discover services dynamically while minimizing the traffic generated by the service discovery process and tolerating the intermittent connectivity of devices.

Device Heterogeneity

A good service discovery protocol has to support a variety of mobile devices since a mobile ad hoc network can consist of devices equipped with heterogeneous interfaces (i.e., IEEE 802.11 WiFi or Bluetooth). Consequently, the service discovery protocol must enable service sharing to with a wide spectrum of devices, regardless of their hardware and software platforms.

Service Description

A service can be identified by a unique id or can be described using a more complex mechanism such as WSDL (Web Service description language). The method used to describe the service influences the service discovery process since it affects the searching, matching, and invocation operations. Using non functional properties for describing a service allows us to differentiate the set of services supported by a network. These properties can be used, for example, to match them against the quality of service requirements of the application.

Bridging Between Infrastructure Based and Ad hoc Networks

A service discovery protocol must be able to bridge MANETs with infrastructure based networks since Intranet/Internet connectivity remains the primary source of services. This enables devices that are located in a MANET to transparently access services offered in the interconnected networks and vice versa.

Scalability

A service discovery protocol should be scalable to operate in large MANETs and to support large class of service applications. Enabling scalability is achieved by addressing two major issues: load balancing and query efficiency. Load balancing is crucial due to the limited resources of the mobile nodes. Nodes being overloaded by registration or queries may delegate their work to other nodes and thus making the most possible use of the available resources. On the other hand, queries have to be efficiently routed through the network in order to minimize the traffic.

OBSERVATIONS AND DISCUSSION

Due to the wide variety of service discovery applications, this topic has become a hot area of research. We have observed that most of the previous research focused on proposing application layer based solutions then diverged to the network layer and cross layer solutions. Although the network layer solutions have many advantages concerning the control message overheads and the communication cost, we believe that they still suffer from several problems that cannot be disregarded.

1. We observe that the control message overhead reduction obtained with the network layer solutions is achieved at the expense of larger routing messages and larger routing tables. Larger messages result usually in more delay and errors.
2. The implementation of the service discovery protocol at the network layer violates

the traditional layering architecture mainly because the services are found at the application layer. We believe that this tight coupling between the routing protocols and the service discovery results in changing the routing protocol whenever the service discovery protocol changes and vice versa. Therefore, the flexibility that is offered by the decoupling of the layers is almost removed.

3. As we all know, the packet loss is very frequent when the network is very congested or when interference exists, thus losing the service discovery and routing aggregated packet will result in retransmission. Nevertheless, in the case of separating both protocols, one only needs to retransmit the lost service discovery or routing packet which implies less overhead than retransmitting an aggregated packet.

FUTURE TRENDS

Service discovery in MANETs is a very important area of research. Current research is headed towards improving the existing protocols or creating new ones to face the challenges of MANETs such as scalability, limited power resources, and device heterogeneity. The scalability and the limited power resources challenges are tightly related since minimizing the energy spent by a mobile node will result in longer network lifetime and may allow the handling of large networks. The cross layer approaches are being studied and adopted mainly to minimize communication by combining service discovery and routing since this may reduce the amount of messages processed by each mobile node which will result in lower energy consumption. In addition, MANETs include the communication between different types of devices. New technologies are constantly emerging with new devices that have wireless support, with different hardware capabilities. Therefore, the need for protocols to bridge the gap seamlessly between the different devices is growing.

CONCLUSION

This chapter presented basic concepts of service discovery and their related issues. It focused on the service discovery approaches that have been proposed for MANETs. These approaches were divided into different comparable categories such as: as broadcast-based, advertisement-based, application-based, network-supported, probability-based, semantic routing-based, directory-based, and directory-less schemes. The chapter described also the service discovery challenges that arise due to the nature and properties of mobile ad hoc networks. The chapter was concluded by some observations and discussion providing guidelines for possible improvements in addition to future trends.

REFERENCES

Arnold, K., Osullivan, B., Scheifler, R., Waldo, J., & Wollrath, A. (1999). *The Jini Specification (The Jini Technology)*. Reading, MA: Addison-Wesley.

Artail, H., Safa, H., Hamze, H., & Mershad, K. (2007). A cluster based service discovery model for mobile ad hoc networks. *3rd IEEE International Conference on Wireless and Mobile Computing, Networking and Communications* (WiMob 2007). New York, USA.

Chakraborty, D., Joshi, A., Yesha, Y., & Fin, T. (2006). Toward Distributed Service Discovery in Pervasive Computing Environments. *IEEE Transactions on Mobile Computing*, 5.

Dhanakoti, N., Gopalan, S., Sridhar, V., & Subramani, S. (2005). A Distributed Service Discovery and Selection Framework in Pervasive Service Environments. In *Proc. Advanced Industrial Conf. on Telecommunications/Service Assurance with partial and Intermittent Resources Telecommunications Workshop*.

Doval, D., & O'Mahony, D. (2002). Nom: Resource location and discovery for ad hoc mobile networks. In *Proceedings of the 1st Annual Mediterranean Ad Hoc Networking Workshop*.

Gao, Z., Wang, L., Yang, M., & Yang, X. (2006). CNPGSDP: An efficient group-based service discovery protocol for MANETs. *Computer Networks: The International Journal of Computer and Telecommunications Networking, 50*(16), 3165–3182.

Gao, Z. G., Yang, X. Z., & Cai, S. B. (2005). Flexible forward probability based service discovery protocol for manets. *Journal of Harbin Institute of Technology, 37*(9), 1256–1260.

Guttman, E., Perkins, C., & Veizades, J. (1997). *RFC 2165. Service Location Protocol.*

Helal, S., Desai, N., & Verma, V. (2003). Konark - A Service Discovery and Delivery Protocol for Ad-hoc Networks. In *Proc. of the Third IEEE Conf. Wireless Communication Networks (WCNC).*

John, R. (1999). *UPnP, Jini and Salutaion—A Look at Some Popular Coordination Frameworks for Future Network Devices* (Technical report). California Software Labs. http://www.cswl.com/whiteppr/tech/upnp.html

Koodli, R., & Perkins, C.E., (2002). Service Discovery in On Demand Ad Hoc Networks. *IETF Internet draft.*

Kozat, U., & Tassiula, L. (2004). Service discovery in mobile adhoc Networks: an Overall perspective on architectural choices and network Layer support issues. *Ad Hoc Networks, 2*(1). doi:10.1016/S1570-8705(03)00044-1

Noh. D., & Shin. H. (2007). SPIZ: an effective service discovery protocol for mobile ad hoc networks. *EURASIP Journal on Wireless Communications and Networking, 2007*(1).

Oh, C. S. Ko., Y. B., & Roh., Y. S. (2005). An integrated approach for efficient routing and service discovery in mobile ad hoc networks. *Consumer Communications and Networking Conference (CCNC)* (pp. 184-189).

Preu, S. (2002). Jesa service discovery protocol. In *Proceedings of the Second international Ifip-Tc6 Networking Conference on Networking Technologies, Services, and Protocols; Performance of Computer and Communication Networks; and Mobile and Wireless Communications* (LNCS 2345, pp. 1196-1201).

Ratsimor, O., Chakraborty, D., Tolia, S., Kushraj, D., Kunjithapatham, A., Gupta, G., Joshi, A., & Finin, T. (2002). *Allia: Alliance-based service discovery for ad-hoc environments.*

Safa, H., Artail, H., Hamze, H., & Mershad, K. (2007). A collaborative service discovery and service sharing framework for mobile ad hoc networks. *IFIP International Conference on Network and Parallel Computing (NPC 2007)* (LNCS). Dalian, China.

Seno, S. A. H., Pahlevanzadeh, B., Wan, T. C., Budiarto, R., & Ghahremani, M. (2008). A new approach CBRP based resource information management in MANETs. *International Journal of Computer Science and Network Security, 8*(5), 301–308.

Toh, C.-K. (2002). *Ad Hoc Mobile Wireless Networks: Protocols and Systems.* New Jersey: Prentice Hall.

Torres, D. A., & Garcia-Macias, J. A. (2005). *Performance analysis of two approaches to service discovery in mobile ad hoc networks* (. *LNCS, 3563*, 368–376.

Varshavsky, A., Reid, B., & De Lara, E. (2005). A cross-layer approach to service discovery and selection in MANETs. In *Proc. IEEE Int'l Conf. on Mobile Ad hoc and Sensor Systems.*

Ververidis, C. N., & Polyzos, G. C. (2005). Routing layer support for service discovery in mobile ad hoc networks. In *Proceedings of the Third IEEE International Conference on Pervasive Computing and Communications Workshops* (pp. 258-262). IEEE Computer Society.

KEY TERMS AND DEFINITIONS

Service Advertisement: is the process that a service provider node may use to make other nodes in the network aware of its services.

Service Description: is the actual representation of a service. The method used to describe the service influences the service discovery process since it affects the searching and matching operations.

Service Discovery: is the process that allows a node to discover services offered by other nodes in the network.

Service Invocation: is the process of actually calling/using a certain service. Once a service has been matched and the service provider has been identified, the requesting node needs now to actually invoke the service from the service provider.

Service Matching: is the process of finding out whether the service requested matches one of the available services at a node or not. If unique IDs are used to represent services, matching will not be considered an issue for the service discovery protocol. However, once the services are assigned functional description then matching the services becomes very critical due to the fact that keywords have to be matched based on a certain criteria. This will lead also to considerations regarding which results have to be returned; the exact matches, the most relevant, the matches with some variations.

Service Selection: is the process of selecting the most optimal service provider from a set of service providers that offer the same service. Service selection was proved to have tremendous effect on the quality of the service discovery and sharing mechanism.

Service Sharing: is the process of making the offered service available for other nodes to use.

Chapter 21
Mobility Prediction in Mobile Ad-Hoc Networks

Damianos Gavalas
University of the Aegean, Greece

Charalampos Konstantopoulos
University of Piraeus, Greece

Basilis Mamalis
Technological Educational Institution of Athens, Greece

Grammati Pantziou
Technological Educational Institution of Athens, Greece

ABSTRACT

A Mobile Ad hoc NETwork (MANET) is a collection of wireless mobile nodes forming a network without using any existing infrastructure. All mobile nodes function as mobile routers that discover and maintain routes to other mobile nodes of the network and therefore, can be connected dynamically in an arbitrary manner. The mobility attribute of MANETs is a very significant one. The mobile nodes may follow different mobility patterns that may affect connectivity, and in turn protocol mechanisms and performance. Mobility prediction may positively affect the service-oriented aspects as well as the application-oriented aspects of ad hoc networking. At the network level, accurate node mobility prediction may be critical to tasks such as call admission control, reservation of network resources, pre-configuration of services and QoS provisioning. At the application level, user mobility prediction in combination with user's profile may provide the user with enhanced location-based wireless services, such as route guidance, local traffic information and on-line advertising. In this chapter we present the most important mobility prediction schemes for MANETs in the literature, focusing on their main design principles and characteristics.

INTRODUCTION

Mobile ad hoc networks are self-organizing and self-configuring multi-hop wireless networks capable of adaptive re-configuration when they are affected by node mobility. A mobile ad hoc network is composed of peer nodes with equal networking capabilities which are able to function as mobile routers i.e., to forward packets and maintain routes.

DOI: 10.4018/978-1-60566-250-3.ch021

Packets can be forwarded in multi-hops from the source nodes to the destination nodes with no need for underlying fixed network infrastructure (e.g. routers and base stations). Therefore, mobile ad hoc networks are not constrained in their deployment by any need for underlying infrastructure and they can be deployed rapidly in situations where wireless access to a backbone is impossible and an infrastructure is difficult to install (e.g., disaster recovery). In addition to the traditional problems of wireless networks (bandwidth optimization and transmission quality enhancement) mobile ad hoc networks introduce new issues such as ad-hoc addressing, increased energy constraints, self-configuration and adaptive reconfiguration, as network topology is affected by node mobility. Furthermore, because of the real time nature of ad hoc network applications (e.g., collaborative mobile computing, battlefield communications, emergency search and rescue operations, disaster recovery), data traffic is routed under timing constraints requiring proactive route construction and maintenance procedures.

Mobility prediction may positively affect the service-oriented aspects (network level) of ad hoc networking as well as the application-oriented aspects (application level). At the network level, accurate mobility prediction may be critical to tasks such as call admission control, congestion control, reservation of network resources, pre-configuration of services and QoS provisioning. At the application level, user mobility prediction in combination with user's profile may provide the user with enhanced location-based wireless services, such as route guidance, local traffic information, tourism services, on-line advertising, etc. Given that 4G and beyond wireless ad hoc and hybrid networks will support real-time multimedia applications, the need for mobility prediction is of great significance.

Because of the importance of mobility prediction in ad hoc networks, there is a significant amount of research work on the topic, while in some cases the proposed techniques follow ideas or approaches used in fixed infrastructure type networks. However, prediction approaches for fixed infrastructure type networks are usually inappropriate in the case of ad hoc networks since: (1) Mobility prediction in fixed wireless networks is based on the use of a static underlying network infrastructure, while in ad hoc networks mobility prediction must be done in a highly dynamic environment, where the network topology is changing and the mobility of other nodes should be taken into consideration. (2) Ad hoc networks are usually applied in emergency operations and military environments, where future node movements cannot be based on a record of previous movements because of the dissimilar requirements of each situation. (3) Since mobility prediction methods for ad hoc networks are executed on the mobile nodes, they should be more light-weight than the methods for fixed wireless networks, typically executed on the base stations.

In this chapter we shall present the most important mobility prediction schemes for mobile ad hoc networks in the literature, focusing on their main design principles and characteristics.

BACKGROUND

Wireless mobile networks can be classified as infrastructure-based networks and ad hoc networks. The former type includes networks with fixed base stations and each mobile node connects to the network by communicating with a base station which is within its communication range. A Mobile Ad hoc NETwork (MANET) is a collection of wireless mobile nodes forming a network without using any existing infrastructure. All mobile nodes function as mobile routers that discover and maintain routes to other mobile nodes of the network and therefore, can be connected dynamically in an arbitrary manner. Therefore, a MANET is self-organized i.e., it is deployed and managed independently of any preexisting infrastructure, while it autonomously determines its own configuration parameters such as position identification, power control, routing and addressing.

Ad hoc network applications vary and their environment may change dynamically and unpredictably. The ad hoc protocols at various layers need to self-tune to adjust to the changes of the environment and to the requirements of the different applications. In addition, MANETs can grow to several thousands of nodes, while their self configuring nature cannot tolerate a physically hierarchical structure to handle the issue of network scalability, as infrastructure type networks do. Therefore, designing scalable MANETs is a very critical challenge.

The mobility attribute of MANETs is a very significant one. The mobile nodes may follow different mobility patterns (e.g., individual random mobility, group mobility, etc.) that may affect connectivity (i.e., links), and in turn protocol mechanisms and performance. Therefore, the mobility model may have an important impact on the selection of a routing scheme and influence the network performance. Several mobility models have been used to model mobile node's movement (Chellapa-Doss et al., 2004). Some of these models are the following: The *Random Walk model,* which defines user movement from one position to the next with memoryless randomly selected speed and direction. The *Random Waypoint model,* which defines user movement as a series of pause and motion periods. The user stays at a particular position for a specified period of time before moving to the next position in a randomly selected direction with speed uniformly distributed in an interval [*0*, *Maxspeed*]. Since, the movement of ad hoc mobile nodes frequently exhibit group mobility characteristics, a number of group mobility models have also been proposed. For example, in the *Reference Point Group Mobility Model (RPGM),* the nodes in the network are organized into mobility groups. Each mobility group has a logical group center, the *reference point*, which defines the movement of the entire group. The member nodes of the group are physically distributed in the vicinity of the *reference point* and the location of each mobile node is described by its physical displacement from the group's reference point.

Many mobile ad hoc networks applications require real-time flow for delivering data packets under strict timing constraints, in environments that are characterized by frequent topology changes. To maintain high-quality service for such applications, the data routing should not only reflect the change of mobile node's location and/or topology changes, as the *reactive or on demand* routing schemes do, but anticipate the movement behavior of mobile nodes employing *proactive* routing procedures. If each mobile node's future location and network topology changes can be predicted accurately enough, then route reconstruction can be done prior to topology changes without overburdening the network with a large number of control packets. Thus, services are pre-connected and pre-assigned at the mobile node's new location before the node moves into the new location. The end user receives immediately services at the new location, almost as efficiently as at the previous location. Therefore, the employment of mobility prediction techniques is very crucial in the design of efficient routing schemes.

It is well known that flat routing schemes do not scale well in terms of performance, while routing on top of a logically hierarchical structures, e.g. a clustered topology, is much more scalable as the flooding traffic is under control. Clustering (McDonald & Znatti, 1999b; Yu & Chong, 2005) is a promising approach for enhancing the scalability of MANETs in the face of frequent topology changes. Clusters are dynamic groupings of mobile nodes. A representative of each cluster is elected as a *cluster head (CH)* and a mobile node, which serves as intermediate for inter-cluster communication, is called a *gateway*. CHs hold routing and topology information while the boundaries of a cluster are defined by the transmission area of its CH. The feasibility of a clustering method is determined by the stability of the cluster structure that it creates, despite network topology changes. Otherwise, frequent reclustering is required thereby, creating a large volume of control messages which in turn consume considerable bandwidth and drain mobile nodes' energy quickly. As the main cause for topology changes in a MANET is node mobility, an efficient clustering method should seriously take the movements of mobile nodes into account in order to form clustering structures resistant to the node mobility. Therefore, employing efficient mobility prediction techniques for cluster formation and maintenance ensures the construction of a relatively stable clustering structure (virtual backbone) that is highly resistant to the topological changes of the ad hoc network.

MOBILITY PREDICTION METHODS AND THEIR APPLICATIONS

We classify the mobility prediction methods for mobile ad hoc networks into three categories as follows (Figure 1):

A. *Movement history based prediction methods*, which predict the "future" location of a mobile user based on his movement history (i.e., previous user movement patterns).
B. *Physical topology based mobility prediction methods,* which base their prediction on the use of the characteristics of MANET's physical topology and therefore, require the use of a Global Positioning System (GPS) to obtain exact node location and mobility information.
C. *Logical topology based mobility prediction methods,* which choose a logical topology of the MANET (e.g. a clustering structure) over which they apply their prediction process. On the contrary to the previous category, they do not require exact location and mobility information and thus they do not make use of a GPS. Estimated values of node location and mobility information may be obtained by other means (e.g, using signal attenuation versus travelled distance to estimate internode distances, or inferring the mobility of each node from how different is the neighbourhood of the node over time).

Figure 1. Classification of the mobility prediction methods

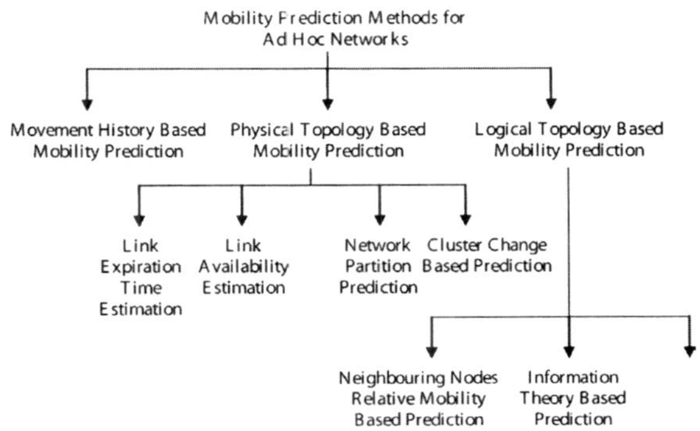

For each one of the above mentioned categories, we present well-known prediction methods, classifying them according to their applications and/or the specific prediction method they are based on.

A. Movement History Based Mobility Prediction

A number of motion prediction algorithms mainly for fixed wireless networks, e.g. (Liu and Maguire, 1996; Erbas et al., 2001) has been proposed which predict the "future" location of a mobile user based on the user's movement history (i.e, previous user movement patterns). The algorithms use different mobility models (e.g., the movement circle model, the movement track model, the Markov chain model) to model the user mobility behavior, exploiting the fact that the movement of a mobile user consists of a random and a regular movement part. The regularity in human movement behavior derives from certain activities that are repeated within a defined period of time (e.g., going to work every day or visiting a family member every week). The proposed schemes consist of regularity-pattern recognition algorithms and motion prediction algorithms and they are very critical for supporting high quality services for mobile applications. For example, by predicting the future location of a mobile user according to his/her movement history, routing may be pre-arranged, recourses may be pre-allocated, services may be pre-assigned at the new location before the user moves into it.

The above methods fail in the case that there are unpredictable changes in user's behavior. Also, there are additional problems when these methods are applied in MANETs because of the nature of those networks applications (military environments, emergency search and rescue operations). Due to dynamic topology and non regular requirements in such applications node mobility prediction based on the movement history is not always feasible and/or efficient.

B. Physical Topology Based Mobility Prediction

1. Link Expiration Time Estimation

By exploiting the fact that in real world situations, usually, a mobile node's movement is not completely random but the node travels in a predictable manner, we can predict the future state of the network topology. By predicting the future state of the network topology, the route reconstruction can be done effectively prior to route breaks and without generating excessive control overhead.

We say that there *exists a wireless link* between two nodes p and q of a MANET if and only if p are q are within the transmission range, say r, of each other, i.e., the distance between their placement is smaller than r. In (Su & Gerla, 1999) a mobility prediction method is presented for estimating the expiration time of the wireless link between two adjacent ad hoc nodes as a way to enhance various unicast and multicast routing protocols. By predicting the link expiration time for any link on a route R, the route's R expiration time is estimated as the least of the link expiration time values of all links on R. Based on this prediction, routes are reconfigured before they disconnect. The estimation of the link expiration time or in other words, the time period T that two ad hoc nodes being within mutual transmission range, remain connected, is done as follows. Let (x_i, y_i) and (x_j, y_j) be the positions of nodes i and j, respectively. Let also v_i and v_j be the speeds, θ_i and θ_j be the moving directions of nodes i and j, respectively, and TX their transmission range. Then, the amount of time T the mobile nodes i and j will stay connected is given by:

$$T = \frac{-(ab + cd) + \sqrt{(a^2 + c^2)TX^2 - (ad - cb)^2}}{(a^2 + c^2)}$$

where $a = v_i \cos \theta_i - v_j \cos \theta_j$, $b = x_i - x_j$, $c = v_i \sin \theta_i - v_j \sin \theta_j$, $d = y_i - y_j$. The exact location and mobility information of each mobile node can be provided by a GPS device. In cases that the use of GPS to compute exact location information is not possible, then approximate values for inter-node distances are computed as follows: Transmission power samples are measured periodically from packets received from a node's neighbour. From this information each node can compute the rate of change for a particular neighbour's power level (Su et al., 2001). Therefore, the time that the power level drops below the acceptable value can be fairly easily estimated. However, the link expiration time method requires exact location and mobility information while using power attenuation for distance estimation is not accurate.

The above mobility prediction mechanism works well in the case of simple node mobility patterns with no sudden changes in the moving directions and speeds, and when the mobility information (obtained from GPS) is accurate. The mechanism was applied to some of the most popular wireless ad hoc routing protocols and simulation results showed that these protocols performed better than their counterparts without mobility prediction (i.e., more data packets were delivered to destinations and the control packets were utilized more efficiently (Su et al., 2001)). Specifically, results show that the Flow Oriented Routing unicast Protocol (FORP) (Lee et al. Gerla, 2001) and Distance Vector (DV) unicast protocol enhanced with mobility prediction offer packet delivery ratios of 0.9 (i.e., 10% of packet loss) for speeds up to 70 km/h. Also, the On-Demand Multicast Routing Protocol (ODMRP) (Lee et al., 2001) with mobility prediction performs better than its counterpart without prediction and offers packet delivery ratio of 0.9 (i.e., 10% of packet loss) for speeds up to 70 km/h (Chellapa-Doss et al., 2004).

2. Link Availability Estimation

A probabilistic link availability model which can predict the future status of a wireless link is proposed in (McDonald & Znabi, 1999a; McDonald & Znabi, 1999b). The *link availability* is defined as the probability that there is an active link between two mobile nodes at time $t+T$ given that there is an active link between them at time t. Note that a link may experience one or more failures and recoveries in the time interval between t and $t+T$. Note that the metric is not practical as a criterion to select a path between two nodes, because if a link fails, then rerouting should immediately take place rather than waiting for the failed link to become available again. But, the link availability criterion is useful during the clustering process as it can be used by mobile nodes to select more reliable neighbours to form more stable clusters.

In (Jiang et al., 2001) a prediction-based link availability estimation and a routing metric in terms of path reliability based on the link availability estimation are presented. The basic idea of this estimation is as follows:

Given an estimation T of the expiration time (i.e., the continuously available time) for an active link $\{v,u\}$ between two nodes v and u at time t (computed e.g., by using the link expiration time algorithm of (Su and Gerla, 1999) presented above), the availability $L(T)$ of link $\{v,u\}$ is defined as

$L(T)=Pr\ \{$ the link $\{v,u\}$ lasts from time t to time $t+T$ ***given that*** the link is available at time $t\}$ which indicates the probability that the link $\{v,u\}$ will be continuously available from time t to time $t+T$. The calculation of $L(T)$ is divided into two parts: the link availability $L_1(T)$ when the speeds and moving directions of the nodes v and u remain unchanged from time t to time $t+T$, and the link availability $L_2(T)$ for the other cases. That is, $L(T) = L_1(T) + L_2(T)$. Assuming that the mobility epoch (a random length interval during which node movement is unchanged) is exponentially distributed with mean λ^{-1}, the nodes' movements are independent of each other and the exponential distribution is 'memoryless', $L_1(T)$ is given by:

$$L_1(T) = [1 - E(T)]^2 = e^{-2\lambda T}$$

An accurate estimation of $L_2(T)$ is not easy due to difficulties in learning the changes in link status caused by changes in nodes' movement. A conservative prediction $L_{min}(T)$ of link availability $L_2(T)$ is given in (Jiang et al., 2001 as follows:

$$L_{min}(T) = (1- e^{-2\lambda T}) /2 \lambda T + \lambda T e^{-2\lambda T}/2$$

Using the estimation of $L(T)$, a routing metric based on $L(T) \times T$ is given in (Jiang et al., 2001) that offers improved network performance, according to simulation results. Note that in highly volatile environments, it is possible that the mobility epoch (during which the mobility of the user is unchanged) is usually very small. This will necessitate a large number of estimations that increase the control overhead.

3. Group Mobility and Network Partition Prediction

Network partition occurs when groups of mobile nodes follow diverse mobility patterns, which cause the separation of the network into disconnected subnetworks. Predicting the occurrence and the timing of network partitioning allows MANET applications to improve their performance by acting in advance and preventing disruptions caused by the partitioning.

A method for network partition prediction which exploits group mobility patterns to compute the remainder time before separation is proposed in (Wang and Li, 2002). In order to describe the basic idea of the method, we consider a simple case of a network consisting of two mobility groups C_i and C_j each moving with velocities $V_i = (vx_i, vy_i)$ and $V_j = (vx_j, vy_j)$ respectively. The relative mobility between them is obtained by fixing one group, say C_i, as stationary. Then the effective velocity V_{ij} at which C_j is moving away from C_i is given by:

$V_{ij} = V_j + (- V_i)$, where $V_{ij} = (vx_{ij}, vy_{ij}) = (vx_j - vx_i, vy_j - vy_i)$

Assume that the two groups cover a circular region of diameter D, wherein the nodes are uniformly distributed. Assume also that the groups are in perfect overlap. Then, in order for the two groups to separate, C_j must move past a distance of the diameter D of C_i's coverage area. Hence, the time taken for the two groups to change from total overlap to complete separation is given by:

$$T_{ij} = D / (vx^2_{ij} + vy^2_{ij})^{1/2}$$

In a network made up of many diverse mobility groups, whose nodes are initially dispersed and inter-mixed, given the mean group velocities, the time of separation can be calculated for any pair of mobility groups. The occurrence of partitioning is predicted as a sequence of the expected time of separations between the different pairs of mobility groups in the network.

The network partition prediction method uses a low-complexity data clustering algorithm that accurately determines the mobility groups and their mobility parameters as well as the group membership of each mobile node. The problem is that the prediction algorithm assumes that group and node velocities are time invariant, which is not a realistic assumption for most MANET applications.

4. Cluster Change Based Prediction

In a clustered ad hoc network each mobile node belongs to a cluster while the position of each mobile node is defined with respect to the cluster head of the cluster it belongs to. A mobile node changes the cluster it belongs to as affected by mobility. The *sectorized* ad hoc mobility prediction algorithm is based on the principle that in order to achieve maximum prediction accuracy the prediction process should be restricted to areas of the network with nodes of high cluster change probability (Chellapa-Doss et al., 2003a; Chellapa-Doss et al., 2003b). The algorithm introduces the *sectorized cluster structure* i.e., the cluster is divided into three regions with respect to the probability of cluster change as follows:

i. The *No-Cluster Change (No-CC)* region of each cluster, which contains the nodes of the cluster that are within communication range of each other and they do not satisfy the requirements for membership to any neighbouring cluster. Thus, for the nodes in the No-CC region cluster change is not possible.
ii. The *Low-Cluster Change (Low-CC)* region of each cluster, which contains the nodes of the cluster that are reachable by all nodes in the No-CC region, either directly, or through

Figure 2. The sectorized cluster structure

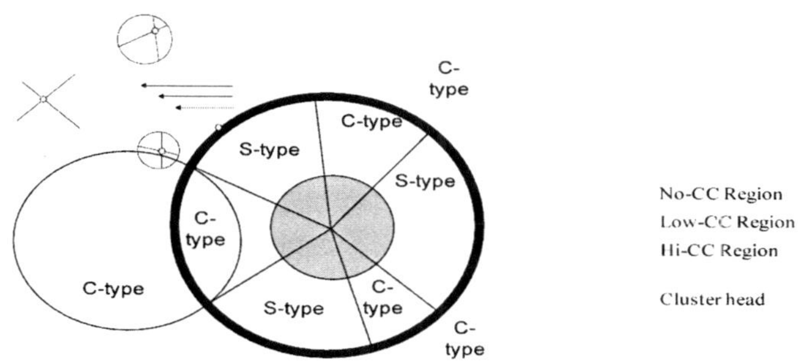

other intermediate nodes belonging to the No-CC region. Thus, for the nodes in Low-CC region the probability of cluster change is fairly low.

iii. The *High-Cluster Change (Hi-CC)* region of each cluster, which contains the nodes of the cluster that are not reachable by any node in the No-CC region, either directly, or through other intermediate nodes belonging to the No-CC region. The nodes in the Hi-CC region are reachable only through the nodes in the Low-CC region and the probability of cluster change for a node in this region is higher than for nodes in the Low-CC region.

The cluster is further divided into *sectors* (Figure 2), (Chellapa-Doss et al., 2003; Chellapa-Doss et al., 2004). Two types of sectors are introduced depending on whether or not the sector is adjacent to a neighbouring cluster. *C-type cluster-sectors* that are adjacent to neighbour clusters, and *S-type cluster-sectors* that are not. It is only from C-type clusters that cluster change is possible. Each C-type cluster-sector is adjacent to only one neighbouring cluster and it is only to this cluster that the node can cluster change to. Nodes in S-type cluster-sectors are not candidates for cluster change as there are no adjacent clusters present.

The cluster sector-numbering scheme (Chellapa-Doss et al., 2003; Chellapa-Doss et al., 2004) is used to predict the next cluster change depending on the mobile node's position in the cluster and its moving direction in the Hi-CC region. The method uses a GPS to build the sectors and locate the mobile nodes positions. Note that, the cluster-sector numbering scheme, as a prediction method, may sit on top of any other ad hoc network clustering algorithm that may be in use.

C. Logical Topology Based Mobility Prediction

1. Neighbouring Nodes Relative Mobility Based Prediction

Many researchers have acknowledged the importance of node mobility estimation for building clustering schemes more stable and less reactive to topological changes of ad-hoc networks. Authors in (McDonald & Znatti, 1999b) propose the *(a,t) clustering scheme*, where mobile nodes form clusters according to a path availability criterion. The network is partitioned into clusters of mobile nodes, that are mutually reachable along cluster internal paths which are expected to be available for a period of time t with a probability of at least a. The parameters of this model are predefined. In addition, it is assumed that the movement of each mobile node is random and entirely independent of the movements of other mobile nodes. However, this random walk model cannot always capture some node mobility patterns occurring in practice in MANETs.

MOBIC in (Basu et al., 2001) elects as CHs the mobile nodes which exhibit the lowest mobility in their neighbourhood. Each node compares the receiving signal strength from its neighbours over the time and uses the variance in these values as an indication of how fast this mobile node is moving in relation to the neighbouring nodes. MOBIC uses only the current mobility to determine the most suitable mobile nodes for CHs.

As an extension of MOBIC, MobDHop (Er & Seah, 2004; Er & Seah, 2006) also uses the variability in receiving signal strength as a hint of neighbourhood mobility and builds variable-diameter clusters. It uses more samples of receiving signal than MOBIC to estimate the predicted mobility, while the prediction model is based on the assumption that the future mobility patterns of mobile nodes will be exactly the same as those of the recent past.

MAPLE (Palit et al., 2006) is another clustering algorithm which also infers node mobility from measurements of the received signal strength. In particular, each mobile node belonging to a cluster can estimate its distance from its CH by using the well-known formula (Schwartz, 2005) of the signal attenuation versus travelled distance. Then, based on past measurements, mobile nodes use a linear model for estimating their future distance from their CH. This helps mobile nodes to proactively join another cluster if they are going to soon leave their current cluster. However, MAPLE does not take node mobility into account during CH election. Specifically, mobile nodes contend for free frames (i.e., time slots) in a single shared broadcast channel during the cluster formation phase and mobile nodes that first reserve the available frames in this phase become CHs. Thus, the election of CHs is mostly a random procedure and it is not based on some CH suitability criteria. In addition, the algorithm sets an upper bound on the number of clusters in the network as well as on the number of mobile nodes per cluster.

2. Information Theory Based Mobility Prediction

Information theory based techniques for node mobility prediction have been first employed in (Bhattacharya & Das, 2002), where the authors focused on the problem of mobile tracking and localization in cellular networks. Later, Sivavakeesar et. al (Sivavakeesar et al., 2004) used the basic technique of (Bhattacharya & Das, 2002) in their cluster formation algorithm for MANETs. In their work they assume that a geographical area is divided into circular-shaped regions (virtual clusters) and each mobile node knows the virtual cluster where it is currently in.

In (Konstantopoulos et al., 2006; Konstantopoulos et al., 2008) a mobility-aware technique for cluster formation and maintenance is proposed. The main idea in the technique is to estimate the future mobility of mobile nodes so as to select cluster heads that will exhibit the lowest predicted mobility in comparison to the other mobile nodes. As a measure of node mobility rate, it is used the probability of a mobile node having the same mobile nodes in its neighbourhood for sufficiently long time. A high probability value for a mobile node indicates either a relatively immobile node or the existence of a group of nodes around this particular node that exhibits the same mobility pattern. Whatever the case is, this mobile node is apparently a good candidate for a CH, because in all probability, it will serve the same neighbours for a long time. For estimating the predicted mobility of a mobile node, it is assumed that the movements of nodes are not random but demonstrate a regular pattern (Camp et al., 2002), which can be predicted, provided that enough "historic" information has been gathered for the movements of each mobile node. For the organization of the historic record and the estimation of future mobility based on this record, prediction techniques from the field of data compression are used. Specifically, the problem of estimating the future neighbourhoods of a mobile node is reduced to that of predicting the next characters in a text given that we have already seen a particular text context. Then, by using context modelling techniques (Bell et al., 1990), the probability of stable neighbourhood around a node can be reliably estimated. The most important characteristic of these methods is the on-line learning of the probability model used for predicting the next character/neighbourhood. This is essential in the case of MANETs because the movements of individual nodes as well as the strong correlation, existing in the movements of these hosts, cannot be easily described by pre-defined random models. The method does not make any use of a fixed geographical partition in contrast to previous work (Bhattacharya & Das, 2002; Sivavakeesar et al., 2004) and thus, the notion of cells is irrelevant to this technique. Also for measuring the mobility, the method does not use any special purpose hardware such as GPS, but the mobility of each node is inferred from how different is the neighbourhood of the node over time.

Besides the stability of the clustering structure, an important objective in cluster creation is keeping the number of elected CHs relatively low, so that the virtual backbone built over these mobile nodes, to be of correspondingly small size and hence routing update protocols can be efficiently ran on this backbone. The well-known highest degree clustering algorithm (Gerla & Tsai, 1995) is a clustering scheme that creates a relatively small number of one-hop clusters and thus a small-size routing backbone. In this technique, each mobile node having the highest degree among all its neighbours is elected as CH. The degree of a mobile node is defined as the number of its one-hop adjacent nodes. The main weakness of the technique is the frequent CH changes due to node mobility. In (Konstantopoulos et al., 2008), a new clustering algorithm (named MobHiD) is proposed which combines the highest degree technique with the mobility prediction scheme, and ensures a relatively small as well as stable virtual backbone despite node mobility.

Note that the mobility prediction technique is of independent interest and may be combined with other clustering algorithms to enhance the stability of the derived clustering structure in the presence of topology changes.

3. Evidence Based Mobility Prediction

The Dempster-Shafer (DS) theory of evidence developed by A. Dempster (Dempster, 1968) and extended by G. Shafer (Shafer, 1975) has attracted considerable attention as an approach that enables combination of different information sources (called *evidences*) to reach decisions in situations characterized by a high degree of uncertainty. In the DS theory, if a probability p is assigned to an event, then $1-p$ represents the confidence not assigned to this event. $1-p$ represents ignorance and uncertainty and it is not necessarily assigned to the opposite event. The main advantage of the DS theory of evidence is its ability to model the narrowing of a hypothesis with the accumulation of evidences and to explicitly represent uncertainty in the form of ignorance or reservation of judgment. The DS theory provides the possibility of giving to different evidences weights according to their relevance and importance in the final decision. It also provides an operation, called the *Dempster's rule of combination,* for combining evidences.

In (Saman & Karmouch, 2005), the DS theory of evidence is used for user mobility prediction based on the use of contextual information. The method can accurately predict user's traveling trajectory using knowledge of mobile user's profile and preferences and analyzed spatial information. Note that knowledge of such contextual information is very important for mobile environments because they may adapt their services according to users' specific demands. Uncertainty of the user's navigation behavior is captured by gathering pieces of evidence concerning different groups of candidate future locations. These groups are then refined to predict the user's future location, when evidence accumulates using the Dempster's combination rule. The method does not require the existence of history of users' movements.

In (Dekar & Kheddouci, 2008) a mobility prediction scheme is proposed that uses the DS theory of evidence to predict the future position of a mobile node based on relevant criteria. The scheme exploits the ability of the DS theory of evidence to explicitly represent uncertainty, which is the main characteristic of the mobility prediction problem. The mobility prediction process is working on a cluster based topology and is performed by a *prediction agent* whose role is to predict the mobile nodes' future clusters before they leave their current clusters. To this end, cluster nodes are organized into the following categories: (a) the *central* nodes, which are either cluster heads or have a link whose strength is greater than a certain value with another central node, (b) the *border* nodes, which have a neighbour that belongs to another cluster, and (c) the *intermediate* nodes, which are neither border nodes nor intermediate nodes. According to this organization, the prediction process is performed only on the border nodes, because only these nodes have neighbours that belong to another cluster, and therefore, can leave their cluster and join another one. The prediction agent uses the following main categories of criteria to predict each mobile node's future cluster:

i. The *physical criteria*, which express the mobile node's physical movement. Such criteria include: *the candidate cluster radius* (if a cluster has the maximum radius it cannot receive new nodes as members), *the links strength with each candidate cluster* (this criterion chooses the candidate cluster with whom the mobile node has the maximum number of strong links) and *the links strength evolution with each candidate* cluster (this criterion chooses the candidate cluster to which the mobile node gets closer).

ii. The *semantic criteria*, which are used to select a cluster when the physical criteria can not do it clearly (e.g., there are several clusters that can be the future cluster according to the physical criteria). Such criteria include: the *user requested services* (this criterion chooses the candidate cluster that contains the most services used by the mobile) and the *user profile* (this criterion chooses the candidate cluster whose profile is the most relevant to the user profile).

By using the DS theory of evidence as a modelling method, the above criteria are weighted according to their importance. Then, the Dempster's rule of combination is applied to combine the different criteria (evidences) and to select the appropriate candidate cluster as the mobile node's future cluster. The method does not require the use of a GPS, as the signal strength may be used to estimate the link strengths and to estimate the distances among the mobile nodes. The mobility prediction scheme has been applied to the Zone Routing Protocol (Haas & Pearlman, 1999) and simulation results have shown its efficiency.

FUTURE TRENDS

So far, ad hoc networks have been deployed for special cases such as battlefield or emergency scenarios. However, this technology is soon expected to be considered as a common place and support our every-day activities and also will coexist with fixed-infrastructure wireless networks (4G networks). The plethora of multimedia services and applications that will run over these networks will pose severe requirements for real-time continuous data flow with guaranteed data rate (QoS requirements). In this new context, predicting the mobile user's future location and the network topology change may significantly improve the QoS parameters by achieving route discovery and route reconstruction prior to the user's movement and to the topology change.

This unification of different wireless technologies does not pose only challenges but also offer opportunities for more accurate mobility prediction methods. The new techniques could be two-tier. The most powerful part could be executed in base stations where advanced machine learning methods could be applied for determining the future positions of mobile nodes at a certain resolution. For more detailed determination of node movements, each node will apply "lightweight" method based only on local information around the node. This also will alleviate the problem incurred in all localization methods in fixed-infrastructure networks where nodes should frequently send position information to the base station in order that the base station keep track of the movements of nodes. Another advantage of the combined centralized and distributed scheme is that base stations collecting information from large node population could reach safer conclusions with regard to node movements as statistical phenomenon. In contrast, each node alone based only on the local view coming from its neighborhood can fail sometimes to determine the behavior of nodes in a larger scale.

The deployment of ad-hoc networks alongside with fixed-infrastructure networks in urban areas also changes some of the basic assumptions about the application field of these networks. Now, the users of mobile devices are not necessarily members of a team that act for a special purpose such as soldiers in a battle-field scenario or rescuers in a emergency scenario but they can be persons not related in an obvious way. So, the correlation of mobile nodes in an urban area is more implicit and hence more difficult to find. Sometimes, it may be possible to infer correlation by taking also into account other restrictions that exist in the movements of persons in a city. For instance, when two nodes are moving along a road, it is sure that the direction of movements of two nodes is the same and cannot change before the next crossroads. Apparently here it is implied that each node is aware of its surrounding topology. This information can be obtained from the base station or from a navigation system now commonly placed in new mobile phones.

In a more general setting, the main issue is the discovery of the relations rising in a large population of individuals. The social behavior observed can be valuable information for inferring correlation among movements of the users of mobile devices. Results from the social networks theory (Da Fontoura et al., 2007) commonly used in the science of sociology could help reveal this kind of relations and thus could be very useful for enhancing predicting capability with regard to the mobile node movements. Also, trace information gathered from the huge number of mobile phones currently in use can provide rich behavioral data that can be studied and then used for deducing safer conclusions about the common behavioral patterns met in large populations (Eagle & Pentland, 2006).

From the discussion above, it becomes evident that although a relatively important number of mobility prediction schemes have already been proposed, there are still issues that should be addressed and improved in the design of such schemes as now more clues about the future behavior of mobile nodes are available and thus they can be used for improving the prediction of mobility of mobile nodes.

CONCLUSION

Mobility prediction, i.e., estimating the trajectory of future positions of the mobile nodes in MANETs, positively affect the service-oriented as well as the application-oriented aspects of ad hoc networking. The employment of mobility prediction techniques at the network level is very crucial in the design of efficient routing schemes, while at the application level, in combination with user's profile may provide the user with enhanced personalized wireless services. This chapter presents well-known mobility prediction schemes classifying them according to the specific prediction methods they employ and/or according to their applications.

REFERENCES

Basu, P., Khan, N., & Little, T. (2001). A Mobility Based Metric for Clustering in Mobile Ad Hoc Networks. In *Proc. of the 21st International Conference on Distributed Computing Systems Workshops (ICDCSW '01)* (pp. 413–418).

Bell, T. C., Cleary, J. G., & Witten, I. H. (1990). *Text Compression*. Prentice Hall.

Bhattacharya, A., & Das, S. K. (2002). LeZi-Update: An Information-Theoretic Framework for Personal Mobility Tracking in PCS Networks. *Wireless Networks*, *8*, 121–135. doi:10.1023/A:1013759724438

Camp, T., Boleng, J., & Davies, V. (2002). A Survey of Mobility Models for Ad Hoc Network Research. [WCMC]. *Wireless Communication & Mobile Computing*, *2*(5), 483–502. doi:10.1002/wcm.72

Chellapa-Doss, R., Jennings, A., & Shenoy, N. (2003a). User Mobility Prediction in Hybrid and Ad Hoc Wireless Networks. In *Proc. of the Australian Telecommunications Networks and Applications Conference (ATNAC'03)*.

Chellapa-Doss, R., Jennings, A., & Shenoy, N. (2003b). A Comparative Study of Mobility Prediction in Cellular and Ad Hoc Wireless Networks. In *Proc. of the IEEE Int'l Conf. on (ICC 2003)*.

Chellapa-Doss, R., Jennings, A., & Shenoy, N. (2004). A Review of Current Mobility Prediction Techniques for Ad Hoc Networks. In *Proc. of the 4th IASTED International Multi-Conference on Wireless and Optical Communications, Canada* (pp. 536-542).

Da Fontoura Costa, L., Oliveira Jr, O. N., Travieso, G., & Rodrigues, F. A. A., Villas Boas, P.R., Antiqueira, L., Viana, M.P. & da Rocha, C. (2007). *Analyzing and Modeling Real-World Phenomena with Complex Networks: A Survey of Applications*. arXiv:0711.3199v1

Dekar, L., & Kheddouci, H., (2008). A Cluster Based Mobility Prediction Scheme for Ad Hoc Networks. *Ad Hoc Networks Journal, 6* (2).

Dempster, A. P. (1968). A Generalization of Bayseian Inference. *Journal of the Royal Statistical Society. Series A (General)*, *30*, 205–247.

Eagle, N., & Pentland, A. (2006). Reality mining: sensing complex social systems. *Personal and Ubiquitous Computing*, *10*(4). doi:10.1007/s00779-005-0046-3

Er, I., & Seah, W. (2004). Mobility-based d-Hop Clustering Algorithm for Mobile Ad Hoc Networks. In *Proc. of the IEEE Wireless Communications and Networking Conference (WCNC 2004)* (Vol. 4, pp. 2359–2364).

Er, I., & Seah, W. (2006). Performance analysis of mobility-based d-hop (MobDHop) clustering algorithm for mobile ad hoc networks. *Computer Networks*, *50*, 3375–3399. doi:10.1016/j.comnet.2005.12.013

Erbas, F., Steuer, J., Kyamakya, K., Eggesieker, D., & Jobmann, K. (2001). A Regular Path Recognition Method and Prediction of User Movements in Wireless Networks. In *Proceedings of the Vehicular Technology Conference (VTC' 2001), IEEE* (pp. 2672-2676).

Gerla, M., & Tsai, J. (1995). Multicluster, Mobile, Multimedia Radio Network. *ACM-Baltzer Journal of Wireless Network*, *1*(3), 255–265. doi:10.1007/BF01200845

Haas, Z.J. & Pearlman, M., (1999, June). *The Zone Routing Protocol (ZRP) for Ad Hoc Networks*. Internet Draft.

Jiang, S., He, D., & Rao, J. (2001). A Prediction-based Link Availability Estimation for Mobile Ad Hoc Networks. *Proceedings - IEEE INFOCOM, 2001*, 1745–1752.

Konstantopoulos, C., Gavalas, D., & Pantziou, G. (2006). A Mobility Aware Technique for Clustering on Mobile Ad-Hoc Networks. In *Proc. International Conference on Distributed Computing and Networking (ICDCN 2006)* (LNCS 4308, pp. 397–408).

Konstantopoulos, C., Gavalas, D., & Pantziou, G. (2008). *Clustering in Mobile Ad-Hoc Networks through Neighbourhood Stability-Based Mobility Prediction. Computer Networks*. Elsevier.

Lee, S.-J., Su, W., & Gerla, M. (2001). Wireless Ad Hoc Multicast Routing with Mobility Prediction. *Mobile Networks and Applications*, *6*, 351–360. doi:10.1023/A:1011478628358

Liu, G. & Maguire Jr. (1996). A Class of Mobile Motion Prediction Algorithms for Wireless Mobile Computing and Communications. *Mobile Networks and Applications Journal, Springer*, *1*(2), 113–121. doi:10.1007/BF01193332

McDonald, A. B., & Znabi, T. F. (1999a). A path availability model for wireless ad hoc networks. In *Proc. IEEE WCNC'1999, New Orleans, USA* (pp. 35-40).

McDonald, A. B., & Znabi, T. F. (1999b). A mobility-based framework for adaptive clustering in wireless ad hoc networks. *IEEE Journal on Selected Areas in Communications*, *17*(8), 1466–1487. doi:10.1109/49.780353

Narendran, B., Agrawal, P., & Anvekar, D. K. (1994). Minimizing Cellular Handover Failures without Channel Utilization Loss. In. *Proceedings of IEEE GLOBECOM, 94*, 1679–1685.

Palit, R., Hossain, E., & Thulasiraman, P. (2006). MAPLE: a Framework for Mobility- Aware Proactive Low Energy Clustering in Ad-hoc Mobile Wireless Networks. *Wireless Communications and Mobile Computing*, *6*(6), 773–789. doi:10.1002/wcm.441

Saman, N., & Karmouch, A. (2005). A Mobility Prediction Architecture Based on Contextual Knowledge and Spatial Conceptual Maps. *IEEE Transactions on Mobile Computing*, *4*(6), 537–551. doi:10.1109/TMC.2005.74

Schwartz, M. (2005). *Mobile Wireless Communications*. Cambridge Univeristy Press.

Shafer, G. (1975). *A Mathematical Theory of Evidence*. Princeton University Press.

Sivavakeesar, S., Pavlou, G., & Liotta, A. (2004). Stable Clustering Through Mobility Prediction for Large-Scale Multihop Ad Hoc Networks. In *Proc. of the IEEE Wireless Communications and Networking Conference (WCNC'2004).*

Su, W., & Gerla, M. (1999). IPv6 Flow Handoff in Ad-Hoc Wireless Networks Using Mobility Prediction. In *Proc. of IEEE GLOBECOM'99.*

Su, W., Lee, S.-J., & Gerla, M. (2001). Mobility Pred iction and Routing in Ad Hoc Wireless Networks. *International Journal of Network Management*, *11*, 3–30. doi:10.1002/nem.386

Wang, K., & Li, B. (2002). Group Mobility and Partition Prediction in Wireless Ad Hoc Networks. In *Proc. of the IEEE International Conference on Communications (ICC'2002).*

Yu, J., & Chong, P. (2005). A Survey of Clustering Schemes for Mobile Ad Hoc Networks. *IEEE Communications Survey*, *7*(1), 32–48. doi:10.1109/COMST.2005.1423333

KEY TERMS AND DEFINITIONS

Ad Hoc Routing Protocol: A convention or standard that specifies how nodes communicate with each other to exchange information that allows them to select routes between any two nodes in a mobile ad-hoc network.

Cluster: Dynamic group of mobile nodes with a cluster head (CH), i.e., a mobile node that holds routing and topology information. The boundaries of a cluster are defined by the transmission area of its CH.

Clustering: The process of partitioning the mobile nodes into dynamic groups, called clusters.

Global Positioning System (GPS): A space-based radionavigation system that provides reliable positioning, navigation, and timing services to users. For anyone with a GPS receiver, the system provides accurate location and time information on a continuous worldwide basis.

Mobile Ad hoc Network (MANET): A self-organizing and self-configuring multi-hop wireless network capable of adaptive re-configuration when it is affected by node mobility, which does not rely on a fixed infrastructure.

Mobility Prediction: The problem of estimating the trajectory of future positions of the nodes in mobile networks.

Network Scalability: The ability of a network to support the continuous increase of the network parameters (such as mobility rate, traffic rate, and network size) without degrading network performance.

Signal Attenuation: A general term that refers to any reduction in the strength of a signal.

Chapter 22
Calling Police Using SMS

Mohammad Shirali-Shahreza
Sharif University of Technology, Iran

M. Hassan Shirali-Shahreza
Amirkabir University of Technology (Tehran Polytechnic), Iran

ABSTRACT

In this chapter, a method is proposed to contact the Police with mobile phones and via SMS (Short Message Service). In this method, when a person wants to contact the Police, he must only press a special key on his mobile phone for a short time to launch a special program. This program sends current location of the person using GPS system and sends it periodically with other useful information such as name, home and work address of that person automatically to the Police Station using SMS. In this method, there is no need to talk and a person can contact the Police easily and tell them the place which crime has occurred. In addition, there is no busy line problem. This method has low cost and can be run on many mobile phones This method is implemented using JavaME (Java Platform Micro Edition) programming language and tested on a Nokia N71 mobile phone by using an 'Evermore BT-R700' GPS receiver.

INTRODUCTION

One of the consequences of urbanization is the rise of crime due to some abnormalities and traumas within the community. The governments and people have always thought of a solution to take care of these problems and lower the crime rates. But it is a necessity to confront criminals and to secure people's safety.

DOI: 10.4018/978-1-60566-250-3.ch022

One of the important factors during the commitment of crimes, especially offenses such as theft, kidnapping and bank robberies, is for one to contact the Police Station and inform them about the crimes in the shortest possible time. For this purpose, different methods have been taken. For instance, in order to prevent the theft of important objects and documents, sirens have been installed on the doors in addition to other devices, which automatically call the Police Station. Moreover, special systems are installed at the telecommunications office to enable Police to trace the location of phone calls. So, if the attacked person cannot speak due to the existing circumstances, he can contact the Police Station, and the policemen can trace the call and locate the victim and go to his location to rescue him.

However, the criminals have also thought of measures to foil these procedures. For instance, they can find the location of the installed protective devices and dismantle them, therefore making it impossible for the Police Station to be informed of the committed crime. Additionally, they disconnect the phone lines, so the victim could not contact the Police.

Nowadays, one of the common devices used by people is the mobile phone. The majority of people always carry their mobile phones. Hence, usage of mobile phones to call the Police Station is appropriate. But its usage also leads to a number of problems. For instance, if the lines are engaged, one cannot call the Police.

One of the initial services provided by the mobile phones is the Short Message Service (SMS). SMS makes it possible to exchange short messages between the mobile phones. SMS has been defined based on GSM. According to GSM03.40 standard, the maximum message should constitute 160 characters and based on the data encoding process in line with the adopted standards, it is saved in 140 bytes (GSM, 2000). These messages can be a combination of numbers and letters or in non-text and binary form.

The exchange of SMS takes place indirectly via a component named SMSC (Short Message Service Center). A number of the SMS advantages are as follows:

- Communication is possible when the network is busy,
- The possibility of exchanging SMS concurrent with the phone call,
- The possibility of sending SMS in the offline mode.

Also one can receive a report from the center about the state of SMS delivery or to determine the validity duration of the SMS (Nokia, 2001).

Regard to the mentioned features of the SMS, one can use this system to contact the police. For this purpose, the following method has been proposed:

In this method, for contacting the Police, one should press a specific key in his mobile phone for a few seconds in order to activate a special program. This program locates the person with the help of GPS (Global Positioning System) and sends other appropriate information such as the individual's name, his place of residence, his workplace and his home number (which have been previously saved in his mobile phone) via SMS in certain continuous intervals to the Police Station. The details of our method have been completely elaborated in the third section.

Usage of other systems including the MMS (Multimedia Messaging Service) is not recommended because these systems have not been activated in numerous countries and not all the mobile phones support these advanced services. On the other hand, the SMS fulfills our needs here and the advanced features of MMS are not useful here. What's more, these services are more expensive than SMS and using them can distorts the network in crises, because the total number of the exchange messages would be huge and the load imposed on the network by MMS is more than the load imposed on the network by the SMS. MMS services may distort the network and/or dismantle it. Also MMS availability is less than SMS and SMS is more reliable in different conditions.

In this method, there is no need to talk and by just pressing a key, all the necessary information including the address of the crime scene is sent to the Police Station. Moreover, the user of SMS will not face problems, such as busy lines, due to the special features of the SMS. On the other hand, if the individual is kidnapped, the person location is sent periodically by SMS, so the Police can locate him in specific intervals via this system. In the section 4, the advantages of this plan are studied.

Obviously, the method proposed in this chapter is not the first method in which the mobile phone is used to contact the Police Station. Such systems have been proposed in the past and are currently in use in some countries including the U.S. But in our method, we have taken advantage of SMS and our method has its differences with other methods. In the section 2, several other systems which are using mobile phones for contacting the Police Station will be studied.

This project has been written in JavaME (Java Platform Micro Edition) programming language and has been tested on a Nokia N71 mobile phone. The 'Evermore BT-R700' GPS Receiver has been used in this project.

BACKGROUND

Mobile phone's locating system has been used in several countries for several years in order to find and rescue individuals. For example, the E911 (enhanced 911) System is used in the U.S. (911 is the telephone number of U.S. Police and Rescue Center). Meanwhile, in the U.S. within the E911 System it has been requested from the companies providing phone services to identify the location of caller with 50-100 meter accuracy and to inform the Police, if the individual calls the Police. In this case, when the caller does not know his whereabouts or cannot speak or his call is disconnected, the Police and rescue forces can locate him.

System 112 (112 is the number of European countries Police and rescue center) is established in Europe alike the E911 system in the U.S., with this difference that there is no defined precision for locating the individuals. Hence, the companies providing mobile phone services in Europe usually use Cell-ID method for identifying individuals, which approximately specifies the position of the mobile phone by using the geographical latitude and longitude. The precision of this method depends on the area of each cell and varies between 100-1000 meters (Warrior, 2003).

In both above-mentioned procedures, a phone contact has been used for getting informed about the incident and to locate individuals. But in the method proposed here, SMS has been firstly used for contacting the Police Station. Hence there is no need to speak. Secondly, in this method, the mobile phone locates the individual (by using a GPS Receiver). The Police are informed about the location so that no necessity would arise for locating the mobile phone by the telecommunications office.

Another similar application of mobile phones and similar wireless devices is in emergency situations such as medical emergency situation. So far, different works have been done in the area of medical care through wireless devices and networks (Wu et al., 2005). But the activities on wireless devices as the PDAs and mobile phones have been less reported due to the limitations of these devices (Pavlovski et al., 2004) and such devices have been mainly used for wireless communication. Especially, the use of SMS has been very limited due to its special limitations (Hassinen and Laitinen, 1998).

One of these applications is the AMBULANCE system presented by Pavlopoulos et al. (Pavlopoulos et al., 1998). This system is intended to establish contact between the emergency medical team and specialist doctors in the hospital during the transfer of the patient to the hospital and for conducting the medical operations needed for the patient until reaching the hospital. For this purpose, some equipment should be placed inside the ambulance and some other equipments in the hospital to establish the communication.

In ambulance, a device must be set up to keep records of patient life conditions and a portable PC to communicate with hospital. The portable PC is fitted with a frame grabber card, a camera and a mobile phone with GSM Modem in order to establish contact with the hospital and transfer information. A workstation in hospital processes input information from ambulance under supervision of a specialized doctor.

OUR SUGGESTED METHOD

In this method, the goal is to contact the Police Station as soon as possible without any additional move and only by just pressing a button, and to identify the location of the incident and present other information automatically, without speaking, by mobile phone via SMS.

In this method, the individual firstly places its personal information including first name, surname, the address of the workplace, the place of residence, house and workplace phone numbers within a special software so that at the time of an incident, the personal details can be sent to the Police. This information is saved in a text file within the mobile phone.

This program is executed by pressing a special button on the mobile phone for several seconds. In case of danger, the individual presses the special button on the mobile phone for a short period to run the special software.

After running, the software initially cuts the voice of the mobile phone and puts it in the silent mode so that the assailants will not detect the mobile phone with the individual. Now, this software identifies the current location of the mobile phone, with the help of the GPS. Some of the new mobile phones like Nokia N95 have built-in GPS. In addition, a great number of phones support Bluetooth GPS modules. The GPS can provide the precise location.

Meanwhile, the GPS is not necessary for this system to work. Although GPS has a high precision in locating, it also has some drawbacks such as not working properly in closed surroundings such as inside the buildings (Bamford, 2006). Additionally, the GPS is not installed in the majority of mobile phones and needs a separate device which is not cheap. If the GPS is not available, the system will use the local information which is handed to the mobile phone by mobile phone network. This information has low accuracy, for example it can provide the location of the mobile phone with 1000 meters accuracy. But, it is acceptable and satisfactory because it can be used to know whether the person is at his home or his office and then the full address of his home or office which is saved in the software will be used.

Afterwards, the software forms a SMS based on the following framework and sends it to the Police Station based on the information previously received from the individual and his current address (Figure 1).

- On the first line, the personal details including his first name and family name are stated.
- In the following lines of the message, other information including his home and workplace addresses are stated.
- The final lines of the message include the current address of the mobile phone (which identifies the person's current location).

In this manner, the Police get informed about the incident and its location, and can immediately take the necessary measures to rescue him from danger.

Figure 1. The SMS sent by the person in need of help

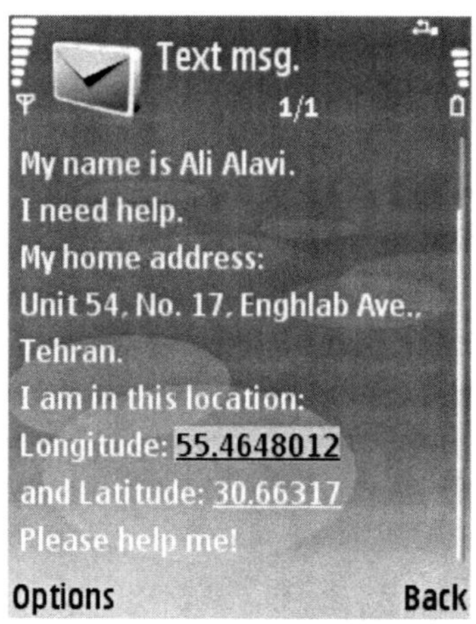

Meanwhile, this SMS is sent to the Police Station continuously in certain intervals so that if the assailants kidnap the person or conditions arise which would force the individual to change his location, the Police would still be able to locate the individual. If the individual once again presses the special button, the program will stop.

The server for this system can be easily set up in Police Stations, because it does not require any special equipment. It only requires a system for receiving SMS messages which is usually available in Police centers which has a SMS number for receiving people reports. This system does not require any special equipment in the network operators.

IMPLEMENTATION

In this section, we will describe the implementation details of the client side program which is run on the mobile phone and will call the Police in the case of an emergency.

The client side program is written by the Java Platform Micro Edition or JavaME (previously named Java 2 Micro Edition or simply J2ME) programming language. This language is an edition – more precisely a subset – of Java platform which is specially designed for small devices such as PDA (Personal Digital Assistant) and mobile phones.

Before describing our system implementation, we have a quick look at the JavaME language and the Java Specification Request (JSR). The Java designers defined a formal process to enable interested parties to be involved in the process of defining future versions of Java platform. This process in established in 1998 and named Java Community Process (JCP). Any organization or individual can become a JCP member and involve in its processes.

The different aspects of Java platform is usually defined by a series of Java Specification Requests (JSR). A JSR is a document which proposes a new specification or a modification to an existing specification. Any JCP member can propose a JSR. A JSR must pass through the process defined by JCP and become final when it is voted by the JCP Executive Committee. The JSR are numbered and usually their numbers are used for calling them. For example, the Java Platform Micro Edition was initially developed as JSR 68.

For the developers who are using Java Platform Standard Edition (JavaSE) or Java Platform Enterprise Edition (JavaEE), the different JSR are not highly important because most of the tools they use – such as compilers or virtual machines – are supporting most of JSR and they have also the ability to update their tools to use new JSR.

But, when we are developing a program for a mobile phone or other devices which have a not-updatable version of Java virtual machine, it is important to know what JSR the device is supporting. Supporting a JSR on a device like mobile phone usually requires special hardware, so each device is supporting a series of JSR and cannot run a program which is using a not supported JSR. So, an important step in designing a program for a mobile phone is deciding which JSR are required by the program.

Among the available JSRs, the two main JSRs for mobile phones are JSR 118 and JSR 139. The JSR 139 is named "Connected Limited Device Configuration 1.1" (CLDC 1.1) and is the revised version of the JSR 30 named "J2ME Connected, Limited Device Configuration" which defined a standard configuration for devices that supports JavaME. The JSR 118 is named "Mobile Information Device Profile 2.0" (MIDP 2) and is the revised version of JSR 37 named "Mobile Information Device Profile for the J2ME Platform" which is defining a profile and extend the JavaME platform to ease the process of application development for the devices which are supporting JavaME. Our client is mainly developed by the APIs provided by CLDC 1.1 and MIDP 2. These JSR are supported by a wide range of mobile phones produced by different manufactures nowadays. So our client program can be run on different mobile phones.

In our system, the client program needs to communicate with the Police center. The communication in our proposed client is done by sending a SMS (Short Message Service). So we need an API to send SMS. The JSR 120 which is named "Wireless Messaging API" defines API for sending different messages such as text messages and binary messages using SMS standard. An advantage of this API is that it can be used in both GSM and CDMA networks. This API is supported by most of the available mobile phones which supports JavaME. This JSR is the main JSR used by our program for communication and hence our client can be used on most of the available mobile phones.

As we described in previous section, the client program tries to find the current location of phone and sends it to the Police Station. We are using the JSR 179 named "Location API for J2ME" for this purpose. This JSR is defined to enable developers to use the current physical location of phone in location-based applications. This JSR requires CLDC 1.1. One of the advantages of this API is that it is independent of the location provider, which means that it can collect the location information from different sources such as GPS or location provided by the network operator. Our client uses the APIs of this JSR for finding the location of user. When the client wants to find the location, it tries to find the location in a defined time period – for example 15 seconds – and send the current location to the Police Station. If the program could not locate the person – for example if the person is inside a building and the GPS cannot receive the required signals from satellites – it will send the help request message without the current location of the person. This feature is provided to enable using the program in situation that the phone cannot find the location or even if the mobile phone is not supporting this JSR.

Our client tries to find the current location of the person to better guide the Police for finding the person. GPS is one the common technologies used to find the location in applications such as finding routes and navigation for use in car. During recent years, a number of mobile phones such as Nokia N95 are produced that have built-in GPS receivers to facilitate the location based applications. But because most of the mobile phones do not have built-in GPS receivers, we must design our client such that it can use external GPS receivers. There are different GPS receiver modules in the market that can be used by mobile phones and PDAs. These modules are usually connected to the device using the Bluetooth technology.

Bluetooth technology can be briefly defined as a standard of short-wave radio frequency for wireless connection of such devices as PDAs, mobile phones, and desktop computers. This technology works in 2.45 Giga Hertz frequency range and covers a scope up to 100-meter distance with the maximum transfer speed of 3 Mega bits per second (Mbps) (in Bluetooth 2 standard). A Bluetooth device can be connected to 8 appliances at the same time (Bluetooth, 2004). It was primarily designed as a cable replacement technology. Bluetooth has become increasingly popular in modern mobile phones with an estimated 24 million Bluetooth enabled phone were in use in 2002 (Garner, 2003). Currently, its main usage is for the connection of hands-free devices to mobile phones or for communicating among devices such as mobile phones or between mobile phone and personal computers. Today, a great number of the mobile phones support Bluetooth technology.

The JSR 82 named "Java APIs for Bluetooth" is defined to enable Java developers to communicate and use external devices in Java applications. Considering the growth of using mobile phone accessories which are connected to phone via Bluetooth, especially hands-free accessories, a great number of mobile phones are supporting this JSR. Our client is using this JSR to communicate with external GPS receivers which are connected to the mobile phone.

Our system is designed for use in emergency situations. So it must be configured in a way that it can be used by the user quickly and easily. Our proposed solution is using programmable key of mobile phones. The mobile phones which are produced today are not sole communication devices, and they have a number of additional functionalities such as receiving FM radio signals. One of these features is playing music files and acting as a music player which is highly welcomed by the users. Today major manufactures are designing and producing special models of their mobile phones for listening to music such as Sony-Ericsson Walkman Series and Nokia N-Series. These mobile phones usually have a number of programmable buttons which are used to run the media player program as default. These buttons can be programmed to run our client program, so the user will be able to run the program easily and by only pressing one button. This can ease the use of our client for elderly people or children. In addition it can enable the user to run the program in situation that he cannot see the screen of the phone because he do not want to bring his phone out of his pocket.

The client code is compiled by standard tools which are provided by the phone manufactures. We test our client on a Nokia N71 mobile phone. A screenshot of the client program running on the Nokia N71 phone is shown in Figure 2. In this project, the Evermore BT-R700 GPS receiver has been used to find the user location, because the Nokia N71, like most of the available mobile phones, do not have built-in GPS receiver. This device communicates with the mobile phone with the use of Bluetooth technology.

ADVANTAGES

Our system has a number of advantages in comparison to available systems:

1. The main advantages of our system is that the location and information of the person are sent to Police Station quickly and accurately. In available systems, the person who calls the Police must say his location or the location is determined with the help of network operator.
2. In our system, in addition to the location, some other information such as the name of person and phone number of his home and office and even the phone number of a relative are sent to the Police. Gathering this information in other systems is not an easy task. For example, founding the name of the person who calls requires contacting network operators. This information can help Police during rescue operations.
3. Today, majority of people even children and the elderly are familiar with the mobile phones and have a personal mobile phone. Hence they can simply benefit from this method to contact the Police. Meanwhile, in the proposed method there is not even a need for familiarity with the mobile phone and it is only necessary to press a button for a short period. Some of the mobile phones such as Nokia N71 have a programmable

Figure 2. A screenshot of the program for calling the police via SMS

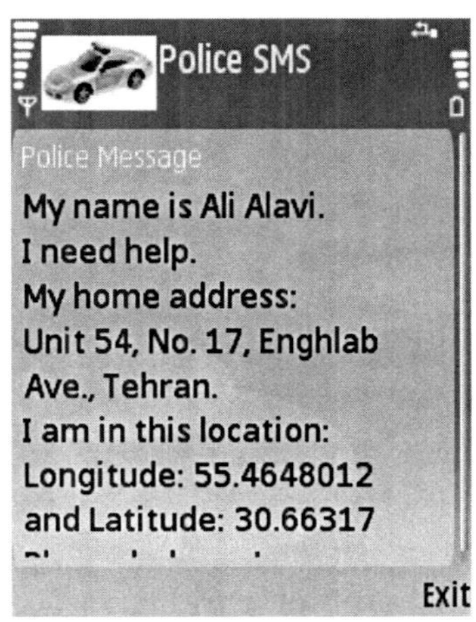

button which can be set to run a special application. In such mobile phones, this button can be programmed to run the program in emergency situation. So elderly people, disabled people or children can use this system easily.

In addition, some other advantages of our method are:

1. All the mobile phones, even the black and white and old models can receive and send SMS. Hence this method covers a wide range of users.
2. The majority of mobile phones support Java language. So this program can be run on the majority of mobile phones.
3. The problem of engagement of phone lines is nonexistent in this method, because the SMS has abilities such as sending short messages in the offline mode and can exchange SMS concurrent with a phone call.
4. If one travels to another country, there would be no language problem because the information can be entered in English. In this case, in whatever country the person is, his information is sent in English to the Police Station of the related country. This can help tourists very much, because they may have problems in speaking in the language of the place they are visiting.
5. There would be no problem in telling the address to the Police because the mobile phone automatically identifies and sends the current address of the mobile phone with the help of the locating system. This can help tourists because they are not familiar with location they are visiting and hence cannot provide their current location to Police Station accurately. This can also help elderly people who may forget their address.
6. The mobile phone network is reliable and the possibility that it is becoming out of order in situations such as public riots is less than other networks including the fixed telephone network.
7. The proposed method can be run in developing countries due to the usage of mobile phones and SMS, which is a public and cheap service on the mobile phones, which is widely available across the globe.
8. Because this system does not require special systems in the Police center, there are no security concerns for the Police center. In addition, the information which are stored in the program on the mobile phone such as name and contact phone numbers are not secret. So there is no risk of revealing secret information even in accidents such as mobile phone lost.

CONCLUSION

In this chapter, a method has been presented with the usage of SMS, in which one can contact the Police at times of danger without any need to speak and only just by pressing a specific button on his mobile phone for a few seconds. In this method, by using the GPS system, the location of the person is automatically sent to the Police Station.

By fining the nuisance individuals, one can prevent the sending of nuisance SMS messages. Also, by installing filters on the SMS receivers at the Police Station, they can separate necessary messages from others.

In addition to the Police Station, this method can be used in the fire stations or hospitals in order to rescue the wounded (Shirali-Shahreza, 2006).

This method can run for the PDA, in addition to the mobile phones. These programs can also be installed on desktop PCs so that if necessary it would automatically send short text messages through the internet.

This method also is useful for the mentally and physically handicapped persons, who can seek help by pressing a special button.

REFERENCES

Bamford, W., Coulton, P., & Edwards, R. (2006, 24-28 April). Location-based Mobile Blogging. In *Proceedings of the 2nd IEEE International Conference on Information & Communication Technologies: from Theory to Applications (ICTTA '06)* (vol. 1, pp. 39-40), Damascus, Syria.

Bluetooth, S. I. G. (2004, November). *Specification of the Bluetooth System.* Version 2, November 2004, http://www.Bluetooth.com/NR/rdonlyres/1F6469BA-6AE7-42B6-B5A1-65148B9DB238/840/Core_v210_EDR.zip

Garner, P. D. (2003, June). Mobile Bluetooth networking: technical considerations and applications. In *Proceedings of the 4th International Conference on 3G Mobile Communication Technologies* (pp. 274-276).

GSM 03.40 v7.4.0 (2000). *Digital cellular telecommunications system (Phase 2+), Technical realization of the Short Message Service (SMS).* ETSI 2000. Retrieved from http://www.etsi.org.

Hassinen, M., & Laitinen, P. (2005). End-to-end Encryption for SMS Messages in the Health Care Domain. *Studies in Health Technology and Informatics, 2005*, 316–321.

Nokia (2001, May). *Sending Content over SMS to Nokia Phones.* Version 1.0, Forum Nokia. Retrieved from http://www.forum.nokia.com.

Pavlopoulos, S., Kyriacou, E., Berler, A., & Koutsouris, D. (1998). A mobile system for emergency health care provision via telematics support-'AMBULANCE. In *Proceedings of the IEEE International Conference on Information Technology Applications in Biomedicine 1998 (ITAB 98)* (pp. 150–154).

Pavlovski, C. HeeSam, K., & Wood, D. (2004). Ubiquitous mobility in clinical healthcare. In *Proceedings of the IDEAS Workshop on Medical Information Systems: The Digital Hospital (IDEAS-DH'04)* (pp. 147-153).

Shirali-Shahreza, M. (2006, October). Emergency SMS. In *Proceedings the SICE-ICASE International Joint Conference 2006 (SICE - ICCAS 2006)* (pp. 1139-1142), Busan, KOREA.

Warrior, J., McHenry, E., & McGee, K. (2003, July). They know where you are [location detection]. *IEEE Spectrum, 40*(7), 20–25. doi:10.1109/MSPEC.2003.1209608

Wu, J., Wang, S., & Lin, L. (2005). What Drives Mobile Health Care? An Empirical Evaluation of Technology Acceptance. *In Proceedings of the 38th Annual Hawaii International Conference on System Sciences (HICSS '05)* (pp.150–159).

KEY TERMS AND DEFINITIONS

Bluetooth: Bluetooth is a standard of short-wave radio frequency for wireless connection among devices such as PDAs, mobile phones, and desktop computers. This technology covers a scope up to 100-meter distance with the maximum transfer speed of 3 Mega bits per second (Mbps). Today, a great number of the mobile phones support Bluetooth technology.

E911: E911 (Enhanced 911) is a location technology advanced by the FCC (Federal Communications Commission) that will enable mobile, or cellular, phones to process 911 emergency calls.

GPS: GPS (Global Positioning System) is a satellite navigation system used to determine ground position and velocity. Though it was created and originally used by the US military, GPS is now available to the general public all over the world.

JavaME: JavaME, or Java Platform Micro Edition – previously known as J2ME or Java 2 Micro Edition – is a collection of Java APIs targeting embedded consumer products such as PDAs, mobile phones and other consumer appliances.

JSR: Java Specification Requests or JSR is a document which proposes a new specification or a modification to an existing aspect of Java platform. A JSR must pass through the process defined by Java Community Process (JCP) and become final when it is voted by the JCP Executive Committee. The JSR are numbered and usually their number is used for calling them.

SMS: SMS or Short Message Service is a service for sending messages of up to 160 characters (224 characters if using a 5-bit mode) to mobile phones that use Global System for Mobile (GSM) communication.

Chapter 23
Vehicular Communications Networks:
Current Trends and Challenges

Moez Jerbi
France Telecom R&D, Core Network Laboratories, France

Sidi-Mohammed Senouci
France Telecom R&D, Core Network Laboratories, France

Yacine Ghamri-Doudane
Networks and Multimedia Systems Research Group, France

Mohamed Cherif
France Telecom R&D, Core Network Laboratories, France

ABSTRACT

Vehicular communications networks (VCNs) are created by vehicles equipped with short and medium range wireless communication technology. They include vehicular ad-hoc networks (VANETs), vehicle-to-vehicle and vehicle-to-infrastructure communications. VCNs enable a plethora of important applications and services, ranging from active safety or safety of life applications to traffic information, music/maps download and multi-hop internet connection. Recently, the promises of wireless communications to support vehicular safety applications have led to several national/international projects around the world. These include the consortia like Vehicle Safety Consortium (US), Car-2-Car Communication Consortium (Europe) and Advanced Safety Vehicle Program (Japan), standardization efforts like IEEE 802.11p (WAVE), and field trials like the large-scale Vehicle Infrastructure Integration Program (VII) in the US. All these efforts have as a main goal to improve safety in vehicular environments by the use of wireless communications, but also consider transport efficiency, comfort and environment. In comparison to other communication networks, VCNs come with unique attractive features: unlimited transmission power, predictable mobility and plethora of potential applications. However, to bring its potency to fruition, VCNs have to cope with formidable challenges that include: rapidly changing topology subject to frequent fragmentations and congestions, lack of connectivity redundancy, and the stringent application requirement on real-time and robust message delivery. In this chapter, we present a detailed description of the state of the art of this fast-moving research area pointing to research, projects and standardiza-

DOI: 10.4018/978-1-60566-250-3.ch023

tion efforts that have been done. We explore the unique features and challenges that characterise these highly dynamic networks as well as their requirements with respect to applications, types of communication, self-organization and security. We discuss various forwarding and routing strategies focussing on position-based techniques including 'anchor-based routing'. We survey various 'intelligent flooding' and information dissemination approaches. Scenarios for highways and cities are taken as example. We conclude by exploring future research directions in this field.

INTRODUCTION

Road safety has been an important concern in the world over the past few years since millions of people die every year because of car accidents and many more are injured. Current statistics show that road traffic accidents in the Member States of the European Union annually claim about 39000 lives and leave more than 1.7 million people injured, representing an estimated cost of 160 billion euros (http //europa.eu.int/comm/transport/care/indexen.htm).

Automated highway systems and intelligent transportation systems (ITS) were introduced to accelerate the development and use of intelligent integrated safety systems that use information and communication technologies as an intelligent solution, in order to increase road safety and reduce the number of accidents in our future roads. In contrast, as mobile wireless devices became an essential part of our lives, and the ubiquitous 'anywhere, anytime' connectivity concept is gaining attraction, Internet access from vehicles is in great demand. The proliferation of cooperated system approach for ITS and the focus on information and communications technologies (ICT) services on one hand and the growing number of communication infrastructure-enabled vehicles on the other hand has opened up new business models and key market segments for many stakeholders in the ITS-market.

Vehicular Communication Networks (VCNs) are a cornerstone of the envisioned Intelligent Transportation Systems (ITS). By enabling vehicles to communicate with each other via Inter-Vehicle Communication (IVC) as well as with roadside base stations via Roadside-to-Vehicle Communication (RVC), vehicular networks could contribute to safer and more efficient roads.

The opportunities and areas of applications of VCNs are growing rapidly, with many vehicle manufacturers and private institutes actively supporting research and development in this field. The integration with on-board sensor systems, and the progressive diffusion of on-board localization systems (GPS) make VCNs suitable for the development of active safety applications, including collision and warning systems, driver assistant and intelligent traffic management systems. On the other hand, inter-vehicular communication (IVC) also fuels the vast opportunities in online vehicle entertainment (such as gaming or file sharing), and enables the integration with Internet services and applications (Nandan et all., 2004).

In this chapter, we present a detailed description of the state of the art of this fast-moving research area pointing to research, projects and standardization efforts that have been done. We explore the unique features and challenges that characterise these highly dynamic networks as well as their requirements, especially in terms of quality of service, market introduction and security. We discuss various forwarding and routing strategies focussing on position-based techniques including 'anchor-based routing'. We survey various 'intelligent flooding' and information dissemination approaches. We conclude by exploring future research directions in this field.

STATE OF THE ART OF VCNS: HISTORY, RECENT PROJECTS & STANDARDIZATION EFFORTS

History

The earliest research into inter-vehicular communications was conducted by JSK (Association of Electronic Technology for Automobile Traffic and Driving) of Japan in the early 1980s (Tsugawa, 2005). This work treated inter-vehicular communications primarily as traffic and driver information systems incorporated in ATMS (Asynchronous Transfer Mode).

From the 1990s through 2000, American PATH (Hedrick et al., 1994) and European "Chaffeur" (Gehring et al., 1997) projects investigated and deployed automated platooning systems through the transmission of data among vehicles.

Recently, the promises of wireless communications to support vehicular safety applications have led to several national/international projects around the world:

Since 2000, many European projects (CarTALK2000, FleetNet, etc.), supported by automobile manufacturers, private companies and research institutes, have been proposed with the common goal to create a communication platform for inter-vehicle communication.

The IST European Project CarTALK2000 (www.cartalk2000.net) was focusing on new driver assistance systems which are based upon inter-vehicle communication. The main objectives were the development of co-operative driver assistance systems and the development of a self-organising ad-hoc radio network as a communication basis with the aim of preparing a future standard;

The FleetNet project in Germany (FleetNet project- Internet on the road, http://www.et2.tuharburg.de/fleetnet), supported by six manufacturers and three universities from the 2000 though 2003, produced important results on several research areas, including the experimental characterization of VANETs, the proposal of novel network protocols (MAC, routing) and the exploration of different wireless technologies.

Ongoing Projects

Many activities in research, development and standardization of vehicular communication are currently ongoing:

In Europe, major R&D projects have been initiated to constitute the basis of a European-wide intelligent transportation system, for example Now, COMeSafety, CVIS, SafeSpot, COOPERS, GST and GeoNet.

- **NoW (Network-on-Wheels, www.network-on-wheels.de):** Network on Wheels is a German project, successor of the project Fleetnet–Internet on the Road, which mainly works on communication aspects for vehicle-to-vehicle and vehicle-to-roadside communication based on WLAN technology. The specific objective of the NoW project is the development of a communication system which integrates both safety (such as extended electronic break light, EEBL) and non-safety applications (such as car-to-home applications).
- **CVIS (Cooperative Vehicle Infrastructure Systems, www.cvisproject.org/):** Aims at developing a communication system that is capable to use a wide range of wireless technologies, including cellular networks (GPRS, UMTS), wireless local area networks (WLAN), short-range microwave beacons (DSRC) and infra-red (IR). Access to these wireless technologies is based on the new international "CALM" standard that allows future vehicular networking implementation to be integrated to the CVIS platform via standardized CALM service access points. A Framework for Open Application Management (FOAM) is defined that con-

nects the in-vehicle systems, roadside infrastructure and back-end infrastructure that is necessary for co-operative transport management.

- **SAFESPOT (Cooperative vehicles and road infrastructure for road safety, www.safespot-eu.org/):** Addresses co-operative systems for road safety, referred to as "smart vehicles on smart roads" to prevent road accidents developing a "safety margin assistant" that detects potentially dangerous situations in advance and extends the drivers' awareness of the surrounding environment "in space and time". This assistant represents an intelligent co-operative system utilizing vehicle-to-vehicle and vehicle-to infrastructure communication based on WLAN technology (IEEE 802.11p).
- **COOPERS (Co-operative Systems for Intelligent Road Safety, www.coopers-ip.eu/):** Focuses on the development of innovative telematics applications on the road infrastructure with the long term goal of a "co-operative traffic management" between vehicle and infrastructure. COOPERS attempts to improve road sensor infrastructure and traffic control applications, develops a communication concept and applications able to cope with the requirements for infrastructure-to-vehicle communication, and demonstrates results at major European motorways with high-density traffic.
- **GST (Global Systems for Telematics, www.gstforum.org/):** Creates an open and standardized end-to-end architecture for automotive telematics services. The project targets at infrastructure-oriented services typically provided by a network operator, such as emergency call services, enhanced floating car data services, safety warning and information services.
- **The EU project GeoNet (www.geonet-project.eu/):** Started in February 2008, implements a reference system for vehicular ad hoc networking using concepts for geographical addressing and routing. Particular focus lies on integration of GeoNetworking with IPv6 and solutions for IP mobility support. In GeoNet, a vehicle is regarded as a mobile network, where the NEMO protocol handles Internet connectivity of the nodes in the mobile network with intermittent access to roadside units. For wide deployment of the project results, it is planned to provide the GeoNet implementations to other R&D projects.

Similar activities can be identified all over the world, such as the Vehicle-Infrastructure Initiative (VII) (www.vehicle-infrastructure.org/) in North America as well as Vehicle Information and Communication System (VICS) (www.vics.or.jp/) in Japan and ITSIndia (Intelligent Transport System India, www.itsindia.org/) in India.

- **The Vehicle Infrastructure Integration (VII) (www.vehicle-infrastructure.org/):** Initiative is a cooperative effort between Federal and State departments of transportation (DOT's) and automobile manufacturers. Together they are evaluating the technical, economic, and social/political feasibility of deploying a communication system that will be used primarily for improving the safety and efficiency of the nation's road transportation system. Additionally, VII will enable the deployment of a variety of applications that support private interests, including those of vehicle manufacturers. It is supported by a radio spectrum at 5.9 GHz, specifically allocated for DSRC (Dedicated Short Range Communications, www.leearmstrong.com/DSRC/DSRCHomeset.htm).

Standardization Efforts

Standardization of vehicular communication is now being pushed by established work groups of major standardization bodies (IEEE, IETF, ETSI, ISO, SAE, ASTM) and consortia such as Vehicular Safety Consortium (VSC), Car-to-Car Communication Consortium (C2C-CC):

- **The Car-to-Car Communication Consortium (C2C-CC, www.car-to-car.org):** Is an industry consortium, initiated by European car makers. It attempts to create and establish an open European industry standard for car-to-car and car-to-infrastructure communication based on wireless LAN components and to guarantee European-wide inter-vehicle operability. The C2C-CC specifies prototypes and demonstrates vehicular communication and pushes the harmonization of vehicular communication standards worldwide.
- Over the last years a working group of the IEEE is defining a new communication standard, which shall be used for future inter-vehicular communication. This so-called IEEE 802.11p or Wireless Access in Vehicular Environments (WAVE, http://grouper.ieee.org/groups/scc32/dsrc/index.html) standard focuses on physical and link layers to provide the required robustness for Vehicular communications and shall be finalized in the near future.
- The ISO Technical Committee 204, Work Group 16 (ISO TC204 WG16) develops a set of standards referred to as 'Continuous Air-interface for Long & Medium range telecommunications (CALM, www.calm.hu/). The CALM concept assumes that systems that future vehicles will be equipped with more than one wireless technology. CALM offers a set of protocol standards that will allow to route data over the most wireless technology that is available at that time, based upon knowledge of the QoS requirements for that application and the real-time performance of the available communications media.

MAIN CHARCACTERISTICS AND CHALLENGES OF VCNS

Characteristics of VCNs

In comparison to other communication networks, VCNs come with unique attractive features (Nekovee, 2005):

- **Sufficient energy and storage.** A common characteristic of nodes in vehicular networks is that nodes have ample energy and computing power (including both storage and processing), since nodes are cars instead of small handheld devices.
- **Higher computational capability.** Operating vehicles can afford significant computing, communication and sensing capabilities.
- **Predictable mobility.** Unlike general mobile ad hoc networks, where it is hard to predict the nodes' mobility, vehicles can have very predictable movement that is (usually) limited to roadways. Roadway information is often available from positioning systems and map-based technologies such as GPS. Given the average speed, current speed, and road trajectory, the future position of a vehicle can be predicted.

However, to bring its potency to fruition, VCNs have to cope with formidable challenges (Blum and al., 2004) that include:

- **Potentially large scale.** Unlike most of the ad hoc networks studied in the literature that usually assume a limited area, VCNs can in principle extend over the entire road network and include many participants.

- **High mobility.** The environment in which vehicular networks operate is extremely dynamic, and includes extreme configurations: in highways, relative speed of up to 300 km/h may occur, while density of nodes may be 1-2 vehicles per kilometer in low busy roads.
- **Highly dynamic topology**. Due to high speed of movement between vehicles, the topology of vehicular networks is always changing. For example, assume that the wireless transmission range of each vehicle is 250 m, so that there is a link between two cars if the distance between them is less than 250 m. In the worst case, if two cars with the speed of 60 mph (25 m/sec) are driving in opposite directions, the link will last only for at most 10 sec.
- **Frequently disconnected network.** Due to the same reason, the connectivity of the vehicular networks could also be changed frequently. Especially when the vehicle density is low, it has higher probability that the network is disconnected. In some applications, such as ubiquitous Internet access, the problem needs to be solved. However, one possible solution is to pre-deploy several relay nodes or access points along the road to keep the connectivity.
- **Various communications environments.** Vehicular networks are usually operated in two typical communications environments. In highway traffic scenarios, the environment is relatively simple and straightforward (e.g., constrained one-dimensional movement); while in city conditions it becomes much more complex. The streets in a city are often separated by buildings, trees and other obstacles. Therefore, there isn't always a direct line of communications in the direction of intended data communication.

Main Challenges of VCNs

VCNs enable a plethora of important applications and services, ranging from active safety or safety of life applications to traffic information, music/maps download and multi-hop internet connection. In spite of these numerous value-added application scenarios, VCNs come with their own set of requirements, especially in the aspects of quality of service, market introduction and security:

- **Quality of service**

Applications envisioned for vehicular networks require fast association and low communication latency between communicating vehicles in order to guarantee: i) service's reliability for safety-related applications while taking into consideration the time-sensitivity during messages' transfer, and ii) the quality and continuity of service for passenger's oriented applications.

- **Market introduction - Technology penetration**

Services and applications which are based on vehicle-to-vehicle communication and do not involve any infrastructure only provide value to the customer in case a sufficient penetration rate of equipped vehicles has been reached. For this reason, car manufacturers have to think about gradual market introduction strategies. Right from the start, car manufactures need to offer attractive infrastructure-based services (e.g. car-to-home data exchange, car-to-garage communications for remote diagnosis or Location Based Services) which provide clear customer benefit and motivate drivers to invest in additional wireless equipment for their vehicles. Eventually, after a longer period of time – it is expected that this process will take up to 10 years– high enough penetration rates can be reached to allow for purely ad-hoc vehicular communication services such as intersection collision warning, local danger warning, and the decentralized dissemination of real-time traffic flow information.

- **Security**

Security is a crucial aspect in VCNs in order to become a reliable and accepted system bringing safety on public roads (Hubaux et al., 2004). Vehicular communication and its services will only be a success and accepted by the customers if a high level of reliability and security can be provided. This includes authenticity, message integrity and source authentication, privacy, and robustness. A model of possible attacks on inter-vehicle communication systems is presented in (Aijaz et al., 2005); the threat model considers four groups of attacks (attacks on car to infrastructure applications, on cars to home applications, on car to car traffic applications, and on car to routing protocols), and defines a set of guidelines which have been used for the proposal of a novel security framework (called VaneSe) (Gerlach, 2005).

Technologically, many different aspects of vehicular communication still need ideas and results from research. They include high performance and efficient physical layer transmission schemes, fair and scalable medium access (MAC) schemes, efficient data dissemination protocols, routing protocols, to name the most critical ones. Some selected research aspects will be presented in the following section.

ROUTING AND INFORMATION DISSEMINATION: CHALLENGES AND SOLUTIONS

Fast and reliable communication between cars (vehicle-to-vehicle) and/or between a car and a road side unit (vehicle-to-infrastructure) are essential for future vehicle alert systems. From a network perspective, this means that messages have to be routed from the information source to one or several destinations without too much administrative overhead and delay.

The unique properties of vehicular networks that we discussed earlier have an impact on designing new network protocols for vehicular communication. In this section, we present the challenges and the most important proposed solutions on unicast routing and data dissemination.

Unicast Routing

As we have shown in the previous section, VCNs are characterized by: (i) a dynamic, rapidly changing topology (due to high mobility); (ii) frequent network fragmentations into isolated clusters; (iii) constrained, largely one/two dimensional movement due to the static topology of streets within the city or highway environment.

These three characteristics significantly degrade the performance of conventional topology based routing protocols designed for mobile ad hoc networks: Neither proactive nor reactive protocols, currently exploited in a MANET environment, are suitable for the VCN, as reported in (Jaap et all., 2005). This is due to packet control overhead (route discovery, route maintenance, etc.) caused by the frequent update of routing information of the whole network, route failures and transient nature of links.

The frequently changing topology suggests that a local routing scheme without the need to keep track of global routing information scales better in VCNs and consumes lower wireless bandwidth. In addition, the popularity of the Global Positioning System (GPS) also makes position-based routing, which maintains only local information about the node's position, a popular routing strategy.

Recently, some position-based routing protocols specific to VCNs have been proposed. In the following, we present the most important ones: GSR, A-STAR, VADD, GVGrid, GYTAR and CAR:

- **'GSR'** (Lochert et all., 2003) (Geographic Source Routing) has been recently proposed as a promising routing strategy for vehicular ad hoc networks in city environments. It combines position-based routing with topological knowledge by computing the sequence of junctions that must be traversed by each packet to reach its destination; this information is then included in the packet in the form of geographic source routing.

- **'A-STAR'** (Seet et all, 2004) (Anchor-based Street and Traffic Aware Routing) is a position-based routing scheme designed specifically for vehicular ad hoc networks in city environments. It features the novel use of city bus route information to identify anchor paths of higher connectivity so that more packets can be delivered to their destinations successfully.
- **'GYTAR'** (Jerbi et all., 2007) (improved GreedY Traffic-Aware Routing) is also an intersection-based routing scheme. GyTAR consists of two modules: (i) dynamic selection of the junctions through which packets must pass to reach their destinations, and (ii) an improved greedy strategy used to forward packets between junctions.
- **'VADD'** (Zhao et all., 2006) (Vehicle-Assisted Data Delivery) is a recent example of vehicular routing that exploits the availability of map information. This routing protocol, aimed at sparsely connected vehicular networks, uses a store and forward technique and approaches the destination by selecting the direction with the lowest estimated delay to the destination. The forwarding algorithm selects the next hop by choosing either the neighbour that is nearest to the destination (which may lead to routing loops), or a neighbour that is approaching the target location.
- **'GVGrid'** (Sun et all., 2006) is a QoS based VANET routing protocol which exploits geographic information. It divides a geographical area into grids and forwards packets along the roads crossing different grids. Nodes toward the same direction are preferred.
- **'CAR'** (Naumov et all., 2007) Like other position-based vehicular routing protocols, the Connectivity-Aware Routing (CAR) protocol finds a route to a destination, but a unique characteristic of CAR is its ability to maintain a cache of successful routes between various source and destination pairs. This characteristic was prompted by observations of other position-based routing protocols and their inability to utilize information gathered about disconnected paths (due to unoccupied streets, for instance) after those disconnections are detected. CAR also predicts positions of destination vehicles, repairs routes as those positions change, and employs geographic marker messages.

Information Dissemination

When a message is disseminated to locations beyond the transmission range, multi-hopping is used. This is, however, a difficult task due to the highly dynamic nature of inter-vehicle networks which results in their frequent fragmentation into disconnected clusters [as mentioned in section 2]. Apart from interference, packet collisions, and hidden nodes which can stop the message dissemination during multi-hop broadcast. Moreover, multi-hop broadcast can consume significant amount of wireless resources because of unnecessary retransmissions. These facts make designing efficient and reliable multi-hop message dissemination a highly complicated task.

There has been considerable research in designing reliable and bandwidth efficient data dissemination protocols in VCNs. The following mechanisms represent general approaches for information dissemination in VCNs with regard to safety applications:

- **Vehicle-to-vehicle location-based broadcast (LBB)** (Xu et all., 2003) is a routing protocol for highway safety communication. It addresses the problem of message lifetime and repeated message broadcasts. The safety application has to assign a reasonable lifetime to a message. The message is only broadcasted during its lifetime and is dropped afterwards.
- **Cluster-based information propagation scheme** (Little et all., 2005) assumes that vehicles tend to travel in consecutive blocks that are separated by gaps. Thus, a directional propagation protocol (DPP) is suggested, which consists of three components: a custody transfer protocol (CTP), an inter-cluster routing protocol that controls communication within a cluster, and an intra-cluster routing protocol that controls the communication between clusters.
- **Optimized dissemination of alarm messages in vehicular ad hoc networks** (Benslimane, 2004) works by restricting rebroadcasts to certain vehicles. In addition, messages are only re-broadcasted in so called risk zones. The approach also suggests an algorithm for calculation of retransmission periods for the message originator that represents the danger (e.g. due to a vehicle breakdown).

The mechanisms below are specially designed for information dissemination in VCNs without special regard to safety applications:

- **UMB** (Korkmaz et al., 2004) is an IEEE 802.11 based Urban Multi-hop Broadcast protocol (UMB) for inter-vehicle communication. This is composed of directional and intersection broadcast. In the directional broadcast, a vehicle source node selects a furthest node on a road segment to be the forwarding vehicle node. This acknowledges in turn a packet received from the source to confirm its role as a next relay vehicle node. When vehicles arrive at an intersection, then intersection broadcast operates to send packets in each road segment constitutes this intersection.
- **SODAD** (Wischhof et al., 2005) is a method for scalable information dissemination in highly mobile VANETs. The method focuses on comfort applications like weather or traffic information systems. Although a variety of optimizations is possible, the basic idea of such a dissemination scheme is that every node maintains a knowledge base, where it stores known information, e. g., on road conditions or parking lot occupancies. The nodes periodically broadcast all or parts of their knowledge to their neighbors. Upon reception, the nodes integrate new or updated information into their knowledge base. Step by step a local overview of the total scenario emerges.
- **Abiding geocast** (Maihöfer et al., 2005) is a special geocast, where the packets need to delivered to all nodes that are sometime during the geocast lifetime (a certain period of time) inside the geocast destination region. The authors provided three solutions: (i) a server is used to store the geocast messages; (ii) an elected node inside the geocast region stores the messages; (iii) each node stores all geocast packets destined for its location and keeps the neighbor information.

CONCLUSION

Vehicular communication is becoming a reality, driven by navigation safety requirements and by the investments of car manufacturers and Public Transport Authorities. Its opportunities and areas of applications are growing rapidly, and include many kinds of service with different goals and requirements. However, it does pose numerous unique and novel challenges from network evolution to event detection and dissemination, making research in this area very attractive.

Many topics in this field are currently under discussion, such as allocation of a protected frequency band for road safety in Europe, potential usage of the IEEE 802.11p / WAVE standard, integration of multiple wireless technologies, data security, congestion control, data transport, and others. These solutions will be harmonized with other emerging worldwide standards that are related to the vehicular communications, such as IEEE 802.11, IEEE 1609, ISO TC 204 WG 16, selected IETF working groups and SAE.

In addition to technical breakthroughs, the phase of market introduction is critical for the success of this new technology. Car manufactures like BMW, Mercedes, Fiat, Ford, Toyota, Nissan, are currently prototyping vehicles equipped with Wi-Fi (802.11a/b/g) and DSRC (802.11p) and Wi-Fi enabled vehicles are expected to be on the road within the next 3-5 years.

REFERENCES

Aijaz A., Bochow, B. & alt (2005). *Attacks on inter vehicle communication systems and analysis* (Technical Report). NOW-Project.

Benslimane, A. (2004). Optimized Dissemination of Alarm Messages in Vehicular Ad-hoc Networks (VANET). In *Proceedings of the 7th IEEE International Conference*, Toulouse, France, June 2004.

Blum, J., Eskandarian, A., & Hoffmman, L. (2004). Challenges of intervehicle ad hoc networks. *IEEE Transactions on Intelligent Transportation Systems*, *5*(4), 347–351. doi:10.1109/TITS.2004.838218

Gehring, O., & Fritz, H. (1997) Practical Results of a Longitudinal Control Concept for Truck Platooning with Vehicle to Vehicle Communication. In *Proc. of the IEEE Conference on Intelligent Transportation Systems*, Boston.

Gerlach, M. (2005) VaneSe: an approach to VANET security. In *Proc. of V2VCOM2005*, July 2005, San Diego, USA.

Hedrick, J., Tomizuka, M., and Varaiya, P. (1994). *Control issues in automated highway systems.*

Hubaux, J. P. (2004). The security and privacy of smart vehicles. *IEEE Security & Privacy*, *4*(3), 49–55. doi:10.1109/MSP.2004.26

Jaap, S., Bechler, M., & Woolf, L. (2005). Evaluation of Routing Protocols for Vehicular Ad Hoc Networks in City Traffic Scenarios. In *Proc. of EUNICE Open European Summer School on Networked Applications*, Colmenarejo, Spain, July 2005

Jerbi, M., Senouci, S., Meraihi, R., & Ghamri-Doudane, Y. (2007). An improved Vehicular Ad Hoc Routing Protocol for City Environments. *IEEE International Conference on Communications (ICC 07)*, Glasgow, Scotland, June 2007.

Korkmaz, G., Ekici, E., Özgüner, F., & Özgüner, Ü. (2004). Urban multihop broadcast protocol for inter-vehicle communication systems. In *VANET '04: Proceedings of the 1st ACM International Workshop on Vehicular Ad Hoc Networks* (pp. 76-85). Philadelphia, PA, USA: ACM Press.

Little, T. D. C., & Agarwal, A. (2005). *An Information Propagation Scheme for VANETs* (MCL Technical Report 07-01-2005). Department of Electrical and Computer Engineering, Boston University, Boston, USA, July 2005.

Lochert, C., Hartenstein, H., Tian, J., Herrmann, D., Füßler, H., & Mauve, M. (2003). A Routing Strategy for Vehicular Ad Hoc Networks in City Environments. *IEEE Intelligent Vehicles Symposium (IV2003)* (pp. 156-161). Columbus, OH, USA, June 2003.

Maihöfer, C., Leinmüller, T., & Schoch, E. (2005) Abiding geocast: time-stable geocast for ad hoc networks. In *Proceedings of the 2nd ACM international workshop on Vehicular ad hoc networks (VANET '05)* (pp. 20–29).

Nandan A., Das S., Pau G., Mario Gerla, & Sanadidi, M.Y. (2004). *Co-operative Downloading in Vehicular Ad-Hoc Wireless Networks* (UCLA CS Technical Report).

Naumov, V., & Gross, T. (2007) Connectivity-aware routing (car) in vehicular ad-hoc networks. In *INFOCOM 2007. 26th IEEE International Conference on Computer Communications* (pp. 1919-1927). IEEE.

Nekovee, M. (2005, May) Sensor networks on the road: the promises and challenges of vehicular adhoc networks and vehicular grids. In *Proc. of the Workshop on Ubiquitous Computing and e-Research*, Edinburgh, UK.

Seet, B.-C., Liu, G., Lee, B.-S., Foh, C. H., Wong, K. J., & Lee, K.-K. (2004). A-STAR: A Mobile Ad Hoc Routing Strategy for Metropolis Vehicular Communications. *NETWORKING, 2004*, 989–999.

Sun, W., Yamaguchi, H., Yukimasa, K., & Kusumoto, S. (2006). GVGrid: A QoS Routing Protocol for Vehicular Ad Hoc Networks. In *Proc. of IEEE International Workshop on Quality of Service 2006.*

Tsugawa, S. (2005). Issues and recent trends in vehicle safety communication systems. IATSS. Research. *Journal of International Association of Traffic and Safety Sciences*, *29*(1).

Wischhof, L., Ebner, A., & Rohling, H. (2005). Information Dissemination in Self-organizing Intervehicle Networks. *IEEE Transactions on Intelligent Transportation Systems*, *6*(1), 90–101. doi:10.1109/TITS.2004.842407

Wu, H., Fujimoto, R., Guensler, R., & Hunter, M. (2004). Analytical models for information propagation in vehicle-to-vehicle networks. In *Proc. IEEE VTC 2004*, Los Angles, USA, September 2004 (pp. 4548-4552).

Xu, Q., Sengupta, R., & Jiang, D. (2003). Design and Analysis of Highway Safety Communication Protocol in 5.9 GHz Dedicated Short Range Communication Spectrum. In *Proceedings of the IEEE Vehicular Technology Conference (VTC 2003-Spring)*, (Vol. 4, pp. 2451– 2455).

Zhao, J., & Cao, G. (2006). VADD: Vehicle-assisted data delivery in vehicular ad hoc networks. In *Proceedings of the 25th Conference on Computer Communications (INFOCOM06).*

KEY TERMS AND DEFINITIONS

Dedicated Short Range Communication (DSRC): is a short to medium range wireless protocol specifically designed for automotive use.

IEEE 802.11: A group of standards that specify the technologies for wireless LANs

Intelligent Transportation Systems (ITS): Program: A worldwide initiative to add information and communications technology to transport infrastructure and vehicles.

Mobile Ad-hoc Network (MANET): is a type of ad hoc network that can change locations and configure itself on the fly.

Vehicle-to-infrastructure communication (V2I): wireless communication between a vehicle and an infrastructure (also termed as vehicle-to-roadside).

Vehicle-to-vehicle communication (V2V): wireless communication between two vehicles equipped with short and medium range wireless communication capabilities (it is also termed as car-to-car communication).

Vehicular Ad-hoc Network (VANET): is a form of Mobile ad-hoc network (MANET), to provide communications among nearby vehicles.

Vehicular Communication Network (VCN): is a form of network created by vehicles equipped with short and medium range wireless communication technology. It includes vehicular ad-hoc networks (VANETs), vehicle-to-vehicle, vehicle-to-infrastructure and vehicle-to-pedestrian communications.

Wireless Local Area Network (WLAN): is one in which a mobile user can connect to a local area network (LAN) through a wireless (radio) connection.

Compilation of References

Abdul, H. S. (2002). *Compressed Video Communications*. John Wiley & Sons, Ltd.

Abolhasan, M., Wysocki, T., & Dutkiewicz, E. (2004). A review of routing protocols for mobile ad hoc networks. *Journal of Ad Hoc Networks, Elsevier*, *2*(1), 1–22. doi:10.1016/S1570-8705(03)00043-X

Abowd, G., Atkeson, C., Hong, J., Long, S., Kooper, R., & Pinkerton, M. (1997). Cyberguide: A mobile context-aware tour guide. *ACM Wireless Networks*, *3*(5), 421–433. doi:10.1023/A:1019194325861

Adkins, D., Lakshminarayanan, K., Perrig, A., & Stoica, I. (2003). *Towards a more functional and secure network infrastructure* (Technical Report UCB/CSD-03-1242). Computer Science Division (EECS), University of California, Berkeley, 2003

Adya, A., Bahl, P., Padhye, P., Wolman, A., & Zhou, L. (2004). A multi-radio unification protocol for IEEE 802.11 wireless networks. *IEEE International Conference on Broadband Communications, Networks and Systems, Broadnets '2004*.

Ahmed, T. (2005). Adaptive *p*acket video streaming over IP networks: A cross-layer approach. *IEEE Journal on Selected Areas in Communications*, *23*(2). doi:10.1109/JSAC.2004.839425

Aittola, M., Ryhänen, T., & Ojala, T. (2003, September 8). *SmartLibrary - Location-aware mobile library service*. Paper presented at the 2003 Fifth International Symposium on Human Computer Interaction with Mobile Devices and Services, Udine, Italy.

Akman, V., & Surav, M. (1997). The use of situation theory in context modeling. *Computational Intelligence*, *13*(3), 427–438. doi:10.1111/0824-7935.00046

Akyildiz, I. F., Su, W., Sankarasubramaniam, Y., & Cayirci, E. (2002). Wireless Sensor Networks: A Survey. *Computer Networks*, *38*(2), 393–422. doi:10.1016/S1389-1286(01)00302-4

Akyildiz, I. F., McNair, J., Ho, J. S. M., Uzunalioglu, H., & Wang, W. (1999). Mobility Management in Next-Generation Wireless Systems. *Proceedings of the IEEE*, *87*(8), 1347–1384. doi:10.1109/5.775420

Akyildiz, I. F., Mohanty, S., & Xie, J. (2005). A Ubiquitous Mobile Communication Architecture for Next-Generation Heterogeneous Wireless Systems. *IEEE Communications Magazine*, *43*(6), 29–36. doi:10.1109/MCOM.2005.1452832

Akyildiz, I. F., Xie, J., & Mohanty, S. (2004). A Survey of Mobility Management in Next-Generation All-IP-Based Wireless Systems. *IEEE Wireless Communications*, *11*(4), 16–28. doi:10.1109/MWC.2004.1325888

Albers, M. C. (1994). The Varese system, hybrid auditory interfaces, and satellite-ground control: Auditory icons and sonification in a complex, dynamic system. In *Proceedings of ICAD '94 - The Second International Conference on Auditory Display, Santa Fe, NM. Santa Fe Institute.*

Ali, A., & Fisal, N. (2008). A real-time routing protocol with load distribution in wireless sensor networks. *Computer Communications*, *31*(14), 3190–3203. doi:10.1016/j.comcom.2008.04.030

Alicherry, M., Bhatia, R., & Li, L. (2005). Joint channel assignment and routing for throughput optimization in multi-radio wireless mesh networks. *ACM International Conference on Mobile Computing and Networking, Mobicom '2005.*

Arkko, J., Devarapalli, V., & Dupont, F. (2004). Using IPsec to Protect Mobile IPv6 Signaling Between Mobile Nodes and Home Agents. IETF RFC 3776.

Arkko, J., Kuijpers, G., Soliman, H., Loughney, J., & Wiljakka, J. (2003). Internet Protocol Version 6 (IPv6) for Some Second and Third Generation Cellular Hosts. IETF RFC 3316.

Arnold, K., Osullivan, B., Scheifler, R., Waldo, J., & Wollrath, A. (1999). *The Jini Specification (The Jini Technology)*. Reading, MA: Addison-Wesley.

Arpinar, I. B., Aleman-Meza, B., Zhang, R., & Maduko, A. (2004). Ontology-driven web services composition. In *Proceedings of IEEE International Conference on E-Commerce Technology (CEC 2004).*

Artail, H., Safa, H., Hamze, H., & Mershad, K. (2007). A cluster based service discovery model for mobile ad hoc networks. *3rd IEEE International Conference on Wireless and Mobile Computing, Networking and Communications* (WiMob 2007). New York, USA.

Ary, B., & Imre, S. (2007). *Advice of Charge in Telecommunication Services. Mobile and Wireless Communications Summit.* Budapest, Hungary: Information Society Technologies.

Ary, B., Debrei, G., & Imre, S. (2005). Real-time charging in third-generation mobile networks. *The 8th International Conference on Telecommunications (ConTEL 2005), 239-245.* Zagreb, Croatia: IEEE.

Aslam, F., Raza, S., Dogar, E. R., Ahmad, I. U., & Uzmi, Z. A. (2005). NPP: A Facility Based Computation Framework for Restoration Routing Using Aggregate Link Usage Information. In *Lecture Notes in Computer Science, Quality of Service in Multiservice IP Networks* (Vol. 3375, pp. 150-163). Springer Berlin.

A-Theotokis, S., & Spinellis, D. (2004). A Survey of Peer-to-Peer Content Distribution Technologies. *ACM Computing Surveys, 36*(4), 335–371. doi:10.1145/1041680.1041681

Bahl, P., Chandra, R., & Dunagan, J. (2004). SSCH: Slotted Seeded Channel Hopping for capacity improvement in IEEE 802.11 ad-hoc wireless networks. *ACM International Conference on Mobile Computing and Networking, Mobicom '2004* (pp. 216–230). Philadelphia, PA, USA.

Bai, Y., & ITO, R.M. (2004). QoS Control for Video and Audio Communication in Conventional and Active Networks: Approaches and Comparison. *IEEE Communications Surveys, 6*(1), 42-49.

Balachandran, A., Voelker, G. M., & Bahl, P. (2005). Wireless hotspots: current challenges and future directions. *Mobile Networks and Applications, 10*(3).

Balaouras, P., & Stavrakakis, I. (2005). Multimedia Transport Protocols for Wireless Networks. In Salkintzis, A. S., & Passas, N. (Eds.), *Emerging Wireless Multimedia Services and Technologies*. John Wiley & Sons, Ltd.

Bamford, W., Coulton, P., & Edwards, R. (2006, 24-28 April). Location-based Mobile Blogging. In *Proceedings of the 2nd IEEE International Conference on Information & Communication Technologies: from Theory to Applications (ICTTA '06)* (vol. 1, pp. 39-40), Damascus, Syria.

Banerjee, N., Acharya, A., & Das, S. K. (2004). SIP-Based Mobility Architecture for Next Generation Wireless Networks. IBM Research Report, September 10, 2004.

Banerjee, S., & Khuller, S. (2001). A Clustering Scheme for Hierarchical Control in Mult-hop Wireless Networks. In *Proceeding of 20th IEEE INFOCOM,* (pp.1028-1037). Anchorage, Alaska.

Bardram, J. (2004). Applications of context-aware computing in hospital work: examples and design principles. *2004 ACM symposium on Applied computing.* Nicosia, Cyprus. (pp. 1574-1579).

Barra, M., & Cillo, T. De Santis, A., Petrillo, U.F., Negro, A., & Scarano, V. (2001). Personal WebMelody: Customized sonification of web servers. In *Proceedings of the International Conference on Auditory Display (ICAD)*, Espoo, Finland.

Barriere, L., Fraigniaud, P., Narayanan, L., & Opatrny, J. (2003). Robust positionbased routing in wireless ad hoc networks with irregular transmission ranges. *Wireless Communications and Mobile Computing*, *3*(2), 141–153. doi:10.1002/wcm.108

Basagni, S., Mastrogiovanni, M., Panconesi, A., & Petrioli, C. (2006). Localized Protocols for Ad Hoc Clustering and Backbone Formation: A Performance Comparison. *IEEE Transactions on Parallel and Distributed Systems*, *17*(4), 292–306. doi:10.1109/TPDS.2006.52

Basagni, S. (1999). Distributed Clustering for Ad Hoc Networks. In *Proceedings of the International Symposium on Parallel Architectures, Algorithms and Networks (ISPAN '99)*, (pp. 310-315). Perth/Fremantle, WA, Australia.

Bassi, A., Denazis, S., Galis, A., Fahy, C., Serrano, M., & Serrat, J. (2007). *Autonomic Internet: A Perspective for Future Internet Services Based on Autonomic Principles*. Paper presented at IEEE 3rd International Week on Management of Networks and Services End-to-End Virtualization of Networks and Services (Manweek 2007) / MACE 2007 2nd IEEE International Workshop on Modelling Autonomic Communications Environments, 29 October – 2 November 2007, San José, USA. Retrieved from http://magellan.tssg.org/2007/mace/mace.php

Basu, P., Khan, N., & Little, T. D. C. (2001). A Mobility Based Metric for Clustering in Mobile Ad Hoc Networks. In *Proceeding of IEEE ICDC*, (pp. 413-418). Phoenix, Arizona, USA.

Baus, J., Cheverst, K., & Kray, C. (2005). A survey of map-based mobile guides. In Meng, L., Reichenbacher, T., & Zipf, A. (Eds.), *Map-based mobile services: Theories, Methods, and Implementations* (pp. 197–216). Springer Berlin Heidelberg. doi:10.1007/3-540-26982-7_13

Becker, C., & Durr, F. (2005). On location models for ubiquitous computing. *Personal and Ubiquitous Computing*, *9*(1), 20–31. doi:10.1007/s00779-004-0270-2

Becker, C., Handte, M., Schiele, G., & Rothermel, K. (2004). PCOM – A Component System for Pervasive Computing. *2nd IEEE Annual Conference on Pervasive Computing and Communications (PERCOM'04)*, Washington, DC, USA.

Belenger, J., & Calafate, C. (2007). A low-cost embedded IDS to monitor and prevent Man-in-the-Middle attack on wired LAN environments.In *Proceedings of t international conference on emerging security information and technologies.*

Bell, T. C., Cleary, J. G., & Witten, I. H. (1990). *Text Compression*. Prentice Hall.

Bettstetter, C., & Konig, S. (2002). On the Message and Time Complexity of a Distributed Mobility-Adaptive Clustering Algorithm in Wireless Ad Hoc Networks. *In Proceeding of the 4th European Wireless* (pp.128-134). Florence, Italy.

Bhattacharya, A., & Das, S. K. (2002). LeZi-Update: An Information-Theoretic Framework for Personal Mobility Tracking in PCS Networks. *Wireless Networks*, *8*, 121–135. doi:10.1023/A:1013759724438

BitTorrent. (2001). BitTorrent webpage. Retrieved from http://bitconjurer.org/BitTorrent

Blattner, M. M., Sumikawa, D. A., & Greenberg, R. M. (1989). Earcons and icons: their structure and common design principles. *Human-Computer Interaction*, *4*(1), 11–44. doi:10.1207/s15327051hci0401_1

Bluetooth, S. I. G. (2004, November). *Specification of the Bluetooth System.* Version 2, November 2004, http://www.Bluetooth.com/NR/rdonlyres/1F6469BA-6AE7-42B6-B5A1-65148B9DB238/840/Core_v210_EDR.zip

Bokor, L. Nováczki, Sz., & Imre, S. (2007). A Complete HIP-based Framework for Secure Micromobility. 5th @ WAS International Conference on Advances in Mobile Computing and Multimedia (MoMM2007) (pp. 111-122). Jakarta, Indonesia, 3-5 December 2007

Borg, J. (2003). *A comparative study of ad hoc & peer to peer networks.* Unpublished Master's thesis, University College London.

Bose, P., Morin, P., Stojmenovic, I., & Urrutia, J. (2001). Routing with guaranteed delivery in ad hoc wireless networks. *Wireless Networks, 7*, 609–616. doi:10.1023/A:1012319418150

Braun, T., Voigt, T., & Dunkels, A. (2007). TCP Support for Sensor Networks. *4th IEEE/IFIP Conference on Wireless On demand Network Systems and Services.* Obergurgl, Austria.

Brewster, S. A., Wright, P. C., & Edwards, A. D. N. (1995). Experimentally derived guidelines for the creation of earcons. *In Adjunct Proceedings of HCI'95*, Huddersfield, UK.

Brewster, S.A. (1997). Using non-speech sound to overcome information overload. *Displays, Special issue on multimedia displays*, 17.

Broch, J., Johnson, D. B., & Maltz, D. A. (1998). *The Dynamic Source Routing Protocol for Mobile Ad Hoc Networks. Ad Hoc Networking* (pp. 139–172). Addison-Wesley.

Broch, J., Maltz, D. A., Johnson, D. B., Hu, Y. C., & Jetcheva, J. (1998). *A performance comparison of multi-hop wireless ad hoc network routing protocols.* Paper presented at the 4th annual ACM/IEEE international conference on Mobile computing and networking Dallas, Texas, United States.

Brown, P. J., Bovey, J., & Chen, X. (1997). Context-aware Applications from the Laboratory to the Marketplace. *IEEE Personal Communications, 4*(5), 58–64. doi:10.1109/98.626984

Brown, L. M., Brewster, S. A., & Ramlol, R. (2003). Design guidelines for audio presentation of graphs and tables. In *Proceedings of the 2003 International Conference on Auditory Display, Boston, MA, USA, July 6-9.*

Brown, P.J. (1996). The stick-e document: a framework for creating context-aware applications, *Electronic publishing, 8*(2-3), 259-272.

Bruning, S., Zapotoczky, J., Ibach, P., & Stantchev, V. (2007). Cooperative Positioning with MagicMap. *Workshop on Positioning, Navigation and Communication 2007 (WPNC'07)*, Hannover, (pp.17-22). Leibniz University of Hannover, Germany.

Buchegger, S., & LeBoudec, J. Y. (2002a). Performance analysis of the CONFIDANT protocol. In *Proceedings of the 3rd ACMS International Symposium on Mobile Ad Hoc Networking Computing and Computing (MobiHoc), Lausanne, Switzerland, June 09–11* (pp. 226–236).

Buchegger, S., & LeBoudec, J. Y. (2002b). Nodes Bearing Grudges: Towards Routing Security, Fairness, and Robustness in Mobile Ad Hoc Networks. In *Proceedings of the Tenth Euromicro Workshop on Parallel, Distributed and Network-based Processing* (pp. 403-410). IEEE Computer Society.

Buettner, M., Yee, G., Anderson, E., & Han, R. (2006). X-MAC: a short preamble MAC protocol for duty-cycled wireless sensor networks. *4th international Conference on Embedded Networked Sensor Systems (SenSys)* (pp. 307-320). Boulder: ACM.

Butler, J., Undercoffer, J., & Pinkston, J. (2003). Hidden processes: the implication for intrusion detection. In *Proceedings of Workshop on Information Assurance, Systems, Man and Cybernetics Society.*

Buttyàn, L., & Hubaux, J. P. (2003). Stimulating cooperation in self-organizing mobile ad hoc networks. *Mobile Networks and Applications, 8*(5), 579–592. doi:10.1023/A:1025146013151

Buttyàn, L., & Hubaux, J. P. (2001). *Nuglets: a virtual currency to stimulate cooperation in self-organized ad hoc networks* (Technical report DSC/2001/001). Swiss Federal Institute of Technology, Lausanne.

Caccamo, M., Zhang, L. Y., Sha, L., & Buttazzo, G. (2002). An Implicit Prioritized Access Protocol for Wireless Sensor Networks. *In Proceedings of the 23rd IEEE Real-Time Systems Symposium (RTSS'2002)* (pp. 39-48). Austin.

Caesar, M., Condie, T., & Kannan, J. (2006). ROFL: Routing on Flat Labels. In *Proceedings of Special Interest Group on Data Communication (SIGCOMM 2006), Pisa, Italy, September 11-15, 2006.*

Camp, T., Boleng, J., & Davies, V. (2002). A Survey of Mobility Models for Ad Hoc Network Research. [WCMC]. *Wireless Communication & Mobile Computing, 2*(5), 483–502. doi:10.1002/wcm.72

Campbell, A. T., Gomez, J., Kim, S., Turányi, Z. R., Valkó, A. G., & Wan, C.-Y. (2002). Internet micromobility. *Journal of High Speed Networks, 11*, 177–198.

Capra, L., Emmerich, W., & Mascolo, C. (2001). Reflective middleware solutions for context-aware applications, *3rd International Conference on Metalevel Architectures and Separation of Crosscutting Concerns* (pp. 126-133).

Carman, M. J., & Knoblock, C. A. (2005). Inducing source descriptions for automated web service composition. In *Proceedings of the AAAI 2005 Workshop on Exploring Planning and Scheduling for Web Services, Grid, and Autonomic Computing.*

Carneiro, G., Ruela, J., & Ricardo, M. (2004). Cross-layer design in 4G wireless terminals. *IEEE Wireless Communications, 11*(2), 7–13. doi:10.1109/MWC.2004.1295732

Casati, F., Ilnicki, S., Jin, L., Krishnamoorthy, V., & Shan, M. (2000). Adaptive and dynamic service composition in eFlow. In *Proceedings of the Conference on Advanced Information Systems Engineering.*

Ceri, S., Gottlob, G., & Tanca, L. (1989). What You Always Wanted to Know About Datalog (And Never Dared to Ask). *IEEE Transactions on Knowledge and Data Engineering, 1*(3), 146–166. doi:10.1109/69.43410

Chakraborty, D., Joshi, A., Yesha, Y., & Fin, T. (2006). Toward Distributed Service Discovery in Pervasive Computing Environments. *IEEE Transactions on Mobile Computing*, 5.

Chan, V., Ray, P. & Parameswaran, N. (2008). Mobile e-Health Monitoring: an agent-based approach. *IET Communications, 2*(2), 223-230. Retrieved June 20, 2008 from IEEE Xplore.

Channa, N., Li, S., Shaikh, A. W., & Fu, X. (2005). Constraint satisfaction in dynamic web service composition. In *Proceedings of the Sixteenth International Workshop on Database and Expert Systems Applications.*

Chappell, D. (2006). Introducing the.Net Framework 3.0. Whitepaper "Introducing the.Net Framework 3.0".

Chatterjee, M., Das, S. K., & Turgut, D. (2002). WCA: A Weighted Clustering Algorithm for Mobile Ad Hoc Networks. *Cluster Computing, 5*(2), 193–204. doi:10.1023/A:1013941929408

Chavez, E., Dobrev, S., Kranakis, E., Opatrny, J., Stacho, L., & Urrutia, J. (2006b). Route discovery with constant memory in oriented planar geometric networks. *Networks, 48*, 7–15. doi:10.1002/net.20114

Chavez, E., Dobrev, S., Kranakis, E., Opatrny, J., Stacho, L., & Urrutia, J. (2006a). Local construction of planar spanners. In *LATIN* (Valdivia, Chile, March 20-24, 2006) (LNCS 3887 pp. 286–297).

Chavez, E., Dobrev, S., Kranakis, E., Opatrny, J., Tejeda, H., & Urrutia, J. (2006c). Half-space proximal: A new local test for extracting a bounded dilation spanner. In *OPODIS 2005* (LNCS 3544).

Chellapa-Doss, R., Jennings, A., & Shenoy, N. (2003a). User Mobility Prediction in Hybrid and Ad Hoc Wireless Networks. In *Proc. of the Australian Telecommunications Networks and Applications Conference (ATNAC'03).*

Chellapa-Doss, R., Jennings, A., & Shenoy, N. (2003b). A Comparative Study of Mobility Prediction in Cellular and Ad Hoc Wireless Networks. In *Proc. of the IEEE Int'l Conf. on (ICC 2003).*

Chellapa-Doss, R., Jennings, A., & Shenoy, N. (2004). A Review of Current Mobility Prediction Techniques for Ad Hoc Networks. In *Proc. of the 4th IASTED International Multi-Conference on Wireless and Optical Communications, Canada* (pp. 536-542).

Chen, J.-C., & Zhang, T. (2004). *IP-Based Next-Generation Wireless Networks: Systems, Architectures, and Protocols.* New York: Wiley-Interscience. doi:10.1002/0471478253

Chen, G., & Kotz, D. (2000). *A Survey of context-aware, mobile computing research* (Dartmouth Computer Science Technical Report TR2000-381). Hanover, NH.

Chen, H. (2004). An Intelligent Broker Architecture for Pervasive Context-Aware Systems, PhD Thesis, University of Maryland, Baltimore County.

Chen, H., Finin, T., & Joshi, A. (2003). *Semantic Web in a Pervasive Context-Aware Architecture.* Paper presented at Workshop on Artificial Intelligence in Mobile Systems, the 5th Annual Conf. on Ubiquitous Computing, Seattle, Washington.

Chen, J., & Zheng, T. (2004). Joint cross-layer design for wireless QoS content delivery. *IEEE International Conference on Communications* (Vol. 7, pp. 4243- 4247).

Chen, J., Zhu, P., & Qi, Z. (2007). PR-MAC: Path-Oriented Real-Time MAC Protocol for Wireless Sensor Network. *In Proceedings of the 2007 International Conference on Embedded Software and System (ICESS'2007)* (pp.530-539). Daegu.

Cheng, J., Wang, C.-Y., Hoh, C. C., & Hwan, R.-H. (2005). *A Scalable Location-Aware Peer-to-Peer System for MANET.* Paper presented at the Mobile Technology, Applications and Systems, 2005 2nd International Conference.

Chereddi, C., Kyasanur, P., So, J., & Vaidya, N. H. (2005). Demo: Multi-Channel Multi-Interface Wireless Mesh Testbed. *ACM International Conference on Mobile Computing and Networking, Mobicom '2005*, Cologne, Germany.

Chetan, S., Al-Muhtadi, J., Campbell, R., & Mickunas, M. D. (2005). Mobile Gaia: A Middleware for Ad-hoc Pervasive Computing. *IEEE Consumer Communications & Networking Conference (CCNC 2005)*, Las Vegas, USA.

Cheverst, K., Davies, N., Mitchell, K., Friday, A., & Efstratiou, C. (2000). Developing a context-aware electronic tourist guide: some issues and experiences. *SIGCHI Conference on Human Factors in Computing Systems* (pp. 17–24), New York.

Chiasserini, C. F., & Rao, R. R. (1999). *Pulsed battery discharge in communication devices.* Paper presented at the 5th annual ACM/IEEE international conference on Mobile computing and networking.

Chinnici, R., Moreau, J.-J., Ryman, A., & Weerawarana, S. (2007). *Web Services Description Language (WSDL) Version 2.0 Part 1: Core Language.* W3C Recommendation 26 June 2007. Retrieved February 13, 2008 from http://www.w3.org/TR/wsdl20/

Chipara, O., He, Z., Xing, G., Chen, Q., Wang, X., Lu, C., et al. (2006). Real-time Power-Aware Routing in Sensor Networks. *In Proceedings of the 14th International Workshop on Quality of Service (IWQoS'2006),* (pp. 83-92). New Haven.

Chlamtac, I., Conti, M., & Liu, J. J. N. (2003). Mobile Ad Hoc Networking: Imperatives and challenges. *Ad Hoc Networks Publication, 1*(1), 13–64. doi:10.1016/S1570-8705(03)00013-1

Chlamtac, I., & Redi, J. (1998). Mobile Computing: Challenges and Opportunities. In Hemmendinger, A. R. D., & Reilly, E. (Eds.), *Encyclopedia of Computer Science* (4th ed.). International Thomson Publishing.

Choi, W., & Woo, M. (2006). A Distributed Weighted Clustering Algorithm for Mobile Ad Hoc Networks. In *Proceedings of the Advanced International Conference on Telecommunications and International Conference on Internet and Web Applications and Services,* (pp. 73-78). Guadeloupe, French Caribbean.

Choudhury, S., & Gibson, J. (2006). Joint PHY/MAC Based Link Adaptation for Wireless LANs with Multipath Fading. *IEEE Wireless Communications and Networking Conference, 2006* (Vol. 2, pp. 757-762).

Clarke, I., Sandberg, O., Wiley, B., & Hong, T. W. (2000). *Freenet: A Distributed Anonymous Information Storage and Retrieval System.* Paper presented at the ICSI Workshop on Design Issues in Anonymity and Unobservability.

Clausen, T., & Jacquet, P. (2003). Optimized Link-State Routing Protocol. *IETF, RFC-3626*, 2003.

Conti, M., Gregori, E., & Turi, G. (2004). *Towards Scalable P2P Computing for Mobile Ad Hoc Networks.* Paper presented at the Second IEEE Annual Conference on Pervasive Computing and Communications Workshops (PERCOMW '04), Orlando, FL, USA.

Conti, M., Gregori, E., & Turi, G. (2005). *A Cross-Layer Optimization of Gnutella for Mobile Ad hoc Networks.* Paper presented at the 6th ACM international symposium on Mobile ad hoc networking and computing, Urbana-Champaign, IL, USA

Corson, M. S., Macker, J. P., & Cirnicione, G. H. (1999). Internet-Based Mobile Ad Hoc Networking. *IEEE Internet Computing, 3*, 63–70. doi:10.1109/4236.780962

Cramer, C., & Fuhrmann, T. (2005c). *Self-Stabilizing Ring Networks on Connected Graphs.* University of Karlsruhe, Germanyo.

Cramer, C., & Fuhrmann, T. (2005a). *ISPRP: A Message-Efficient Protocol for Initializing P2P Networks.* Paper presented at the 24th IEEE International, Performance, Computing, and Communications Conference (IPCCC), Phoenix, AZ, USA.

Cramer, C., & Fuhrmann, T. (2005b). *Proximity Neighbor Selection for a DHT in Wireless Multi-Hop Networks.* Paper presented at the Fifth IEEE International Conference on Peer-to-Peer Computing (P2P'05)

Cramer, C., & Fuhrmann, T. (2006). *Performance evaluation of chord in mobile ad hoc networks.* Paper presented at the 1st international workshop on Decentralized resource sharing in mobile computing and networking, Los Angeles, California.

Crowley, J. L., Hall, D., & Emonet, R. (2007). Autonomic Computer Vision Systems. *International Conference on Computer Vision Systems (ICVS'07)*, Bielefeld, Germany.

Czyzowicz, J., Dobrev, S., Fevens, T., Gonzalez-Aguilar, H., Kranakis, E., Opatrny, J., & Urrutia, J. (2008a). Local algorithms for dominating and connected dominating sets of unit disk graphs. In *LATIN 2008*, April 07-11, Buzios, Brazil.

Czyzowicz, J., Dobrev, S., Gonzalez-Aguilar, H., Kranakis, E., Opatrny, J., & Urrutia, J. (2008b). Local 7-coloring for planar subgraphs of unit disk graph. In *TAMC 2008*, April 25-29, Xi'an, China (LNCS 4978, pp. 170-181).

Czyzowicz, J., Dobrev, S., Kranakis, E., Opatrny, J., & Urrutia, J. (2007). Local edge colouring of Yao-like subgraphs of unit disk graph. In *14th SIROCCO* (June 6-8, 2007, Livorno, Italy) (LNCS 4474).

Da Fontoura Costa, L., Oliveira Jr, O. N., Travieso, G., & Rodrigues, F. A. A., Villas Boas, P.R., Antiqueira, L., Viana, M.P. & da Rocha, C. (2007). *Analyzing and Modeling Real-World Phenomena with Complex Networks: A Survey of Applications.* arXiv:0711.3199v1

Dam, T. V., & Langendoen, K. (2003). An Adaptive Energy-Efficient MAC Protocol for Wireless Sensor Networks. *In Proceedings of the first ACM Conference on Embedded Networked Sensor Systems (SenSys'2003)* (pp. 171-180). Los Angeles.

Das, S. R., Perkins, C. E., & Royer, E. M. (2000). *Performance comparison of two on-demand routing protocols for ad hoc networks.* Paper presented at IEEE INFOCOM, Nineteenth Annual Joint Conference of the IEEE Computer and Communications Societies.

Datta, A. (2003). *MobiGrid: Peer-to-Peer Overlay and Mobile Ad-Hoc Network Rendezvous - a Data Management Perspective.* Paper presented at CAiSE 2003 Doctoral Symposium, Klagenfurt, Austria.

David, S., Man, L., & Renaud, C. (Eds.). (2006). *QoS and QoE Management in UMTS Cellular Systems.* Wiley.

De Couto, D. S. J., Aguayo, D., Bicket, J., & Morris, R. (2003, September), A High Throughput Path Metric for Multihop Wireless Routing. In *Proceedings of the Annual International Conference on Mobile Computing and Networking (ACM MOBICOM)*, San Diego, California, USA.

Deering, S., & Hinden, R. (1998). Internet Protocol, Version 6 (IPv6) Specification. IETF RFC 2460.

Deguang, L., Xiaoming, F., & Hogrefe, D. (2006). A review of mobility support paradigms for the internet. *IEEE Communications Surveys & Tutorials*, *8*(1), 38–51. doi:10.1109/COMST.2006.323441

Dekar, L., & Kheddouci, H., (2008). A Cluster Based Mobility Prediction Scheme for Ad Hoc Networks. *Ad Hoc Networks Journal, 6* (2).

Delmastro, F. (2005). *From Pastry to CrossROAD: CROSS-layer Ring Overlay for AD hoc networks.* Paper presented at the Workshop on Mobile Peer-to-Peer Computing - MP2P '05, Kauai, Hawaii.

Demirkol, A. (2006, April). MAC Protocols for Wireless Sensor Networks. *IEEE Communications Magazine*, 115. doi:10.1109/MCOM.2006.1632658

Dempster, A. P. (1968). A Generalization of Bayseian Inference. *Journal of the Royal Statistical Society. Series A (General)*, *30*, 205–247.

Devarapalli, V., Wakikawa, R., Petrescu, A., & Thubert, P. (2005, January). *Network Mobility (NEMO) Basic Support Protocol.* RFC 3963.

Dey, A., Salber, D., & Abowd, G. (2001). A conceptual framework and a toolkit supporting the rapid prototyping of context-aware applications. *Human-Computer Interaction*, *16*(2-4), 97–166. doi:10.1207/S15327051HCI16234_02

Dey, A., & Abowd, G. (2000). Towards a better understanding of context and context-awareness, *Workshop on the What, Who, Where, When and How of Context-Awareness.* New York: ACM Press.

Dhanakoti, N., Gopalan, S., Sridhar, V., & Subramani, S. (2005). A Distributed Service Discovery and Selection Framework in Pervasive Service Environments. In *Proc. Advanced Industrial Conf. on Telecommunications/Service Assurance with partial and Intermittent Resources Telecommunications Workshop.*

Diaz, A., Merino, P., & Panizo, L. (2007). Experimental Analysis of Peer-to-Peer Streaming in Cellular Networks. *IEEE International Conference on Advanced Information Netowrking and Applications* (pp. 784-791).

Diffie, W., & Helman, M. (1976, November). New directions in cryptography. *IEEE Transactions on Information Society*, *22*(6), 644–654. doi:10.1109/TIT.1976.1055638

Dimic, G., Sidiropoulos, N. D., & Zhang, R. (2004). Medium Access Control — Physical Cross-Layer Design. *IEEE Sig. Proc.*, *21*(5), 40–50. doi:10.1109/MSP.2004.1328087

Dimopoulou, L., Leoleis, G., & Venieris, I. S. (2005). Fast Handover Support in a WLAN Environment: Challenges and Perspectives. *IEEE Network*, *19*(3), 14–20. doi:10.1109/MNET.2005.1453394

Ding, G., & Bhargava, B. (2004). *Peer-to-peer File-sharing over Mobile Ad hoc Networks.* Paper presented at the Second IEEE Annual Conference on Pervasive Computing and Communications Workshops (PERCOMW '04), Orlando, FL, USA.

Docter, J., Licciardi, C., & Marchetti, M. (2007). The Telecom Industry and Context Awareness. *International Conference on the Management of Mobile Business* (pp. 35-42), Toronto, Ontario.

Dong, J., Sun, Y., & Yang, S. (2006). Owl-s ontology framework extension for dynamic web service composition. In *Proceedings of the International Conference on Software Engineering and Knowledge Engineering (SEKE 2006).*

Dourish, P. (2004). What we talk about when we talk about context. *Personal and Ubiquitous Computing*, *8*(1), 19–30. doi:10.1007/s00779-003-0253-8

Doval, D., & O'Mahony, D. (2002). Nom: Resource location and discovery for ad hoc mobile networks. In *Proceedings of the 1st Annual Mediterranean Ad Hoc Networking Workshop.*

Draves, R., Padhye, J., & Zill, B. (2004). Routing in Multi-Radio, Multi-Hop Wireless Mesh Networks. *ACM International Conference on Mobile Computing and Networking, Mobicom '2004.*

Droms, R., Bound, J., Volz, B., Lemon, T., Perkins, C., & Carney, M. (2003). Dynamic Host Configuration Protocol for IPv6 (DHCPv6). IETF RFC 3315.

Du, S., & Khan, A. PalChaudhuri, S., Post, A., Saha, A. K., Druschel, P., et al. (2004). *Self-Organizing Hierarchical Routing for Scalable Ad Hoc Networking*. Department of Computer Science, Rice University, Houston, TX, USA.

Dubnicki, C., Ungureanu, C., & Kilian, W. *FPN: A Distributed Hash Table for Commercial Applications* Paper presented at the 13th IEEE International Symposium on High Performance Distributed Computing (HPDC-13 '04)

Dugas, C. (2005). Configuring and managing a large-scale monitoring network: solving real world challenges for ultra-low-powered and long-range wireless mesh networks. *International Journal of Network Management*, *15*(4), 269–282. doi:10.1002/nem.573

Durocher, S., Kirckpatrick, D., & Narayanan, L. (2008). *On routing with guaranteed delivery in three dimensional ad hoc wireless networks* (pp. 546–557). ICDCN.

Eagle, N., & Pentland, A. (2006). Reality mining: sensing complex social systems. *Personal and Ubiquitous Computing*, *10*(4). doi:10.1007/s00779-005-0046-3

Earle, A. E. (2006). *Wireless security handbook*. Boca Raton, FL: Auerbach Publications.

Eberspacher, J., Schollmeier, R., Zols, S., & Kunzmann, G. (2004). *Structured P2P Networks in Mobile and Fixed Environments*. Paper presented at the International Working Conference on Performance Modeling and Evaluation of Heterogeneous Networks (HET-NETs '04), Ilkley, West Yorkshire.

EBU. (2003). BPN 056:SAMVIQ - Subjective Assessment Methodology for Video Quality.

ElBatt, T. A., Krishnamurthy, S. V., Connors, D., & Dao, S. (2000). Power management for throughput enhancement in wireless ad-hoc networks. *IEEE International Conference on Communications* (pp. 1506–1513).

El-Moussa, F., Linge, N., & Hope, N. (2007). *Active router approach to defeating denial-of-service attacks in networks*. IET Communications white paper.

Enz. (August 2004). WiseNET: An Ultralow-Power Wireless Sensor Network Solution. *IEEE Computer*, 62-70.

Er, I., & Seah, W. (2006). Performance analysis of mobility-based d-hop (MobDHop) clustering algorithm for mobile ad hoc networks. *Computer Networks*, *50*, 3375–3399. doi:10.1016/j.comnet.2005.12.013

Er, I., & Seah, W. (2004). Mobility-based d-Hop Clustering Algorithm for Mobile Ad Hoc Networks. In *Proc. of the IEEE Wireless Communications and Networking Conference (WCNC 2004)* (Vol. 4, pp. 2359–2364).

ER. I., & Seah, W. (2005). Clustering Overhead and Convergence Time Analysis of the Mobility-based Multi-hop Clustering Algorithm for Mobile Ad Hoc Networks. In *Proceedings of the 11th International Conference on Parallel and Distributed System,* (pp. 1144-1155). Fuduoka, Japan.

Erbas, F., Steuer, J., Kyamakya, K., Eggesieker, D., & Jobmann, K. (2001). A Regular Path Recognition Method and Prediction of User Movements in Wireless Networks. In *Proceedings of the Vehicular Technology Conference (VTC' 2001), IEEE* (pp. 2672-2676).

Ergen, M., Lee, D., Sengupta, R., & Varaiya, P. (2004). WTRP - Wireless token ring protocol. *IEEE Transactions on Vehicular Technology*, *53*(6), 1863–1881. doi:10.1109/TVT.2004.836928

Ergen, S. C., & Varaiya, P. (2006). PEDAMACS: Power Efficient and Delay Aware Medium Access Protocol for Sensor Networks. *IEEE Transactions on Mobile Computing*, *7*(5), 920–930. doi:10.1109/TMC.2006.100

Ericcson (2005). Prepaid Postpaid Convergent Charging. *Whitepaper.*

Ernst, T., & Lach, H-Y. (2007, July). *Network Mobility Support Terminology*. RFC 4885.

Exposito, E., Van Wambeke, N., Chassot, C., & Drira, K. (2008). Introducing a cross-layer interpreter for multimedia streams, Computer Networks. *The International Journal of Computer and Telecommunications Networking*, *52*(6), 1125–1141.

Farrel, A. (2004). *The Internet and its Protocol: A Comparative Approach*. Morgan Kaufmann.

Feeney, L. M., & Nilsson, M. (2001). *Investigating the Energy Consumption of a Wireless Network Interface in an Ad Hoc Networking Environment*. IEEE INFOCOM.

Félegyházi, M., Hubaux, J.-P., & Buttyán, L. (2006). Nash Equilibria of Packet Forwarding Strategies in Wireless Ad Hoc Networks. *IEEE Transactions on Mobile Computing*, *5*(5), 463–476. doi:10.1109/TMC.2006.68

Fèlegyhàzi, M., Buttyàn, L., & Hubaux, J. P. (2003a). *Equilibrium Analysis of Packet Forwarding Strategies in Wireless Ad Hoc Networks – The Dynamic Case (extended version)* (Technical report Nr. IC/2003/68). Laboratory of to Computer Communications and Applications, Swiss Federal Institute of Technology, Lausanne.

Fèlegyhàzi, M., Buttyàn, L., & Hubaux, J. P. (2003b). *Equilibrium Analysis of Packet Forwarding Strategies in Wireless Ad Hoc Networks – The Static Case (extended version)* (Technical report Nr. IC/2003/33). Laboratory of to Computer Communications and Applications, Swiss Federal Institute of Technology, Lausanne.

Felemban, E., Lee, C. G., & Ekici, E. (2006). MMSPEED: Multipath Multi-SPEED Protocol for QoS Guarantee of Reliability and Timeliness in Wireless Sensor Networks. *IEEE Transactions on Mobile Computing*, *5*(6), 738–754. doi:10.1109/TMC.2006.79

Fels, S., Sumi, Y., Etani, Y., Simonet, N., Kobayashi, K., & Mase, K. (1998). Progress of C-Map: a Context-aware Mobile Assistant. *AAAI Spring Symposium on Intelligent Environments* (pp. 60-67), Palo Alto, California.

Forman, G. H., & Zahorjan, J. (1999). The challenges of mobile computing source. In Milojicic, D., Douglis, F., & Wheeler, R. (Eds.), *Mobility: processes, computers, and agents* (pp. 269–285). New York: ACM Press/Addison-Wesley Publishing Co.

Franciscani, F. P., Vasconcelos, M. A., Couto, R. P., & Loureiro, A. A. F. (2005). (Re)configuration algorithms for peer-to-peer over ad hoc networks. *Journal of Parallel and Distributed Computing*, *65*(2), 234–245. doi:10.1016/j.jpdc.2004.09.007

Franciscani, F. P., Vasconcelos, M. A., Couto, R. P., & Loureiro, A. A. F. (2003). *Peer-to-Peer over Ad-hoc Networks: (Re)Configuration Algorithms*. Paper presented at the International Parallel and Distributed Processing Symposium (IPDPS'03) Nice, France

Fraser, M. J. (2007). Local routing on tori. In *Proceedings of 6th ADHONOW* (LNCS 4686, pp. 1–14).

FreeNet. (2000). Freenet Homepage. Retrieved from http://freenetproject.org/.

Freepastry. Freepastry Homepage. Retrieved from http://freepastry.rice.edu/FreePastry/

Frey, H., & Stojmenovic, I. (2006). On Delivery Guarantees of Face and Combined Greedy-FaceRouting in Ad Hoc and Sensor Networks. *In Proceedings of the Twelfth ACM Annual International Conference on Mobile Computing and Networking (MobiCom'2006)* (pp.390-401), Los Angeles.

Fujii, K., & Suda, T. (2006). Semantics-based dynamic web service composition. *International Journal of Cooperative Information Systems (IJCIS), special issue on Service-Oriented Computing.*

Galis, A., Denazis, S., Brou, C., & Klein, C. (Eds.). (2004). *Programmable Networks for IP Service Deployment*. Artech House Books.

Ganesan, D., Estrin, D., Woo, A., Culler, D., Krishnamachari, B., & Wicker, S. (2002). *Complex Behavior at Scale: An Experimental Study of Low-Power Wireless Sensor Networks* (Technical Report CSD-TR 02-0013). University of California at Los Angeles (UCLA).

Gao, Z., Wang, L., Yang, M., & Yang, X. (2006). CNP-GSDP: An efficient group-based service discovery protocol for MANETs. *Computer Networks: The International Journal of Computer and Telecommunications Networking*, *50*(16), 3165–3182.

Gao, Z. G., Yang, X. Z., & Cai, S. B. (2005). Flexible forward probability based service discovery protocol for manets. *Journal of Harbin Institute of Technology*, *37*(9), 1256–1260.

Garcia-Ruiz, M. A., Edwards, A., Aquino-Santos, R., Vargas Martin, M., & Mendoza Gonzalez, R. (2007). Using sonification to teach network intrusion detection: A preliminary usability study. In *Proceedings of ED-Media*, Vancouver, Canada.

Garifo, C. (2001). Increasing Flexibility: Billing Systems Stretch to Handle Varieties of IP-based Services. *Xchange magazine*. Retrieved August 26, 2008, from http://www.xchangemag.com/articles/192back.html

Garlan, D., Siewiorek, D., Smailagic, A., & Steenkiste, P. (2002). Project Aura: Towards distraction-free pervasive computing. *IEEE Pervasive Computing / IEEE Computer Society [and] IEEE Communications Society*, *21*(2), 22–31. doi:10.1109/MPRV.2002.1012334

Garner, P. D. (2003, June). Mobile Bluetooth networking: technical considerations and applications. In *Proceedings of the 4th International Conference on 3G Mobile Communication Technologies* (pp. 274-276).

Gaver, W. W. (1986). Auditory icons: Using sound in computer interfaces. *Human-Computer Interaction*, *2*, 167–177. doi:10.1207/s15327051hci0202_3

Gerla, M., & Tsai, J. (1995). Multicluster, Mobile, Multimedia Radio Network. *ACM-Baltzer Journal of Wireless Network*, *1*(3), 255–265. doi:10.1007/BF01200845

Gerla, M., Lee, Y.-Z., Park, J.-S., & Yi, Y. (2005, December). On Demand Multicast Routing with Unidirectional Links. In *Proceedings of the Wireless Communications and Networking Conference (WCNC)*, Volume 4, New Orleans, LA, USA.

Ghanbari, M. (2003). *Standard Codecs: Image Compression to Advanced Video Coding*. Institution of Electrical Engineers.

Ghidini, C., & Giunchiglia, F. (2001). Local models semantics, or contextual reasoning = locality + compatibility. *Artificial Intelligence*, *127*(2), 221–259. doi:10.1016/S0004-3702(01)00064-9

Gilfix, M., & Couch, A. (2000). Peep (The network auralizer): Monitoring your network with sound. In *Proceedings of 14th System Administration Conference*, New Orleans, USA.

Giordano, S. (2001). Mobile Ad-hoc Networks. In Stojmenovic, I. (Ed.), *Handbook of Wireless Network and Mobile Computing*. John Wiley & Sons.

Gnutella. (2000). The Gnutella Protocol specification. Retrieved from http://dss.clip2.com/GnutellaProtocol04.pdf.

Goldsmith, J., & Wicker, S. (2002). Design challenges for energy-constrained ad hoc wireless networks. *IEEE Wireless Communications Magazine*, *9*, 8–27. doi:10.1109/MWC.2002.1028874

Gopinath, M. C. (2004). *Auralization of intrusion detection system using Jlisten*. Unpublished thesis, Birla Institute of Technology and Science, India.

GPP (2000). Charging and Accounting Mechanisms. *Technical report (TR 122 924).*

GPP (2004). 3rd Generation Partnership Project; Technical Specification Group Services and System Aspects; Telecommunication management; Charging management; Charging architecture and principles (Release 6). *Technical report (TS 32.240).*

GPP. (n.d.). 3rd Generation Partnership Project. Retrieved from http://www.3gpp.org/

Grimm, R. (2004). One.World: Experiences with a Pervasive Computing Architecture. *IEEE Pervasive Computing / IEEE Computer Society [and] IEEE Communications Society*, *3*(3), 22–30. doi:10.1109/MPRV.2004.1321024

GSM 03.40 v7.4.0 (2000). *Digital cellular telecommunications system (Phase 2+), Technical realization of the Short Message Service (SMS).* ETSI 2000. Retrieved from http://www.etsi.org.

Gu, T., Pung, H., & Zhang, D. (2005). A service-oriented middleware for building context-aware services. *Journal of Network and Computer Applications*, *28*(1), 1–18. doi:10.1016/j.jnca.2004.06.002

Gudgin, M., Hadley, M., Mendelsohn, N., Moreau, J.-J., Nielsen, H. F., Karmarkar, A., & Lafon, Y. (2007). *SOAP Version 1.2 Part 1: Messaging Framework* (2nd ed.). *W3C Recommendation 27 April 2007*. W3C. Retrieved February 13, 2008 from http://www.w3.org/TR/soap12

Gummadi, K., Gummadi, R., Gribble, S., Ratnasamy, S., Shenker, S., & Stoica, I. (2003). *The Impact of DHT Routing Geometry on Resilience and Proximity.* Paper presented at the ACM SIGCOMM 2003 conference on Applications, technologies, architectures, and protocols for computer communications Karlsruhe, Germany.

Gundavelli, S., Leung, K., Devarapalli, V., Chowdhury, K., & Patil, B. (2007). Proxy Mobile IPv6. IETF Draft, draft-ietf-netlmm-proxymip6-07.txt.

Gustafsson, E., & Jonsson, A. (2003). Always Best Connected. *IEEE Wireless Communications*, *10*(1), 49–55. doi:10.1109/MWC.2003.1182111

Guttman, E., Perkins, C., & Veizades, J. (1997). *RFC 2165: Service Location Protocol.*

Gwizdka, J. (2000). What's in the context? Workshop on The What, Who, Where, When, and How of Context-Awareness, Conference on Human Factors in Computing Systems (CHI 2000), April 2000.

Haas, Z.J. & Pearlman, M., (1999, June). *The Zone Routing Protocol (ZRP) for Ad Hoc Networks*. Internet Draft.

Han, P., Zhou, X., Li, Y., Chen, X., & Gao, C. S. (2007). An Adaptive Real-Time Routing Scheme for Wireless Sensor Networks. *In Proceedings of the 21st International Conference on Advanced Information Networking and Applications Workshops (AINAW'2007)*, (pp. 918-922). Ontario.

Handley, M. et al, (2003). *TCP Friendly Rate Control (TFRC)*. IETF RFC 3448.

Hansen, M. D., Charp, E., Lodha, S., Meads, D., & Pang, A. (1999). PROMUSE: A system for multi-media data presentation of protein structural alignments. In *Proceedings of the Pacific Symposium on Biocomputing*, Hawaii, EUA.

Haseeb, S., & Ismail, A. F. (2007). Handoff Latency Analysis of Mobile IPv6 Protocol Variations. *Computer Communications*, *30*(4), 849–855. doi:10.1016/j.comcom.2006.10.010

Hass, Z. (1997, October). A New Routing Protocol for the Reconfigurable Wireless Networks, *In Proceedings of the 6th IEEE International Conference on Universal Personal Communications (IEEE ICUPC)*, San Diego, CA, USA.

Hassinen, M., & Laitinen, P. (2005). End-to-end Encryption for SMS Messages in the Health Care Domain. *Studies in Health Technology and Informatics*, *2005*, 316–321.

He, T., Stankovic, J. A., Lu, C., & Abdelzaher, T. F. (2003). SPEED: a stateless protocol forreal-time communication in sensor networks. *In Proceedings of the International Conference on Distribued Computing Systems* (ICDCS'2003). Rhode Island.

He, T., Vicaire, P., Yan, T., Luo, L., Gu, L., Zhou, G., et al. (2007). Achieving Real-Time Target Tracking Using Wireless Sensor Networks. *In Proceedings of the 12th IEEE Real-Time and Embedded Technology and Applications Symposium* (RTAS'2007), (pp.37-48). Washington, DC.

Heer, T. M. (2006, August). *LHIP – Lightweight Authentication for the Host Identity Protocol.* Diploma Thesis.

Heer, T. S, G., S, R., & K, W. (2006). *Adapting distributed hash tables for mobile ad hoc networks.* Paper presented at the Pervasive Computing and Communications Workshops (PerCom Workshops 2006), Fourth Annual IEEE International Conference.

Heinzelman, W., Chandrakasan, A., & Balakrishnan, H. (October 2002). An Application Specific Protocol Architecture for Wireless Microsensor Networks. *IEEE Transactions on Wireless Communications.*

Helal, S., Desai, N., & Verma, V. (2003). Konark - A Service Discovery and Delivery Protocol for Ad-hoc Networks. In *Proc. of the Third IEEE Conf. Wireless Communication Networks (WCNC).*

Henderson, T. R., Ahrenholz, J. M., & Kim, J. H. (2003, March). Experience with the Host Identity Protocol for Secure Host Mobility and Multihoming. *IEEE Wireless Communications and Networking*, *3*, 2120–2125.

Henderson, T. (2007, March). *End-Host Mobility and Multihoming with the Host Identity Protocol*. IETF Internet Draft <draft-ietf-hip-mm-05>.

Henricksen, K., & Indulska, J. (2006). Developing context-aware pervasive computing applications: Models and approach. *Pervasive and Mobile Computing*, *2*(1), 37–64. doi:10.1016/j.pmcj.2005.07.003

Henricksen, K., Indulska, J., & Rakotonirainy, A. (2002) Modeling context information in pervasive computing systems. *1st International Conference on Pervasive Computing*, Zurich, Switzerland (pp. 167–180).

Herborn, S., Huber, A., Boreli, R., & Seneviratne, A. (2007). *A Scheme for Host Identity Delegation.* Paper presented at the IEEE/Create-Net/ICST International Conference on COMmunication System softWAre and MiddlewaRE (COMSWARE), Bangalore, India.

Herborn, S., Huber, A., Boreli, R., & Seneviratne, A. (2007, January). Secure Host Identity Delegation for Mobility. In *COMSWARE'07* (pp. 1-9).

Hiertz, G. R., Sebastian, M., Rui, Z., Denteneer, D., & Berlemann, L. (2007). Principles of IEEE 802.11s. In *Proceedings of 16th International Conference on Computer Communications and Networks*, ICCCN 2007.

Hightower, J., & Borriello, G. (2001). Location Systems for Ubiquitous Computing. *Journal of IEEE Computer*, *34*(8), 57–66.

Hirsch, F., Kemp, J., & Ilkka, J. (2006). *Mobile Web Services: Architecture and Implementation*. West Sussex: John Wiley & Sons. doi:10.1002/9780470017982

Hoesel, V., & Havinga. (2004). A Lightweight Medium Access Protocol (LMAC) for Wireless Sensor Networks: Reducing Preamble Transmissions and Transceiver State Switches. *INSS.*

Hoesel, V., & Havinga. (2004). A TDMA-based MAC Protocol for WSNs. *ACM SenSys.*

Hofmann, A., Horeis, T., & Sick, B. (2002). Feature selection for intrusion detection: An evolutionary approach. In *Proceedings of the IEEE International Joint Conference on Neural Networks*, Budapest, Hungary (Vol. 2).

Hong, W., Xu, K., & Gerla, M. (2002). Scalable Routing Protocols for Mobile Ad Hoc Networks. *IEEE Network*, *16*(4), 11–21. doi:10.1109/MNET.2002.1020231

Horrocks, I., Patel-Schneider, P., Boley, H., Tabet, S., Grosof, B., & Dean, M. (2004). *SWRL: A Semantic Web Rule Language Combining OWL and RuleML.* W3C. Retrieved February 13, 2008 from http://www.w3.org/Submission/SWRL/

Hsieh, R., Zhou, Z. G., & Seneviratne, A. (2003). S-MIP: A Seamless Handoff Architecture for Mobile IP. *22nd Annual Joint Conference of the IEEE Computer and Communications Societies (INFOCOM 2003)*, (pp. 1774-1784). San Francisco, CA, USA.

Hu, W.-C., Lee, C.-W., & Jen, J.-H. (2004). Mobile Commerce Systems. In Nansi, S. (Ed.), *Mobile Commerce Applications* (pp. 1–23). Hershey, PA: Idea Group Publishing.

Hu, Y. C., Das, S. M., & Pucha, H. (2003). *Exploiting the Synergy between Peer-to- Peer and Mobile Ad Hoc Networks.* Paper presented at the HotOS IX Workshop, Lihue, HI, USA.

Huber, J. F. (2004). Mobile next-generation networks. *IEEE MultiMedia*, *11*(1), 72–83. doi:10.1109/MMUL.2004.1261110

Hui, S. Y., & Yeung, K. H. (2003). Challenges in the Migration to 4G Mobile Systems. *IEEE Communications Magazine*, *41*(12), 54–59. doi:10.1109/MCOM.2003.1252799

Hurni, P., & Braun, T. (2008). Evaluation of WiseMAC on Sensor Nodes. *10th IFIP International Conference on Mobile and Wireless Communications Networks (MWCN2008).* Toulouse: IFIP.

Hurni, P., Braun, T., & Feeney, L. M. (2006). Simulation and Evaluation of Unsynchronized Power Saving Mechanisms in Wireless Sensor Networks. *4th International Conference on Wired/Wireless Internet Communications* (LNCS 3970, pp. 311-324).

Hwang, J., Huang, C., & Chang, C. (2008). Cross-Layer End-to-End QoS for Scalable Video over Mobile WiMAX. In Chen, K., Roberto, J., & Marca, B. (Eds.), *Mobile WiMAX.* John Wiley & Sons, Ltd. doi:10.1002/9780470723937.ch13

Iannaccone, G., Chuah, C.-N., Bhattacharyya, S., & Diot, C. (2004). Feasibility of IP Restoration in a Tier-1 Backbone. *IEEE Network, 18*(2), 13–19. doi:10.1109/MNET.2004.1276606

IEEE Std.802.15.4(2003) Wireless Medium Access Control(MAC) and Physical Layer(PHY) Specifications for Low-Rate Wireless Personal Area Networks(LR-WPANs).

IEEE. (2003). *Specific requirements Part 15.4: Wireless Medium Access Control (MAC) and Physical Layer (PHY) Specifications for Low-Rate Wireless Personal Area Networks (LR-WPANs).* 802.15.4 IEEE Standard for Information technology - Telecommunications and information exchange between systems — Local and metropolitan area networks.

IEEE-draft. (1999). *IEEE p802.11/D10* o. IETF-manet. IETF MANET working Group. Retrieved from http://www.ietf.org/html.charters/manet-charter.html

Ilyas, M., & Mahgoub, I. (2005). *Handbook of Sensor Networks: Compact Wireless and Wired Sensing Systems.* CRC Press.

Indulska, J., Robinson, R., Rakotonirainy, A., & Henricksen, K. (2003). Experiences in using CC/PP in context-aware systems. *4th International Conference on Mobile Data Management* (pp. 247–261).

Indulska, K., & Sutton, P. (2003). Location Management in Pervasive Systems. *Australasian Information Security Workshop Conf. ACSW Frontiers*, Adelaide, Australia (pp. 143–151).

ISO 11898:1993 (1993), Road Vehicles – Interchange of Digital Information – Controller Area Network (CAN) for High-Speed Communication.

ITU-R. (2002). *Recommendation BT500.* Methodology for the Subjective Assessment of the Quality of Television Pictures.

ITU-R. BT.1683 (2004). Objective Perceptual Video Quality Measurement Techniques for Standard Definition Broadcast Television in the Presence of a Full Reference.

ITU-T. (1998). *Recommendation* (p. 911). Subjective Audiovisual Quality Assessment Methods for Multimedia Applications.

ITU-T. (1999). *Recommendation* (p. 910). Subjective Video Quality Assessment Methods for Multimedia Applications.

ITU-T. J.144 (2004). Objective Perceptual Video Quality Measurement Techniques for Digital Cable Television in the Presence of a Full Reference.

Iyer, A., & Rosenberg, C. (2006). Understanding the Key Performance Issues with MAC Protocols for Multi-hop Wireless Networks. *Wireless Communications and Mobile Computing, 6*(6).

Jain, N., Das, S., & Nasipuri, A. (2001). A Multichannel CSMA MAC Protocol with Receiver-Based Channel Selection for Multihop Wireless Networks. *IEEE International Conference on Computer Communications and Networks, IC3N '2001.*

Jiang, S., Liu, Y., Jiang, Y., & Yin, Q. (2004). Provisioning of adaptability to variable topologies for routing schemes in MANETs. *IEEE Journal on Selected Areas in Communications, 22*(7), 1347–1356. doi:10.1109/JSAC.2004.829352

Jiang, S., He, D., & Rao, J. (2001). A Prediction-based Link Availability Estimation for Mobile Ad Hoc Networks. *Proceedings - IEEE INFOCOM, 2001*, 1745–1752.

Jiang, M., Ji, J. & Tay, Y.C. (1999). *Cluster based routing protocol.* Internet Draft, draft-ietf-manet-cbrp-spec-01.txt, work in progress, 1999.

Jin, H. & Wu, H. (2005). Semantic-enabled specification for web services agreement. *The International Journal of Web Services Practices.*

John, R. (1999). *UPnP, Jini and Salutaion—A Look at Some Popular Coordination Frameworks for Future Network Devices* (Technical report). California Software Labs. http://www.cswl.com/whiteppr/tech/upnp.html

Johnson, D. B., & Maltz, D. A. (1996). Dynamic Source Routing (DSR) in AdHoc Wireless Networks. In Imielinski, K. (Ed.), *Mobile Computing*. Kluwer Academic Publishers. doi:10.1007/978-0-585-29603-6_5

Johnson, B. (2005). Telecom Billing Systems - Build, Buy, Outsource. Retrieved December 11, 2007, from http://job.wowak.com/introduction-telecom-billing.html.

Johnson, D., Perkins, C., & Arkko, J. (2004). Mobility Support in IPv6. IETF RFC-3775.

Johnsson, M., Rembarz, R., Schieder, A., Surtees, A., Ahmed, K., & Schoo, P. (2008). *Ambient Networks – An Extensible Architecture for Self-Managing and Composable Systems in Heterogeneous and Dynamic Environments*. IEEE Communication Magazine.

Johnsson, M., et al. (n.d.). *Draft System Description*. IST Project Ambient Networks D7. Retrieved from http://www.ambient-networks.org/Files/deliverables/D7-A.2_PU.pdf

Jokela, P., Melen, J., & Ylitalo, J. (2006, June). *HIP Service Discovery*. IETF Internet Draft <draft-jokela-hip-service-discovery-00>.

Jokela, P., Moskowitz, R., & Nikander, P. (2007,. June). *Using ESP transport format with HIP*. IETF Internet Draft <draft-ietf-hip-esp-06>.

Jokela, P., Rinta-aho, T., Jokikyyny, T., Wall, J., Kuparinen, M., Mahkonen, H., & Melen, J. &Kauppinen, T. (2005, November 13-14). *HIPernet: a decentralized security infrastructure for large scale grid environments*. Paper presented at 6th IEEE/ACM International Workshop on Grid Computing'05.

Joly, A., Nathalie, M., & Marcel, B. (2001). *Audio-Visual Quality and Interactions between Television Audio and Video*. Paper presented at the International Symposium on Signal Processing and its Applications, Malaysia.

Jubin, J., & Tornow, J. D. (1987). *The DARPA Packet Radio Network Protocols*. Paper presented at the Proc. of the IEEE.

Jung, H. Y., Kim, E. A., Yi, J. W., & Lee, H. H. (2005). A Scheme for Supporting Fast Handover in Hierarchical Mobile IPv6 Networks. *ETRI Journal*, *27*(6), 798–801. doi:10.4218/etrij.05.0205.0009

Jurca, D., Chakareski, J., Wagner, J. P., & Frossard, P. (2007). Enabling Adaptive Video Streaming in P2P Systems. *IEEE Communications Magazine*, *45*(6), 108–114. doi:10.1109/MCOM.2007.374427

Kappler, C., Poyhonen, P., Johnsson, M., & Schmid,S. (2007). Dynamic network composition for beyond 3G networks: a 3GPP viewpoint. *Network IEEE, 21*(1).

Karl, H., & Willig, A. (2006). *Protocols and Architectures for Wireless Sensor Networks*. Wiley.

Karp, B., & Kung, H. T. (2000, August), Greedy Perimeter Stateless Routing for Wireless Networks. In *Proceedings of the Sixth Annual ACM/IEEE International Conference on Mobile Computing and Networking (MobiCom)*, Boston, MA.

Karygiannis, T., & Owens, L. (2002). *Wireless network security: 802.11, Bluetooth, and Handheld Devices*. National Institute of Standards and Technology white paper.

KaZaA. (2002). KaZaA Homepage. Retrieved from http://www.kazaa.com.

Kempf, J., Wood, J., & Fu, G. (2003). Fast Mobile IPv6 Handover Packet Loss Performance: Measurements for Emulated Real Time Traffic. *IEEE Wireless Communications and Networking Conference (WCNC 2003)* (pp. 1230-1235). New Orleans, LA, USA.

Khurri, A., Vorobyeva, E., Gurtov, A. (2007). Performance of Host Identity Protocol on Lightweight Hardware. In *Proceedings of The 2nd ACM International Workshop on Mobility in the Evolving Internet Architecture (MobiArch 2007)* August 27, 2007, Kyoto, Japan

Kim, D., Toh, C. K., & Choi, Y. (2003, April). GAHA and GAPA: Two Link-Level Approaches for Supporting Link Asymmetry in Mobile Ad-hoc Networks. *IEICE Transactions on Communications. E (Norwalk, Conn.), 86-B*(4), 1297–1303.

Kim, K. (2006). A Novel Factor for Robust Clustering in Mobile Ad Hoc Networks. *IEICE Transactions on Communications. E (Norwalk, Conn.), 89-B*(4), 1436–1439.

Kim, D., Toh, C. K., & Choi, Y. (2000, October). RODA: A New Dynamic Routing Protocol using Dual Paths to Support Asymmetric Links in Mobile Ad-hoc Networks. In *Proceedings of the 9th International Conference on Computer Communications and Networks (IEEE ICCCN)*, Las Vegas, NV, USA.

Kimoto, M., & Ohno, H. (2002). Design and implementation of Stetho - network sonification system. In *Proceeding of ICMC2002* (pp 273-279).

Klemm, A., Lindemann, C., & Waldhorst, O. (2003). *A special-purpose peer-to-peer file sharing system for mobile ad hoc networks.* Paper presented at the IEEE Vehicular Technology Conference (VTC 2003).

Klyne, G., & Carroll, J. (2004). Resource Description Framework (RDF): Concepts and Abstract Syntax. W3C. Retrieved September 20, 2008 from http://www.w3.org/TR/rdf-concepts/

Kofler, I., Timmerer, C., Hellwagner, H., & Ahmed, T. (2007). Towards MPEG-21-Based Cross-Layer Multimedia Content Adaptation. In *Proceedings of the Second International Workshop on Semantic Media Adaptation and Personalization* (pp. 3-8).

Koh, S. J., Chang, M. J., & Lee, M. (2004, March). mSCTP for soft handover in transport layer. *IEEE Communications Letters, 8*(3), 189–191. doi:10.1109/LCOMM.2004.823432

Konstantopoulos, C., Gavalas, D., & Pantziou, G. (2008). *Clustering in Mobile Ad-Hoc Networks through Neighbourhood Stability-Based Mobility Prediction. Computer Networks*. Elsevier.

Konstantopoulos, C., Gavalas, D., & Pantziou, G. (2006). A Mobility Aware Technique for Clustering on Mobile Ad-Hoc Networks. In *Proc. International Conference on Distributed Computing and Networking (ICDCN 2006)* (LNCS 4308, pp. 397–408).

Koodli, R. (2005). Fast Handovers for Mobile IPv6. IETF RFC 4068.

Koodli, R. (2007). Mobile IPv6 Fast Handovers. IETF Draft, draft-ietf-mipshop-fmipv6-rfc4068bis-03.txt.

Koodli, R., & Perkins, C.E., (2002). Service Discovery in On Demand Ad Hoc Networks. *IETF Internet draft.*

Kortuem, G., & Schneider, J. (2001). *An Application Platform for Mobile Ad-hoc Networks.* Paper presented at the Workshop on Application Models and Programming Tools for Ubiquitous Computing (UBICOMP 2001), Atlanta, Georgia.

Kovacshazi, Z., & Vida, R. (2007). Host Identity Specific Multicast. *International Conference on Networking and Services (ICNS '07)* (p. 1).

Kozat, U., & Tassiula, L. (2004). Service discovery in mobile adhoc Networks: an Overall perspective on architectural choices and network Layer support issues. *Ad Hoc Networks, 2*(1). doi:10.1016/S1570-8705(03)00044-1

Kramer, G. (Ed.). (1994). *Auditory display: Sonification, audification, and auditory interfaces. Santa Fe Institute Studies in the Sciences of Complexity, Proc* (*Vol. XVIII*). Reading, MA: Addison-Wesley.

Kramer, G., Walker, B., Bonebright, T., Cook, P., Flowers, J. H., Miner, N., et al. (1997). *Sonification report: status of the field and research agenda.* Available at: http://www.icad.org/websiteV2.0/References/nsf.html

Kranakis, E., Mott, T., & Stacho, L. (2008). Online routing in quasi-planar and quasi-polyhedral graphs. *Discrete Applied Mathematics*, *156*(18), 3430–3442. doi:10.1016/j.dam.2008.01.027

Kranakis, E., Singh, H., & Urrutia, J. (1999). Compass routing on geometric networks. In *Proceedings of the Canadian Conference on Computational Geometry (CCCG-99)* (pp. 51–54).

Krause, A., Smailagic, A., & Siewiorek, D. (2006). Context-aware mobile computing: learning context-dependent personal preferences from a wearable sensor array. *IEEE Transactions on Mobile Computing*, *5*(2), 113–127. doi:10.1109/TMC.2006.18

Kreller, B., Carrega, D., Shankar, J., Salmon, P., Bottger, S., & Kassing, T. (1998). *A Mobile-Aware City Guide Application* (pp. 60–65). Rhodos, Greece: ACTS Mobile Communication Summit.

Krüger, A., Butz, A., Müller, C., Stahl, C., Wasinger, R., Steinberg, K. E., & Dirschl, A. (2004, January 13). *The connected user interface: realizing a personal situated navigation service*. Paper presented at 9th International Conference on Intelligent User Interfaces, Funchal, Madeira, Portugal. Retrieved September 20, 2007 from ACM.

Krumm, J., & Horvitz, E. (2007, July 25). *Driver Destination Models*. Paper presented at 11th International Conference on User Modeling, Corfu, Greece. Retrieved October 27 from Springer Link.

Kuflik, T., Sheidin, J., Jbara, S., Goren-Bar, D., Soffer, P., Stock, O., & Zancanaro, M. (2007, January 28). *Supporting small groups in the museum by context-aware communication services*. Paper presented at 12th International Conference on Intelligent User Interfaces, Hawaii, USA. Retrieved March 3, from ACM.

Kumar, D. A., Alazemi, H., Vijaykumar, R., & Roy, S. (2005). Optimization Models for Fixed Channel Assignment in Wireless Mesh Networks with Multiple Radios. *IEEE Communications Society Conference on Sensor, Mesh and Ad Hoc Communications and Networks, SECON '2005*, Santa Clara, USA.

Kumaran, S., I. (2002). JINI Technology an Overview. *Prentice Hall PTR*.

Kyasanur, P., & Vaidya, N. H. (2005). Routing and Interface Assignment in Multi-Channel Multi-Interface Wireless Networks. *IEEE Wireless Communications and Networking Conference, WCNC '2005*, New Orleans, LA.

Laganier, J., & Eggert, L. (2006). Host Identity Protocol (HIP) Rendezvous Extension. IETF Internet Draft <draft-ietf-hip-rvs-05>

Laganier, J., Koponen, T., &Eggert, L. (2006, June). Host Identity Protocol (HIP) Registration Extension. IETF Internet Draft <draft-ietf-hip-registration-02>

Laliwala, Z., Khosla, R., Majumdar, P., & Chaudhary, S. (2006). Semantic and rules based event-driven dynamic web services composition for automation of business processes. In *Proceedings of the IEEE Services Computing Workshops (SCW'06)*.

Lau. G, J. M., & Ravindran. G. (2005). *RAON: A P2P Network for MANET.* Paper presented at the Wireless and Optical Communications Networks (WOCN 2005) Second IFIP International Conference.

Lawrence, J. (2002). *LEAP into Ad-Hoc Networks.* Paper presented at the Workshop on Ubiquitous Agents on Embedded, Wearable, and Mobile Devices, Bologna, Italy.

Lecue, F., & Leger, A. (2006). A formal model for semantic web service composition. In *Proceedings of ISWC*.

Lee, S., Gerla, M., & Toh, C. (1999, August). A Simulation Study of Table-driven and On-demand Routing Protocols for Mobile Ad-hoc Networks. *IEEE Network*, *13*(4), 48–54. doi:10.1109/65.777441

Lee, S.-J., Su, W., & Gerla, M. (2001). Wireless Ad Hoc Multicast Routing with Mobility Prediction. *Mobile Networks and Applications*, *6*, 351–360. doi:10.1023/A:1011478628358

Leonhardt, U., Magee, J., & Dias, P. (1996). Location service in mobile computing environments. *Computers & Graphics*, *20*(5), 627–632. doi:10.1016/S0097-8493(96)00036-2

Leung, M. F., & Chan, G. (2007). Broadcast-Based Peer-to-Peer Collaborative Video Streaming among Mobiles. *IEEE Transactions on Broadcasting*, *53*(1), 350–361. doi:10.1109/TBC.2006.889093

Li, X.-Y., Wang, Y., & Song, W.-Z. (2004). Applications of k-Local MST for topology control and broadcasting in wireless ad hoc networks. In *INFOCOM- 04*, Piscataway, NJ, USA, Mar. 7–11.

Lin, P., Qiao, C., & Wang, X. (2004). Medium Access Control with A Dynamic Duty Cycle for Sensor Networks. *In Proceedings of the IEEE Wireless Communications and Networking Conference* (WCNC'2004) (pp. 1534-1539). Atlanta.

Lindemann, C., & Waldhorst, O. (2002). *A Distributed Search Service for Peer -to-Peer File Sharing in Mobile Applications*. Paper presented at the 2nd IEEE Conference on Peer-to-Peer Computing (P2P 2002), Linkoping, Sweden.

Linial, N. (1992). Locality in distributed graph algorithms. *SIAM Journal on Computing*, *21*(1), 193–201. doi:10.1137/0221015

Liu, Q., Zhou, S., & Giannakis, G. B. (2004). Cross-Layer Combining of Adaptive Modulation and Coding with Truncated ARQ Over Wireless Links. *IEEE Transactions on Wireless Communications*, *3*(5), 1746–1755. doi:10.1109/TWC.2004.833474

Liu, G. & Maguire Jr. (1996). A Class of Mobile Motion Prediction Algorithms for Wireless Mobile Computing and Communications. *Mobile Networks and Applications Journal, Springer*, *1*(2), 113–121. doi:10.1007/BF01193332

Liu, J., Sailhan, F., Sacchetti, D., & Issarny, V. (2005). Group Management for Mobile Ad Hoc Networks: Design, Implementation and Experiment. In *Proceedings of the 6th international conference on Mobile Data Management*, (pp.192-199). Ayia Napa, Cyprus.

Liu, R. P., Rosberg, Z., Collings, I. B., Wilson, C., Dong, A. Y., & Jha, S. (2008, September). Overcoming Radio Link Asymmetry in Wireless Sensor Networks, In *Proceedings of the International Symposium on Personal, Indoor and Mobile Radio Communications (PIMRC)*.

Liu, S., Fan, K.-W., & Sinha, P. (2007). CMAC: An Energy Efficient MAC Layer Protocol Using Convergent Packet Forwarding for Wireless Sensor Networks. *4th Annual IEEE Communications Society Conference on Sensor, Mesh and Ad Hoc Communications and Networks (SECON)* (pp. 11-20). IEEE.

Liu, X., Doermann, D., & Li, H. (2006, December 4). *Imaging as an alternative data channel for camera phones*. Paper presented at 5th International Conference on Mobile and Ubiquitous Multimedia, Stanford, California. Retrieved May 15, 2007 from ACM.

Lu, C., Blum, B. M., Abdelzaher, T. F., Stankovic, J. A., & He, T. (2002). Rap: A real-time communication architecture for large-scale wireless sensor networks. *In Proceedings of the 8th IEEE Real-Time and Embedded Technology and Applications Symposium* (RTAS'2002). San Jose.

Lua, E. K., Crowcroft, J., Pias, M., Sharma, R., & Lim, S. (2005). *A Survey and Comparison of Peer-to-Peer Overlay Network Schemes*. IEEE Communications Surveys & Tutorials.

Lucas, D. M. (2004). Where Should Rating be Implemented. *Billing World and OSS today*.

Mahapatra, A., Anand, K., & Agrawal, D. P. (2006). QoS and energy aware routing for real-time traffic in wireless sensor networks. *Computer Communications*, *29*(4), 437–445. doi:10.1016/j.comcom.2004.12.028

Majithia, S., Walker, D. W., & Gray, W. A. (2004). Automated web service composition using semantic web technologies. In *Proceedings of the International Conference on Autonomic Computing*.

Malandrino, D., Mea, D., Negro, A., Palmieri, G., & Scarano, V. (2003). Nemos: Network monitoring with sound. In *Proceedings of the 2003 International Conference on Auditory Display (ICAD)*, Boston, MA, USA, July 6-9.

Mangharam, R., Rowe, A., & Rajkumar, R. (2007). FireFly: A Cross-Layer Platform for Real-Time Embedded Wireless Networks. *Real-Time Systems*, *37*(2), 183–231. doi:10.1007/s11241-007-9028-z

Manner, J., &Kojo, M. (2004, June). *Mobility Related Terminology.* IETF RFC 3753

Marina, M. K., & Das, S. R. (2002, June). Routing Performance in the Presence of Unidirectional Links in Multihop Wireless Networks. In *Proceedings of the International Symposium on Mobile Ad Hoc Networking and Computing (MobiHoc)*, EFPL Lausanne, Switzerland.

Marti, S., Giuli, T. J., Lai, K., & Baker, M. (2000). Mitigating routing misbehaviour in mobile ad hoc networks. In *Proceedings of the Sixth annual ACM/IEEE International Conference on Mobile Computing and Networking* (pp. 255–265).

Martin, D., Burstein, M., Hobbs, J., Lassila, O., McDermott, D., McIlraith, S., et al. (2004). OWL-S: Semantic Markup for Web Services. *W3C Member Submission 22 November 2004.* Retrieved on February 13, 2008 from http://www.w3.org/Submission/OWL-S/

Mascolo, C., Hailes, S., Lymberopoulos, L., Picco, G.P., Costa, P., Blair, G., Okanda, P., Sivaharan, T., Fritsche, W., Karl, M., Rnai, M.A., Fodor, K., & Boulis, A. (2005). *Survey of middleware for networked embedded systems.* (Technical Report D.5.1).

Mathieu, B., Galis, A., Jean, K., Ocampo, R., Lai, Z., Stiemerling, M., et al. (2007). *Dynamic Adaptable Overlay Networks for Personalised Service Delivery.* Paper presented at IEEE 7th International Symposium on Communications and Information Technologies - http://www.elec.uow.edu.au/ISCIT2007/ M2NM 2007- First Ambient Networks Workshop on Mobility, Multiaccess, and Network Management; 16-19 Oct 2007, Sydney, Australia; and http://nicta.com.au/research/projects/ambient_networks/m2nm-2007

Mathieu, B., Song, M., Galis, A., Cheng, L., Jean, K., Ocampo, R., et al. (2007). *Autonomic Management of Context-Aware Ambient Overlay Networks.* Paper presented at IEEE International Conference on Communications and Networking in China (IEEE CHINACOM 2007), 15 - 17 August 2007, Shanghai. Retrieved from www.chinacom.org/2007

Mathieu, B., Song, M., Galis, A., Cheng, L., Jean, K., Ocampo, R., et al. (2007). *Self-Management of Context-Aware Overlay Ambient Networks.* Paper presented at 10th IFIP/IEEE International Symposium on Integrated Management (IM 2007); 21-25 May 2007, Munich, Germany. Retrieved from http://www.im2007.org

McCann, P. (2005, November). *Mobile IPv6 Fast Handovers for 802.11 Networks.* IETF RFC 4260.

McDermott, D. (2002). Estimated-regression planning for interactions with web services. In *Proceedings of the 6th International Conference on AI Planning and Scheduling.*

McDonald, A. B., & Znabi, T. F. (1999b). A mobility-based framework for adaptive clustering in wireless ad hoc networks. *IEEE Journal on Selected Areas in Communications, 17*(8), 1466–1487. doi:10.1109/49.780353

McDonald, A. B., & Znabi, T. F. (1999a). A path availability model for wireless ad hoc networks. In *Proc. IEEE WCNC'1999, New Orleans, USA* (pp. 35-40).

McIlraith, S. A., Son, T. C., & Zeng, H. (2001). Semantic Web Services. *IEEE Intelligent Systems, 16*(1), 46–53. doi:10.1109/5254.920599

McNair, J., & Zhu, F. (2004). Vertical Handoffs in Fourth-Generation Multinetwork Environments. *IEEE Wireless Communications, 11*(3), 8–15. doi:10.1109/MWC.2004.1308935

Michiardi, P., & Molva, R. (2002). Core: A Collaborative Reputation Mechanism To Enforce Node Cooperation In Mobile AD HOC Networks. In *Proceedings of the 6th IFIP Communications and Multimedia Security Conference, Portorosz, Slovenia, September.*

Michiardi, P., & Molva, R. (2003). A game theoretical approach to evaluate cooperation enforcement mechanisms in mobile ad hoc networks. In *Proceedings of WiOpt 2003: Modeling Optimization and Mobile in, Ad Hoc and Wireless Networks, INRIA Sophia-Antipolis, France, March 3-5.*

Microsoft. (2003), *Business Process Execution Language for Web Services Specification.* Retrieved from http://msdn2.microsoft.com/en-us/library/aa479359.aspx

Minar, N. (2001). *Distributed Systems Topologies*. Paper presented at the O'Reilly P2P and Web Services Conference

Misra, R., & Mandal, C. R. (2005, January), Performance Comparison of AODV/DSR On-Demand Routing Protocols for Ad-Hoc Networks in Constrained Situation. In *Proceedings of IEEE International Conference on Personal Wireless Communications (ICPWC)*, New Delhi, India.

MobileIn. (2006). What is MVNO. *MobileIn.* Retrieved August 26, 2008, from http://www.mobilein.com/what_is_a_mvno.htm

Moore, N. (2006). Optimistic Duplicate Address Detection (DAD) for IPv6. IETF RFC 4429.

Morino, H., Miyoshi, T., & Ogawa, M. (2005, April). Ad Hoc Unidirectional Routing Protocol Based on Relay Control of Route Requests. In *Proceedings of the IEEE Autonomous Decentralized Systems (ISADS)*, Chengdu, China.

Morino, H., Miyoshi, T., & Ogawa, M. (2007, March). Unidirectional Ad-Hoc Routing Protocol with Area-Controlled Flooding Using Overheard N Node Information. In *Proceedings of the IEEE Autonomous Decentralized Systems (ISADS)*, Sedona, AZ, USA.

Moskowitz, R., & Nikander, P. (2006, May). *Host Identity Protocol (HIP) Architecture*. IETF RFC 4423.

Moskowitz, R., Nikander, P., Jokela, P., & Henderson, T. (2007, October 30). *Host Identity Protocol*. IETF Internet Draft (draft-ietf-hip-base-10).

Muqattash, A., & Krunz, M. (2005). POWMAC: A single-channel power-control protocol for throughput enhancement in wireless ad hoc networks. *IEEE Journal on Selected Areas in Communications*, *23*(5), 1067. doi:10.1109/JSAC.2005.845422

Murthy, C. S. R., & Manoj, B. S. (2004). *Ad Hoc wireless Networks, Architecture and Protocols*. Prentice Hall.

Naor, M., & Stockmeyer, L. (1995). What can be computed locally? *SIAM Journal on Computing*, 24.

Napster. (1999). Napster webpage. Retrieved from http://www.napster.com

Narendran, B., Agrawal, P., & Anvekar, D. K. (1994). Minimizing Cellular Handover Failures without Channel Utilization Loss. In. *Proceedings of IEEE GLOBECOM*, *94*, 1679–1685.

Narten, T., Nordmark, E., Simpson, W., & Soliman, H. (2007). Neighbor Discovery for IP version 6 (IPv6). IETF RFC 4861.

Nasser, N., Hasswa, A., & Hassanein, H. (2006). Handoffs in Fourth Generation Heterogeneous Networks. *IEEE Communications Magazine*, *44*(10), 96–103. doi:10.1109/MCOM.2006.1710420

Nesargi, S., & Prakash, R. (2000, April). A Tunneling Approach to Routing with Unidirectional Links in Mobile Ad-hoc Networks. In *Proceeding of the 9th International Conference on Computer Communications and Networks (ICCCN),* Las Vegas, NV, USA.

Newsome, J., Karp, B., & Song, D. (2005). Polygraph: Automatically generating signatures for polymorphic worms. In *Proceedings of the 2005 IEEE Symposium on Security and Privacy*, Oakland, USA.

Niebert, N., & Flinck, H. Hancock, R. Karl, H. & Prehofer, C. (2004). *Ambient Networks – Research for Communication Networks Beyond 3G.* Paper presented at 13th IST Mobile and Wireless Communications-Summit 2004, 27-30 June 2004, Lyon. Retrieved from www.mobile-summit2004.org

Niemela, E., & Latvakoski, J. (2004). Survey of requirements and solutions for ubiquitous software. *3rd international conference on Mobile and ubiquitous multimedia (MUM '04)* (pp. 71-78). New York.

Nikander, P., & Arkko, J. (2004). Delegation of Signaling Rights. In *Security Protocols* (LNCS 2845, pp 203-214).

Nikander, P., & Laganier, J. (2007, April). *Host Identity Protocol (HIP) Domain Name System (DNS) Extensions*. IETF Internet Draft <draft-ietf-hip-dns-09>.

Nikander, P., Arkko, J., & Ohlman, B. (2004). *Host Identity Indirection Infrastructure (Hi3).* Paper presented at the Second Swedish National Computer Networking Workshop 2004 (SNCNW2004).

Noh. D., & Shin. H. (2007). SPIZ: an effective service discovery protocol for mobile ad hoc networks. *EURASIP Journal on Wireless Communications and Networking, 2007*(1).

Nokia (2001, May). *Sending Content over SMS to Nokia Phones.* Version 1.0, Forum Nokia. Retrieved from http://www.forum.nokia.com.

Nováczki, S., Bokor, L., Jeney, G., & Imre, S. (2008). Design and Evaluation of a Novel HIP-Based Network Mobility Protocol. *Journal ofNetworks*, *3*(1). doi:10.4304/jnw.3.1.10-24

Nováczki, S., Bokor, L., Jeney, G., & Imre, S. (2007). A HIP based Network Mobility Protocol. In *Proceedings of the 2007 International Symposium on Applications and the Internet Workshops* (SAINTW'07)

O'Hara, B., & Petrick, A. (1999). *The IEEE 802.11 handbook: A designer's companion. Standards Information Network.* Los Alamitos, CA: IEEE Press.

OASIS. (2008), OASIS UDDI. *Specification TC.* Retrieved February 13, 2008 from http://www.oasis-open.org/committees/tc_home.php?wg_abbrev=uddi-spec

OGSI. (2007). *About the OSGi Service Platform* (Technical White paper). Retrieved from www.osgi.org

Oh, C. S. Ko., Y. B., & Roh., Y. S. (2005). An integrated approach for efficient routing and service discovery in mobile ad hoc networks. *Consumer Communications and Networking Conference (CCNC)* (pp. 184-189).

Olaf, L., Stefan, G., & Klaus, W. (2006). *Towards Scalable Mobility in Distributed Hash Tables.* Paper presented at the Sixth IEEE International Conference on Peer-to-Peer Computing.

Oliveira, L. B., Siqueira, I. G., & Loureiro, A. A. F. (2005). On the performance of ad hoc routing protocols under a peer-to-peer Application. *Journal of Parallel and Distributed Computing*, *65*(11). doi:10.1016/j.jpdc.2005.05.023

Oliveira, L. B., Siqueira, I. G., & Loureiro, A. A. (2003). *Evaluation of ad-hoc routing protocols under a peer-to- peer application.* Paper presented at IEEE Wireless Communications and Networking (WCNC 2003)

OSGi Alliance. (2007). OSGi Service Platform. *Core Specification Release 4.1.* Retrieved from http://www.osgi.org/Specifications/HomePage

Otzturk, P., & Aamodt, A. (1997). Towards a model of context for case-based diagnostic problem solving. *Interdisciplinary conference on modeling and using context* (pp. 198–208). Rio de Janeiro, Brazil.

Palit, R., Hossain, E., & Thulasiraman, P. (2006). MAPLE: a Framework for Mobility-Aware Pro-active Low Energy Clustering in Ad-hoc Mobile Wireless Networks. *Wireless Communications and Mobile Computing*, *6*(6), 773–789. doi:10.1002/wcm.441

Papadopouli, M., & Schulzrinne, H. (2001). *Effects of Power Conservation, Wireless Coverage and Cooperation on Data Dissemination among Mobile Devices.* Paper presented at the ACM Symposium on Mobile Ad Hoc Networking and Computing (MOBIHOC 2001), Long Beach, CA.

Papadopoulos, C., Kyriakakis, C., Sawchuk, A., & He, X. (2004). CyberSeer: 3D audio-visual immersion for network security and management. In *Proceedings of the 2004 ACM Workshop on Visualization and Data Mining For Computer Security,* Washington DC, USA.

Park, K., & Lee, H. (2001). On the effectiveness of route-based packet filtering for distributed DoS attack prevention in power-law internet. In *Proceedings of Special Interest Group on Data Communication.* San Diego, USA: SIGCOMM.

Pavlopoulos, S., Kyriacou, E., Berler, A., & Koutsouris, D. (1998). A mobile system for emergency health care provision via telematics support-'AMBULANCE. In *Proceedings of the IEEE International Conference on Information Technology Applications in Biomedicine 1998 (ITAB 98)* (pp. 150–154).

Pavlovski, C. HeeSam, K., & Wood, D. (2004). Ubiquitous mobility in clinical healthcare. In *Proceedings of the IDEAS Workshop on Medical Information Systems: The Digital Hospital (IDEAS-DH'04)* (pp. 147-153).

Peleg, D. (2000). *Distributed Computing: A Locality-Sensitive Approach. SIAM, 2000.* Monographs in Discrete Mathematics and Applications.

Perera, E., Boreli, R., Herborn, S., Georgiades, M., Eisl, J., & Hepworth, E. (2008). A mobility toolbox architecture for all-IP networks: an ambient networks approach. *Wireless Communications IEEE, 15*(2).

Perera, E., Sivaraman, V., & Seneviratne, A. (2004). Survey on network mobility support. *ACM SIGMOBILE Mobile Computing and Communications Review, 8*(2).

Perkins, C. (2002). IP Mobility Support for IPv4. IETF RFC-3344.

Perkins, C. (2007). IP Mobility Support for IPv4, Revised. IETF Draft. draft-ietf-mip4-rfc3344bis-05.

Perkins, C. E., & Bhagwat, P. (1994). *Highly dynamic destination-sequenced distance-vector routing (DSDV) for mobile computers.* Paper presented at the ACM SIGCOMM's conference on Communications architectures, protocols and applications London, UK.

Perkins, C. E., & Royer, E. M. (1999). *Ad hoc on-demand distance vector routing(AODV).* Paper presented at the 2nd IEEE Workshop on Mobile Computing Systems and Applications, New Orleans, LA.

Pister, K. (2008). Technical Overview of Time Synchronized Mesh Protocol (TSMP) [White paper]. Retrieved from http://www.dustnetworks.com

Polastre, J., Hill, J., & Culler, D. (2004). Versatile low power media access for wireless sensor networks. *2nd international Conference on Embedded Networked Sensor Systems (SenSys)* (pp. 95-107). Baltimore.

Ponnekanti, S. R., & Fox, A. (2002). Sword: A developer toolkit for web service composition. In *Proceedings of the 11th World Wide Web Conference.*

Prakash, R. (1999, August), Unidirectional Links Prove Costly in Wireless Ad-Hoc Networks. In *Proceedings of the 3rd International Workshop on Discrete Algorithms and Methods for Mobile computing and Communications (DIAL M)*, Seattle, WA, USA.

Prengel, F. (2006). *Net Micro Framework: Programming Small Connected Devices with.* Net, Core Specification. [White paper]

Preu, S. (2002). Jesa service discovery protocol. In *Proceedings of the Second international Ifip-Tc6 Networking Conference on Networking Technologies, Services, and Protocols; Performance of Computer and Communication Networks; and Mobile and Wireless Communications* (LNCS 2345, pp. 1196-1201).

Pucha, H., Das, S. M., & Hu, Y. C. (2004). *Ekta: An Efficient DHT Substrate for Distributed Applications in Mobile Ad Hoc Networks.* Paper presented at the 6th IEEE Workshop on Mobile Computing Systems and Applications (WMCSA 2004), English Lake District, UK.

Qi, L., Vargas Martin, M., Kapralos, B., Green, M., & Garcia-Ruiz, M. A. (2007). Toward sound-assisted intrusion detection systems. In *Proceedings of the Information Security Symposium. LNCS 4804*, Algarve, Portugal.

Qu, Y., Lung, C.-H., & Srinivasan, A. (2006). Multi-channel OLSR with Dedicated Control Interface. *International Symposium on Performance Evaluation of Computer and Telecommunication Systems, SPECTS '2006,* Calgary, Alberta (pp. 155–162).

Quintero, A., Garcia, O., & Pierre, S. (2004). An Alternative Strategy for Location Update and Paging in Mobile Networks. *Computer Communications, 27*(15), 1509–1523. doi:10.1016/j.comcom.2004.06.001

Radha, H., Van der Schaar, M., & Chen, Y. (2001). The MPEG-4 Fine-Grained Scalable video coding method for multimedia streaming over IP. *IEEE Transactions on Multimedia, 3*(1). doi:10.1109/6046.909594

Raento, M., Oulasvirta, A., Petit, R., & Toivonen, H. (2005). ContextPhone: a prototyping platform for context-aware mobile applications. *Pervasive Computing, 4*(2), 51–59. doi:10.1109/MPRV.2005.29

Raento, M., Oulasvirta, A., Petit, R., & Toivonen, H., (2005). ContextPhone: A Prototyping Platform for Context-Aware Mobile Applications. *IEEE Pervasive Computing, 4*(2), 51-59. Retrieved December 9, 2006 from IEEE Xplore.

Raghavendra, C., Sivalingam, K., & Znati, T. (2004). *Wireless Sensor Networks*. Kluwer. doi:10.1007/b117506

Raisinghani, V. T., & Iyer, S. (2004). Cross-Layer Design Optimizations in Wireless Protocol Stacks. *IEEE Comp. Commun., 27*, 720–724. doi:10.1016/j.comcom.2003.10.011

Rajagopalan, S. S., & Chien-Chung. (2006). *A Cross-layer Decentralized BitTorrent for Mobile Ad hoc Networks*. Paper presented at the Mobile and Ubiquitous Systems - Workshops, 2006. 3rd Annual International Conference.

Ramasubramanian, V., & Mosse, D. (2008, February). BRA: A Bidirectional Routing Abstraction for Asymmetric Mobile Ad Hoc Networks. *IEEE/ACM Transactions on Networking, 16*(1), 116-129.

Raniwala, A., Gopalan, K., & Chiueh, T.-C. (2004). Centralized Channel Assignment and Routing Algorithms for Multi-Channel Wireless Mesh Networks. *Mobile Computing and Communications Review, 8*(2), 50–65. doi:10.1145/997122.997130

Raniwala, A., & Chiueh, T.-C. (2005). Architecture and Algorithms for an IEEE 802.11-Based Multi-Channel Wireless Mesh Network. *IEEE Conference on Computer Communications INFOCOM '2005*.

Ratnasamy, P. Francis, M.Handley, R.Karp, & S.Schenker. (2001). *CAN: A scalable content-addressable network*. Paper presented at ACM SIGCOMM.

Ratsimor, O., Chakraborty, D., Tolia, S., Kushraj, D., Kunjithapatham, A., Gupta, G., Joshi, A., & Finin, T. (2002). *Allia: Alliance-based service discovery for ad-hoc environments*.

Reinbold, P., & Bonaventure, O. (2003). *IP Micro-Mobility Protocols* (pp. 40–57). IEEE Communications Surveys & Tutorials Third Quarter.

Reuters. (2007). *Global cell phone use at 50 percent*. Retrieved November 30, 2007, from http://news.yahoo.com/s/nm/20071129/tc_nm/cellphones_world_dc

RFC 2210 Wroclawski, J. (1997). The use of RSVP With IETF Integrated Services.

RFC 2211 Wroclawski, J. (1997). Specification of the Controlled-Load Network Element Service.

RFC 2212 Shenker, S., Partridge,C., & Guerin, R. (1997). Specification of Guaranteed Quality of Service.

RFC 2597 Heinanen, J., Baker, F., Weiss, W., & Wroclawski, J. (1999). Assured Forwarding PHB Group.

RFC 2707 Bergman, R., Hastings, T., Isaacson, S., & Lewis, H. (1999). Job Monitoring MIB-V1.

RFC 3031 Rosen, E., Viswanathan, A., & Callon, R. (2001). Multiprotocol Label Switching Architecture.

RFC 3246 Davie, B., Charny, A., Bennett, J.C.R., Benson, K., Le Boudec, J.Y., Courtney, W., Davari, S., Firoiu, V., & Stiliadis, D. (2002). An Expedited Forwarding PHB (Per-Hop Behaviour).

RFC 3945 Mannie, E (2004). Generalized Multi-Protocol Label Switching (GMPLS) Architecture.

Rhee, Warrier, Aia, & Min. (2005). Z-MAC: A Hybrid MAC for Wireless Sensor Networks. *ACM Sensys*.

Richardson, I. E. G. (2003). *H.264 and MPEG-4 Video Compression: Video Coding for Next Generation Multimedia*. Wiley. doi:10.1002/0470869615

Ringwald, M., & Römer, K. (2005). Bitmac: a deterministic, collision-free, and robust mac protocol for sensor networks. *In Proceedings of the second European Conference on Wireless Sensor Networks* (EWSN'2005) (pp. 57-69). Istanbul.

Rodd, M. G., Dimyati, K., & Motus, L. (1998). The design and analysis of low-cost real-time fieldbus systems. *Control Engineering Practice, 6*(1), 83–91. doi:10.1016/S0967-0661(97)10053-3

Roedig, U., Barroso, A., & Sreenan, C. J. (2006). f-MAC: A Deterministic Media Access Control Protocol without Time Synchronization. *In Proceedings of the 3rd European Conference on Wireless Sensor Networks (EWSN'2006)* (pp. 276-291). Zurich.

Roosta, T., Menzo, M., & Sastry, S. (2005, May). Probabilistic Geographic Routing in Ad Hoc and Sensor Networks. In *Proceedings of the International Workshop on Wireless Ad Hoc Networks (IWWAN)*, London, UK.

Rowstron, A., & Druschel, P. (2001). *Pastry: scalable, distributed object location and routing for large-scale peer-to-peer systems.* Paper presented at Middleware, Heidelberg, Germany.

Royer, E. M., & Toh, C.-K. (1999). A Review of Current Routing Protocols Ad Hoc Mobile Wireless Networks. *IEEE Personal Communications*, *6*, 46–55. doi:10.1109/98.760423

Rudolph, L. (2001). Project Oxygen: Pervasive, Human-Centric Computing - An Initial Experience. *Advanced Information Systems Engineering, 13th International Conference (CaiSE2001)* (LNCS 2068, pp. 765-780) Interlaken, Switzerland.

Rukzio, E. (2006). *Physical mobile interactions: mobile devices as pervasive mediators for interactions with the real world.* PhD thesis. Faculty for Mathematics, Computer Science and Statistics. University of Munich.

Safa, H., Artail, H., Hamze, H., & Mershad, K. (2007). A collaborative service discovery and service sharing framework for mobile ad hoc networks. *IFIP International Conference on Network and Parallel Computing (NPC 2007)* (LNCS). Dalian, China.

Saha, D., Mukherjee, A., Misra, I. S., Chakraborty, M., & Subhash, N. (2004). Mobility support in IP: a survey of related protocols. *IEEE Network*, *18*(6), 34–40. doi:10.1109/MNET.2004.1355033

Salminen, T. (2005). *Lightweight middleware architecture for mobile phones*. Diploma Thesis, University of Oulu, Department of Electrical and Information Engineering.

Saman, N., & Karmouch, A. (2005). A Mobility Prediction Architecture Based on Contextual Knowledge and Spatial Conceptual Maps. *IEEE Transactions on Mobile Computing*, *4*(6), 537–551. doi:10.1109/TMC.2005.74

Samulowitz, M., Michahelles, F., & Linnhoff-Popien, C. (2001). Capeus: An architecture for context-aware selection and execution of services. *New developments in distributed applications and interoperable systems.* Krakow, Poland (pp. 23–39).

Sanders, M. S., & McCormick, E. J. (1987). *Human factors in engineering and design*. New York: McGraw-Hill.

Sang, L., Arora, A., & Zhang, H. (2007, September), On Exploiting Asymmetric Wireless Links via One-way Estimation. In *Proceedings of the ACM International Symposium on Mobile Ad Hoc Networking and Computing (MobiHoc)*, Montreal, QC, Canada.

Särelä, M., & Nikander, P. (2004). Applying host identity protocol to tactical networks. In *Proceedings of IEEE Milcom 2004, Monterey, California, USA, Nov 2004.*

Satyanarayanan, M. (2001). Pervasive Computing: Vision and Challenges. *IEEE Personal Communication, 8*(4), 10-17.

Schilit, B., Adams, N., & Want, R. (1994). Context-aware computing applications. *IEEE Workshop on Mobile Computing Systems and Applications*, Santa Cruz, CA.

Schiller, J., Liers, A., Ritter, H., Winter, R., & Voigt, T. (2005). ScatterWeb - Low Power Sensor Nodes and Energy Aware Routing. *Hawaii International Conference On System Sciences (HICSS 2005).*

Schmidt, A., & van Laerhoven, K. (2001). How to Build Smart Appliances. *IEEE Personal Communications*, *8*(4), 66–71. doi:10.1109/98.944006

Schollmeier, R., Gruber, I., & Finkenzeller, M. (2002). *Routing in Mobile Ad Hoc and Peer-to-Peer Networks, A Comparison.* Paper presented at the Revised Papers from the NETWORKING 2002 Workshops on Web Engineering and Peer-to-Peer Computing.

Schollmeier, R., Gruber, I., & Niethammer, F. (2003). *Protocol for peer-to-peer networking in mobile environments.* Paper presented at the IEEE 12th International Conference on Computer Communications and Networks (ICCCN).

Schulzrinne, H. et al (2002). *SIP: Session Initiation Protocol.* IETF RFC 3261.

Schulzrinne, H. et al (2003). *RTP: A Transport Protocol for Real-Time Applications.* IETF RFC 3550.

Schulzrinne, H., Rao, A., & Lanphier, R. (1998). *Real Time Streaming Protocol (RTSP).* IETF RFC 2326.

Schurgers, T. G. (2002, January). Optimizing Sensor Networks in the Energy-Latency-Density Design Space. *IEEE Transactions on Mobile Computing.*

Schwartz, L. (2005). *Bridging the Gap: Linking CRM and Business Support Systems.* Billing World & OSS Today.

Schwartz, S. (2003). *Next-Gen Rating: It Will Be Only As Good as the Network.* Billing World & OSS Today.

Schwartz, M. (2005). *Mobile Wireless Communications.* Cambridge Univeristy Press.

Seno, S. A. H., Pahlevanzadeh, B., Wan, T. C., Budiarto, R., & Ghahremani, M. (2008). A new approach CBRP based resource information management in MANETs. *International Journal of Computer Science and Network Security, 8*(5), 301–308.

Sentinelli, A., Marfia, G., Gerla, M., & Kleinrock, L. (2007). Will IPTV Ride the Peer-to-Peer Stream? *IEEE Communications Magazine, 45*(6), 86–92. doi:10.1109/MCOM.2007.374424

Shadbolt, N. (2003). Ambient Intelligence. *IEEE Intelligent Systems, 18*(1), 2–3. doi:10.1109/MIS.2003.1200718

Shafer, G. (1975). *A Mathematical Theory of Evidence.* Princeton University Press.

Shakkottai, S. Rappaport & T.S. Karlsson, P.C. (2003). Cross-layer design for wireless Networks. *IEEE Communications Magazine, 41*(10). doi:10.1109/MCOM.2003.1235598

Shen, Z., & Su, J. (2007). *On automated composition for web services. Poster Paper on World Wide Web (WWW).* Conference.

Sheng, Q. Z., & Benatallah, B. (2005) ContextUML: a UML-based modeling language for model-driven development of context-aware web services, *International Conference on Mobile Business* (pp. 206–212).

Sheu, P., & Wang, C. (2006). A Stable Clustering Algorithm Based on Battery Power for Mobile Ad Hoc Networks. *Tamkang Journal of Science and Engineering, 9*(3), 233–242.

Shiflet, C. F., Belding-Royer, E. M., & Perkins, C. E. (2004). Address Aggregation in Mobile Ad-hoc Networks. In *Proceedings of the IEEE International Conference on Communications*, (pp. 3734-3738). Paris, France.

Shirali-Shahreza, M. (2006, October). Emergency SMS. In *Proceedings the SICE-ICASE International Joint Conference 2006 (SICE - ICCAS 2006)* (pp. 1139-1142), Busan, KOREA.

Shneiderman, B., & Plaisant, C. (2004). *Designing the user interface: Strategies for effective human-computer interaction* (4th ed.). Boston, MA: Addison-Wesley.

Singh, S., Estan, C., Varghese, G., & Savage, S. (2003). *The EarlyBird system for realtime detection of unknown worms* (Technical Report CS2003-0761). University of California, San Diego, USA.

Sirin, E., Hendler, J., & Parsia, B. (2002). Semi-automatic composition of web services using semantic descriptions. In *Proceedings of Web Services: Modeling, Architecture and Infrastructure workshop in conjunction with ICEIS2003.*

Sivavakeesar, S., Pavlou, G., & Liotta, A. (2004). Stable Clustering Through Mobility Prediction for Large-Scale Multihop Ad Hoc Networks. In *Proc. of the IEEE Wireless Communications and Networking Conference (WCNC'2004).*

So, J. Y. H., Jidong, W., & Jones, D. (2005). SHIP mobility management hybrid SIP-HIP scheme. *6th International Conference of Software Engineering, Artificial Intelligence, Networking and Parallel. Distributed Computing, 05*, 226–230.

So, J. Y. H., & Jidong, W. (2006). *HIP Based Mobility Management for UMTS/WLAN Integrated Networks.* Paper presented at the Australian Telecommunication Networks and Applications Conference 2006

So, J., & Vaidya, N. H. (2004). Multi-channel MAC for Ad Hoc Networks: Handling Multi-Channel Hidden Terminals using a Single Transceiver. *ACM International Symposium on Mobile Ad Hoc Networking and Computing, Mobihoc '2004.*

Sobrinho, J., & Krishnakumar, A. (1999). Quality-of-Service in Ad Hoc Carrier Sense Multiple Access Wireless Networks. *IEEE Journal on Selected Areas in Communications, 17*(8), 1353–1368. doi:10.1109/49.779919

Soliman, H., Castelluccia, C., El Malki, K., & Bellier, L. (2005). Hierarchical Mobile IPv6 Mobility Management (HMIPv6). IETF RFC 4140.

Soliman, H., Castelluccia, C., El Malki, K., & Bellier, L. (2007). Hierarchical Mobile IPv6 Mobility Management (HMIPv6). IETF Draft, draft-ietf-mipshop-4140bis-00.txt.

Srinivasan, V., Nuggehalli, P., Chiasserini, C. F., & Rao, R. R. (2005). An analytical approach to the study of cooperation in wireless ad hoc networks. *IEEE Transactions on Wireless Communications, 4*(2), 722–733. doi:10.1109/TWC.2004.842950

Srinivasan, K., & Levis, P. (2006, May). RSSI is Under Appreciated. In *Proceedings of the Third Workshop on Embedded Networked Sensors (EmNets)*, Boston, MA.

Srivastava, V., & Motani, M. (2005). Cross-layer design: a survey and the road ahead. *IEEE Communications Magazine, 43*(12), 112–119. doi:10.1109/MCOM.2005.1561928

Staub, T., Bernoulli, T., Anwander, M., Wälchli, M., & Braun, T. (2006). Experimental Lifetime Evaluation for MAC Protocols on Real Sensor Hardware. *ACM Workshop on Real-World Wireless Sensor Networks (REALWSN'06)* (pp. 25-29). Uppsala, Sweden: ACM Press.

Stoica, I., Morris, R., Karger, D., Kaashoek, M. F., & Balakrishnan, H. (2001). *Chord: A Scalable peer-to-peer lookup service for internet applications.* Paper presented at the ACM SIGCOM, California, USA.

Strang, T., & Linnhoff-Popien, C. (2004). A context modeling survey. *Workshop on Advanced Context Modeling, Reasoning and Management, The 6th International Conference on Ubiquitous Computing*, Nottingham, England.

Su, W., Lee, S.-J., & Gerla, M. (2001). Mobility Prediction and Routing in Ad Hoc Wireless Networks. *International Journal of Network Management, 11*, 3–30. doi:10.1002/nem.386

Su, W., & Gerla, M. (1999). IPv6 Flow Handoff in Ad-Hoc Wireless Networks Using Mobility Prediction. In *Proc. of IEEE GLOBECOM'99.*

Subramanian, A. P., Gupta, H., & Das, S. (2007). Minimum Interference Channel Assignment in Multi-Radio Wireless Mesh Networks. *IEEE Communications Society Conference on Sensor, Mesh and Ad Hoc Communications Networks, SECON '07*, San Diego California, USA.

Sucec, J., & Marsic, I. (2004). Hierarchical Routing Overhead in Mobile Ad Hoc Netorks. *IEEE Transactions on Mobile Computing, 3*(1), 46–56. doi:10.1109/TMC.2004.1261816

Sudame, P., & Badrinath, B. (2001). On providing support for protocol adaptation in mobile wireless networks. *Mobile Networks and Applications, 6*(1), 43–55. doi:10.1023/A:1009861720398

Suh, C., & Ko, Y.-B. (2005). A traffic aware, energy efficient MAC protocol for wireless sensor networks. *International Symposium on Circuits and Systems (ISCAS'05).*

Suh, J.-J., Quan, S. G., Park, J. D., Park, J. H., & Kim, Y. Y. (2007, January). Asymmetric Link Dynamic Routing Algorithm for Ubiquitous Multimedia Services. In *Proceeding of the 13th International Multimedia Modeling Conference (MMM)*, Singapore.

Tam, W.-H., & Tseng, Y.-C. (2007). Joint multi-channel link layer and multi-path routing design for wireless mesh networks. *IEEE Conference on Computer Communications INFOCOM 2007.*

Tanenbaum, A. S. (Ed.) (2002). *Computer Networks* (4th ed.). Prentice Hall.

Tang, Y., Luo, J.-G., Zhang, Q., Zhang, M., & Yang, S.-Q. (2007). Deploying P2P Networks for Large-Scale Live Video Streaming Service. *IEEE Communications Magazine*, *45*(6), 100–106. doi:10.1109/MCOM.2007.374426

Théoleyre, F., & Valois, F. (2005). Virtual structure routing in ad hoc networks. *In Proceedings of the IEEE International Conference on Communications (ICC'2005)* (pp.3078-3082), Seoul.

Thomesse, J. P. (1998). A review of the fieldbuses. *Annual Reviews in Control*, *22*, 35–45. doi:10.1016/S1367-5788(98)00003-0

Thomson, S., Narten, T., & Jinmei, T. (2007). IPv6 Stateless Address Autoconfiguration. IETF RFC 4862.

Tobagi, F., & KleinRock, L. (1975). Packet Switching in Radio Channels: Part II-the Hidden Terminal Problem in Carrier Sense Multiple-Access Modes and the Busy-Tone solution. *IEEE Transactions on Communications*, *23*(12). doi:10.1109/TCOM.1975.1092767

Toh, C.-K. (1997, March). Associativity Based Routing for Ad-hoc Mobile Networks, *Wireless Personal Communications Journal. Special Issue on Mobile Networking and Computing Systems*, *4*(2), 103–139.

Toh, C.-K. (2002). *Ad Hoc Mobile Wireless Networks: Protocols and Systems*. New Jersey: Prentice Hall.

Tong, L., Naware, V., & Venkitasubramaniam, P. (2004). Signal Processing in Random Access. *IEEE Sig. Proc.*, *21*(5), 29–39. doi:10.1109/MSP.2004.1328086

Torres, D. A., & Garcia-Macias, J. A. (2005). *Performance analysis of two approaches to service discovery in mobile ad hoc networks* (. *LNCS*, *3563*, 368–376.

Tsakountakis, A., Kambourakis, G., & Gritzalis, S. (2007). Towards effective wireless intrusion detection in IEEE 802.11i. In *Proceedings of Third International Workshop on Security Privacy and Trust in Pervasive and Ubiquitous Computing*, IEEE.

Twycross, J., & Williamson, M. M. (2003). Implementing and testing a virus throttle. In *Proceedings of the 12th USENIX Security Symposium*, Washington, USA.

U.S. Department of Defense DISA. (2008). *Net-Enabled Command Capability (NECC) Overview.* Retrieved September 20, 2008 from http://www.disa.mil/necc/

UMTS Forum (2002). Charging, Billing and Payment Views on 3G Business Models. *Technical report (Report No 21).*

Unoki, H., & Higaki, H. (2004, March). Loop-Based Source Routing Protocol for Mobile Ad-hoc Networks. In *Proceeding of the IEEE International Parallel & Distributed Processing Symposium (IPDPS)*, Long Beach, CA, USA.

UPnP forum (2006). *UPnP Technology – the simple, seamless home network* [White paper].

Urpi, A., Bonuccelli, M., & Giordano, S. (2003). Modelling cooperation in mobile ad hoc networks: A formal description of selfishness. In *Proceeding of WiOpt 2003: Modeling Optimization and Mobile in, Ad Hoc and Wireless Networks, INRIA Sophia-Antipolis, France, March 3-5.*

Urrutia, J. (2007). Local Solutions for Global Problems in Wireless Networks. *Journal of Discrete Algorithms*, *5*(3), 395–407. doi:10.1016/j.jda.2006.05.004

Vaidya, N. H. (2004). *Mobile Ad Hoc Networks: Routing, MAC and Transport Issues*. Paper presented at the INFOCOM 2004 (IEEE International Conference on Computer Communication).

Van der Schaar, M., Krishnamachari, S., & Choi, S. (2003). Adaptive Cross-Layer Protection Strategies for Robust Scalable Video Transmission Over 802.11 WLANs. *IEEE Journal on Selected Areas in Communications*, *21*(10). doi:10.1109/JSAC.2003.815231

Van der Schaar, M., & Sai Shankar, D. (2005). Cross-layer wireless multimedia transmission: Challenges, principles and new paradigms. *IEEE Wireless Communications*, *12*(4), 50–58. doi:10.1109/MWC.2005.1497858

Van der Schaar, M., & Turaga, D. S. (2007). Cross-Layer Packetization and Retransmission Strategies for Delay-Sensitive Wireless Multimedia Transmission. *IEEE Transactions on Multimedia*, *9*(1), 185–197. doi:10.1109/TMM.2006.886384

Van Oorschot, P. C., Robert, J.-M., & Vargas Martin, M. (2006). A monitoring system for detecting repeated packets with applications to computer worms. *International Journal of Information Security*, *5*(3), 186–199. doi:10.1007/s10207-006-0081-8

Varner, P. E., & Knight, J. C. (2001). Security monitoring, visualization, and system survivability. In *Proceedings of IEEE/SEI. Information Survivability Workshop (ISW)*.

Varshavsky, A., Reid, B., & De Lara, E. (2005). A cross-layer approach to service discovery and selection in MANETs. In *Proc. IEEE Int'l Conf. on Mobile Ad hoc and Sensor Systems*.

Verikouris, C., Alonso, L., & Giamalis, T. (2005). *Cross-layer optimization for wireless systems: A European research key challenge*. Global Communications Newsletter.

Veronikis, S., Zoutsou, K., & Papatheodorou, C. (2007, October 1). *An evaluation methodology of ubiquitous access in physical and digital information spaces*. Paper presented at 16th Panhellenic Conference of Academic Libraries, Piraeus, Greece.

Ververidis, C. N., & Polyzos, G. C. (2005). Routing layer support for service discovery in mobile ad hoc networks. In *Proceedings of the Third IEEE International Conference on Pervasive Computing and Communications Workshops* (pp. 258-262). IEEE Computer Society.

Walker, J. (2001). *Security event correlation: Where are we now? NetIQ whitepaper.* Retrieved from http://www.net-. iq.com/products/sm/whitepapers.asp

Wang Q. & Abu-Rgheff, M.A., (2003). Cross-Layer Signalling for Next-Generation. *Wireless Systems, IEEE WCNC 2003* (Vol. 2, pp. 1084–89).

Wang, K., & Li, B. (2002). Group Mobility and Partition Prediction in Wireless Ad Hoc Networks. In *Proc. of the IEEE International Conference on Communications (ICC'2002).*

Wang, X., Zhang, D., Gu, T., & Pung, H. (2004). Ontology Based Context Modeling and Reasoning using OWL. *2nd IEEE Conference on Pervasive Computing and Communications* (pp. 18-22), Orlando, Florida.

Wang, Y., & Kim, M. S. (2007). Bandwidth-adaptive Clustering for Mobile Ad Hoc Networks. *International Conference on Computer Communications and Networks,* (pp. 103-108). Honolulu, Hawaii, USA.

Wang, Z., Lu, J., & Tang, J. (2007, March). Neighbor Monitoring Mechanism to Solve Unidirectional Link Problem in MANET. In *Proceedings of the 3rd International Conference on Wireless and Mobile Communications (ICWMC)*, Guadeloupe, French Caribbean.

Want, R., Hopper, A., Falcao, V., & Gibbons, J. (1992). The active badge location system. *ACM Transactions on Information Systems*, *10*(1), 91–102. doi:10.1145/128756.128759

Want, R., Fishkin, K. P., Gujar, A., & Harrison, B. L. (1999, May 15). *Bridging physical and virtual worlds with electronic tags*. Paper presented at SIGCHI Conference on Human Factors In Computing Systems: the CHI is the limit, Pennsylvania, USA. Retrieved February 8, 2007 from ACM.

Warrior, J., McHenry, E., & McGee, K. (2003, July). They know where you are [location detection]. *IEEE Spectrum*, *40*(7), 20–25. doi:10.1109/MSPEC.2003.1209608

Watteyne, T., Dohler, M., & Augé-Blum, I. (2009). Beyond Localization: Communicating Using Virtual Coordinates. In Mao, G., & Fidan, B. (Eds.), *Localization Algorithms and Strategies for Wireless Sensor Networks: Monitoring and Surveillance Techniques for Target Tracking* (pp. 446–467). Hershey, PA: IGI Global.

Watteyne, T., Augé-Blum, I., & Ubeda, S. (2006). Dual-Mode Real-Time MAC Protocol for Wireless Sensor Networks: A Validation/Simulation Approach. *In Proceedings of the First International Conference on Integrated Internet Ad hoc and Sensor Networks (InterSense'2006)* (pp. 2-8). Nice.

Weiser, M. (1991). The computer for the 21st century. *Scientific American*, *265*(3), 66–75. doi:10.1038/scientificamerican0991-94

Whitehouse, K. Zhao, F. & Liu, J. (2005). *Semantic Streams: a framework for declarative queries and automatic data interpretation* (Technical Report MSR-TR-2005-45). Microsoft Research.

Wiegand, T., Schwarz, H., Joch, A., Kossentini, F., & Sullivan, G. J. (2003). Rate-Constrained Coder Control and Comparison of Video Coding Standards. *IEEE Transactions on Circuits and Systems for Video Technology*, *13*(7), 688–703. doi:10.1109/TCSVT.2003.815168

Wiese, A., & Kranakis, E. (2008). Local PTAS for dominating and connected dominating set in location aware unit disk graphs. In *6th Workshop on Approximation and Online Algorithms* (LNCS).

Wu, S.-L., Tseng, Y.-C., Lin, C.-Y., & Sheu, J.-P. (2002). A Multi-Channel MAC Protocol with Power Control for Multi-Hop Mobile Ad Hoc Networks. [SCI]. *The Computer Journal*, *45*(1), 101–110. doi:10.1093/comjnl/45.1.101

Wu, J. (Ed.). (2006). *Handbook on Theoritical and Algorithmic Aspects of Sensor, Ad Hoc Wireless, and Peer-to-Peer Networks*. Auerbach Publications.

Wu, D., Sirin, E., Hendler, J., Nau, D., & Parsia, B. (2003). Automatic web services composition using shop2. In *Proceedings of the International Conference on Automated Planning and Scheduling*.

Wu, J., Wang, S., & Lin, L. (2005). What Drives Mobile Health Care? An Empirical Evaluation of Technology Acceptance. *In Proceedings of the 38th Annual Hawaii International Conference on System Sciences (HICSS '05)* (pp.150–159).

Wu, S.-L., Lin, C.-Y., Tseng, Y.-C., & Sheu, J.-P. (2000). A New Multi-Channel MAC Protocol with On-Demand Channel Assignment for Multi-Hop Mobile Ad Hoc Networks. *International Symposium on Parallel Architectures, Algorithms and Networks, ISPAN '00.*

Wu, S.-L., Tseng, Y.-C., Ni, S.-Y., & Sheu, J.-P. (2000, January). Route Maintenance in a Wireless Mobile Ad Hoc Network. In *Proceedings of the 33rd Hawaii International Conference on System Sciences (HICSS)*, Maui, Hawaii, USA.

WWI-AN. (n.d.). Ambient Networks Project / Reports. Retrieved from www.ambient-networks.org

Xiuyuan, L., Xiaojuan, W., Wenming, L., & Xiaogang, W. (2008). An Adaptive Cross-Layer Scheduling Algorithm for Multimedia Networks. *International Conference on Intelligent Information Hiding and Multimedia Signal Processing* (pp. 52-55).

Xue, M. ER, I., & Seah, W. K. G. (2006). Analysis of Clustering and Routing Overhead for Clustered Mobile Ad Hoc Networks. In *Proceedings of the 26th IEEE international Conference on Distributed Computing Systems*, (pp. 46-53). Lisboa, Portugal.

Yan, L., Sere, K., & Zhou, X. (2004). *Towards an Integrated Architecture for Peer-to- Peer and Ad Hoc Overlay Network Applications.* Paper presented at the 10th IEEE International Workshop on Future Trends of Distributed Computing Systems (FTDCS '04), Suzhou, China.

Yang, H., Xie, L., & Sun, J. (2004). Intrusion detection solution to WLANs. In *Proceedings of IEEE Circuit and System Symposium on Emerging Technologies: Frontiers of Mobile and Wireless Communications*. IEEE.

Yatani, K., Onuma, M., Sugimoto, M., & Kusunoki, F. (2004). Musex: A system for supporting children's collaborative learning in a museum with PDAs. *Systems and Computers in Japan 35*(14), 54-63. Retrieved February 20, 2007 from IEEE Xplore.

Ye, F., Zhong, G., Lu, S., & Zhang, L. (2005). GRAdient Broadcast: A robust data delivery protocol for large scale sensor networks. *Wireless Networks*, *11*(3), 285–298. doi:10.1007/s11276-005-6612-9

Ye, W., & Heidemann, J. (2003). *Medium Access Control in Wireless Sensor Networks* (USC/ISI Technical Report ISI-TR-580).

Ye, W., Heideman, J., & Estrin, D. (June 2004). Medium Access Control With Coordinated Adaptive Sleeping for Wireless Sensor Networks. *IEEE/ACM Transactions on Networking*, 496.

Ye, W., Heidemann, J., & Estrin, D. (2004). Medium Access Control with Coordinated Adaptive Sleeping for Wireless Sensor Networks. *IEEE/ACM Transactions on Networking, 3*(12), 493-506.

Yu, J. Y., & Chong, P. H. J. (2005). A Survey of Clustering Schemes for Mobile Ad Hoc Networks. *IEEE Communications Survey &Tutorials*, *7*(1), 32–48. doi:10.1109/COMST.2005.1423333

Yuan, L. F., Cheng, W. Q., & Du, X. (2007). An energy-efficient real-time routing protocol for sensor networks. *Computer Communications*, *30*(10), 2274–2283. doi:10.1016/j.comcom.2007.06.002

Zahn, T., Winter, R., & Schiller, J. (2004). *Simple, efficient peer-to-peer overlay clustering in mobile, ad hoc networks.* Paper presented at the 12th IEEE International Conference on Networks (ICON 2004), Singapore.

Zhang, Q., Guo, C., Guo, Z., & Zhu, W. (2003). Efficient Mobility Management for Vertical Handoff between WWAN and WLAN. *IEEE Communications Magazine*, *41*(11), 102–108. doi:10.1109/MCOM.2003.1244929

Zhao, L., Kan, B. Q., Xu, Y. J., & Li, X. W. (2007). FT-SPEED: A Fault-Tolerant, Real-Time Routing Protocol for Wireless Sensor Networks. *In Proceedings of the International Conference on Wireless Communications, Networking and Mobile Computing (WiCom'2007)* (pp. 2531-2534). ShangHai.

Zhong, S. G. (2001). An Ultra-Low Power and Distributed Access Protocol For Broadband Wireless Sensor Networks.

Zhong, S., Chen, J., & Yang, Y. R. (2003). Sprite: A Simple Cheat-Proof Credit-Based System for Mobile Ad-Hoc Networks. In *Proc. of the 22nd Annual Joint Conference of the IEEE Computer and Communications Societies (INFOCOM'03), 3*(30), 1987-1997.

Zhou, G., Krishnamurthy, S., & Stankovic, J. A. (2004, June). Impact of Radio Irregularity on Wireless Sensor Networks. In *Proceedings of the 2nd International Conference on Mobile Systems, Applications, and Services (MobiSys)*, Boston, MA, USA.

Zhou, Y. F. Ngai, Edith C.-H., Lyu, Michael R., & Liu, J. (2007). POWER-SPEED: A Power-Controlled Real-Time Data Transport Protocol for Wireless Sensor-Actuator Networks. *In Proceedings of the IEEE Wireless Communications and Networking Conference (WCNC'2007)* (pp. 3739-3743), Hong Kong.

Zhu, X., & Atwood, W. J. (2007). A Web Database Security Model Using the Host Identity Protocol. *11th International Database Engineering and Applications Symposium'07* (pp. 278-284), 6-8 Sept. 2007

Zidek, K., Saloky, T., & Polanecka, I. (2006). Usability of GPS Systems for Mobile Robots Navigation. *4th Sloyakian-Hungarian Joint Symposium on Applied Machine Intelligence*. Herlany, Slovakia.

Zuniga Zamalloa, M., & Krishnamachari, B. (2007). An Analysis on Unreliability and Asymmetry in Low-Power Wireless Links. *ACM Transactions on Sensor Networks (TOSN), 3*(2).

Zupeng Li, X. Y. Peiyang Yao, & Jinnan Huang. (2006). *Implementation of P2P Computing in Design of MANET Routing Protocol.* Paper presented at the First International Multi-Symposiums on Computer and Computational Sciences (IMSCCS'06), Washington, DC, USA

About the Contributors

Samuel Pierre received the B.Eng. degree in civil engineering in 1981 from École Polytechnique de Montréal, Québec, the BSc and MSc degrees in mathematics and computer science in 1984 and 1985, respectively, from the UQAM, Montréal, the MSc degree in economics in 1987 from the Université de Montréal, and the Ph.D. degree in Electrical Engineering in 1991 from École Polytechnique de Montréal. From 1987 to 1998, he was a Professor at the Université du Québec à Trois-Rivières prior to joining the Télé-Université of Québec, an Adjunct Professor at Université Laval, Ste-Foy, Québec, an Invited Professor at the Swiss Federal Institute of Technology, Lausanne, Switzerland, then the Université Paris 7, France. He is currently a Professor of Computer Engineering at École Polytechnique de Montréal where he is Director of the Mobile Computing and Networking Research Laboratory (LARIM), chairholder of the NSERC - ERICSSON Chair in Next Generations Mobile Networking Systems and Director of Mobile Computing and Networking Research Group (GRIM). His research interests include wireline and wireless networks, mobile computing, artificial intelligence, and telelearning. He is a senior member of IEEE and a member of ACM. He is a Regional Editor of the Journal of Computer Science, an Associate Editor of IEEE Communications Letters, IEEE Canadian Journal of Electrical and Computer Engineering and IEEE Canadian Review, and serves on the editorial board of Telematics and Informatics edited by Elsevier Science. He has received many distinctions, such as the Prix Poly 1873 for excellence in teaching (2001 and 2005), Fellow of Engineering Institute of Canada (2003), among others. He is a Fellow of the Canadian Academy of Engineering (2008). In 2009, he has been awarded Chevalier de l'Ordre national du Québec, "the highest distinction given by the Québec government."

* * *

Abdelhalim Baaziz is professor of computer science at the computer science department of the University of Annaba (Algeria). He holds Engineer Diploma in Computer Engineering (1990) of the University of Annaba, the Master's degree in Computer System (1994) from the same university and Philosophiae Doctor's Degree in Computer Engineering (2008) of the polytechnic School of Montreal (Canada). His research interests are mainly mobile ad hoc networks, but he is also interested in web technologies and grid computing.

Alex Galis is a visiting professor in the Networks and Services Research Laboratory of the Department of Electronic and Electrical Engineering, University College London (UCL). He has co-authored and published (www.ee.ucl.ac.uk/~agalis) 4 books and more than 135 scientific papers on various networking topics, including network and service management, intelligent services, programmable and active networks, grid systems and ambient networks.

Bertrand Mathieu joined France Telecom, Division R&D in 1994. Until 1999, he had worked on network management including interfaces, protocols and platforms. From 1999 to 2001, he had worked on distributed computing (Corba, DCOM, RMI, …) for the Telecommunications domain and contributed to collaborative projects. Since 2001, he is working on programmable networks, ad hoc networks and he is currently focusing its research activity on dynamic overlay networks and P2P networks. Content adaptation and context-awareness are the main subjects he is interested in and he is studying how they can be improved using those networks approaches.

François Jan is a research engineer in the field of the wireless networks at Orange R&D located in Lannion (France). His main interests are the survey of wireless mesh and ad hoc networks in real environments, the survey and the evaluation of ad hoc routing protocols, the performances improvement of these protocols and architectures survey and designing.

Carine Toham got her PhD on December 2008 from the Institut National Polytechnique de Grenoble (FRANCE). She worked for 3 years at Orange Labs Lannion (France) to achieve that PhD thesis. She also got a Master Degree in Computer Sciences in 2004 from ESEO-Angers (France). She is interested in wireless networks and IP architectures designing for Next Generation Networks.

Christos Bouras obtained his Diploma and PhD from the Department Of Computer Engineering and Informatics of Patras University (Greece). He is currently a Professor in the above department. Also he is a scientific advisor of Research Unit 6 in Research Academic Computer Technology Institute (CTI), Patras, Greece. His research interests include Analysis of Performance of Networking and Computer Systems, Computer Networks and Protocols, Telematics and New Services, QoS and Pricing for Networks and Services, e-Learning Networked Virtual Environments and WWW Issues. He has extended professional experience in Design and Analysis of Networks, Protocols, Telematics and New Services. He has published 300 papers in various well-known refereed conferences and journals. He is a co-author of 8 books in Greek. He has been a PC member and referee in various international journals and conferences. He has participated in R&D projects such as RACE, ESPRIT, TELEMATICS, EDUCATIONAL MULTIMEDIA, ISPO, EMPLOYMENT, ADAPT, STRIDE, EUROFORM, IST, GROWTH and others. Also he is member of experts in the Greek Research and Technology Network (GRNET), Advisory Committee Member to the World Wide Web Consortium (W3C), IEEE - CS Technical Committee on Learning Technologies, IEEE ComSoc Radio Communications Committee, IASTED Technical Committee on Education WG6.4 Internet Applications Engineering of IFIP, ACM, IEEE, EDEN, AACE, New York Academy of Sciences and Technical Chamber of Greece.

Apostolos Gkamas obtained his Diploma, Master Degree and PhD from the Computer Engineering and Informatics Department of Patras University (Greece). He is currently an R&D Computer Engineer at the Research Unit 6 of the Research Academic Computer Technology Institute, Patras, Greece. He is also visiting lecturer in the Hellenic Open University. His research interests include Computer Networks, Telematics, Distributed Systems, Multimedia and Hypermedia. More particular he is engaged in transmission of multimedia data over networks and multicast congestion control. He has published more than 40 papers in international Journals and well-known refereed conferences. He is also co-author of three books (one with subject Multimedia and Computer Networks one with subject Special Network Issues and one with subject IPv6). He has participated in various R&D project (in both EU and national) such as IST, FP6, eLearning, PENED, EPEAEK, Information Society.

Georgios Kioumourtzis is a PhD candidate in the Computer Engineering and Informatics Department at the University of Patras, Greece. He has received my B.S from the Hellenic MIlitary Academy and graduated also from the School of Telecommunications for Signal Officers. In 2005, He received the Master of Science in System's Engineering and the Master of Science in Computer Science from the Naval Postgraduate School California, USA. His thesis work was related to Mobile AdHoc Wireless Networks (MANETs). His current research interests include Multimedia transmission over heterogeneous networks, transmission protocols, cross-layer optimization in wireless networks and IEEE 802.11x technology

Sándor Imre was born in Budapest in 1969. He received the M.Sc. degree in Electronic Engineering from the Budapest University of Technology (BUTE) in 1993. Next he started his PhD studies at BUTE and obtained dr. univ. degree in 1996, PhD degree in 1999 and DSc degree in 2007. Currently he is carrying his teaching activities as Professor at Dept. of Telecommunications of BUTE. He was invited to join the Mobile Innovation Centre of BUTE as R&D director in 2005. His research interest includes mobile and wireless systems. Especially he has contributions on different wireless access technologies, mobility protocols and reconfigutable systems.

Bálint Dávid Ary was born in Pécs in 1981. He received the MSc degree in IT from the Budapest University of Technology (BUTE) in 2000. While starting his PhD studies at BUTE at the same year, he is started to working for an international software company, which is focusing on billing and customer management systems especially for telecommunication companies.

Jorge Urrutia is Professor of Mathematics at the Instituto de Matematicas, Universidad Nacional Autonoma de Mexico. He has published numerous articles in Discrete and Computational Geometry. He was co-founding editor, and co-editor-in-chief (until Dec. 1999) of the journal Computational Geometry, Theory and Applications published by Elsevier. Currently he is a member of the editorial boards of Computational Geometry, Theory and Applications, Graphs and Combinatorics, and Boletin de la Sociedad Matematica Mexicana.

Evangelos Kranakis received a B.Sc. in Mathematics from the University of Athens in 1973 and a Ph.D. Mathematical Logic from the University of Minnesota, USA, in 1980. From 1980 to 1982 he was at the Mathematics Department of Purdue University, USA, from 1982 to 1983 at the mathematisches institut of the University of Heidelberg, Germany, from 1983 to 1985 at the Computer Science Department of Yale University, USA, from August to December of 1985 at the Computer Science Department of the Universiteit van Amsterdam, and from 1986 to 1991 at the Centrum voor Wiskunde en Informatica in Amsterdam, The Netherlands. He joined the faculty of the School of Computer Science of Carleton University in the Fall of 1991 and was director from 1994 to 2000. He has been in the Research Management Committee and Communications, Networks, and Security Theme Leader of MITACS (Mathematics of Information Technology and Complex Systems) since 1998. He is currently serving in the NSERC Grant Selection Committee since fall 2007. He has published three research monographs and several articles in the analysis of algorithms, bioinformatics, communication and data (ad hoc and wireless) networks, computational and combinatorial geometry, distributed computing, and network security. He became Carleton University Chancellor's Professor in 2006.

Florence Agboma is a PhD candidate in the Department of Computing and Electronic Systems at the University of Essex, UK. Her research interests include Quality of Experience of mobile services, multimedia applications, Quality of Service management and peer-to-peer streaming. Florence received the MSc degree in Computer Information and Networks from the University of Essex and her BSc degree in Electronic Communication Systems from the University of Plymouth, UK. She is a student member of the IEEE and IET society.

Antonio Liotta holds the Chair of Communication Networks at the Eindhoven University of Technology (The Netherlands) where he leads the Autonomic Networks Team. He was previously a Reader at the University of Essex (UK), coordinating the Pervasive Services Team. Antonio is a Fellow of the U.K. Higher Education Academy and serves a number of senior advisory boards including Research Councils, Journals and international conferences. He has over 100 publications to his credit covering the broad area of advanced networking, service management and P2P systems.

Isabelle Augé-Blum obtained the Master degree in 1996 in Industrial Computer Sciences and the Ph.D. degree in 2000 about Formal Validation of Real-Time Fieldbus, both from LAAS laboratory, University Toulouse III. Since September 2000, she is an associate professor at the INSA of Lyon in the Telecommunications department. She is also a member of the Center for Innovations in Telecommunication and Services integration (CITI Lab) and a member of the INRIA Project named ARES for Architecture of Networks of Services. Her research interests include self-organization, real time and reliability in Wireless Sensor and Mesh Networks.

Fei Yang received the B.S. degree in electronic engineering from HuBei Polytechnical University, Wuhan, China, in 2004 and the M.S degree in control science and engineering from NorthWestern Polytechnical University, XiAn China, in 2007. He is working toward the Ph.D degree at the CITI lab, INSA lyon, France. His research interest is in real-time communications in Wireless Sensor Networks.

Thomas Watteyne holds an MSc degree, specialized in Telecommunications and Networking from INSA Lyon, France, 2005. He is currently finishing his PhD studies at the CITI Laboratory and Orange Labs. His research has facilitated him a four months stay at the University of California at Berkeley, working with Prof. Kris Pister. His research interests include wireless sensor networks, self-organization principles and energy-efficiency. He is member of the IEEE Region 8 Student Activities Committee, has published several conference papers, holds two patents, has been TPC member and member of the organizing committees of various conferences.

Li Jun Zhang received a Ph. D and M.Sc.A degree in Computer Engineering (2007) from the Ecole Polytechnique de Montreal, Montreal, Canada and a B.Sc in Mathematics (1993) from Jilin University, Changchun, China. From July 1993 to May 2001, she worked as Software Engineer, System Analyst, System Administrator, System Integrator, Network Administrator and Database Engineer at

China Construction Bank, Dalian, China. Since September 2003, she has been working as Graduate Research Assistant at Mobile Computing and Networking Research Laboratory (LARIM) and Intern Research Assistant at Ericsson Research Canada, since January 2004. She works as Teaching Assistant at the Department of Computer Engineering and Software Engineering of the Ecole Polytechnique de Montreal. She served as a reviewer for several conference and journal papers. She is currently working as R&D Project Manager, Computing and Networking in R&D division at Geninov Group Inc. Her current research interests include mobile and wireless networking architectures, mobility management and wireless network security for next-generation heterogeneous networks, 3G and 4G wireless network design and planning, wireless number portability, RF design, etc. She has applied three seamless handoff related patents in USA, two book chapters, and has published several peer-reviewed journal and conference papers.

Miguel A. Garcia-Ruiz graduated in Computer Systems engineering and obtained his MSc in Computer Science from the University of Colima, Mexico. He received his PhD in Computer Science and Artificial Intelligence from the University of Sussex, England. Miguel has been a visiting professor at the University of Ontario, Institute of Technology, Canada. Miguel is currently an Assistant Professor at the University of Colima, where teaches CS courses and do research mainly on virtual reality. He has published more than twenty scientific papers and chapters, including a book, and has directed a video documentary about an introduction to virtual reality.

Miguel Vargas Martin, PhD., PEng., is an Assistant Professor at the University of Ontario Institute of Technology (Oshawa, Canada), and Chief Technology Officer of Hoper Inc., an Oshawa-based research and development company that offers innovative Web tools. He is a licensed Professional Engineer in Ontario, Canada. He holds a Ph.D. in Computer Science (Carleton University, Canada), a Master's degree in Electrical Engineering (CINVESTAV, Mexico), and a Bachelor of Computer Science (UAA, Mexico). His current research interests include computer forensics, mitigation of denial-of-service attacks, security and human computer interaction, hidden communication channels, and Web modeling and optimization.

Raúl Aquino Santos graduated from the University of Colima with a BE in Electrical Engineering, received his MS degree in Telecommunications from the Centre for Scientific Research and Higher Education in Ensenada, Mexico in 1990. He holds a PhD from the Department of Electrical and Electronic Engineering of the University of Sheffield, England. Since 2005, he has been with the College of Telematics, at the University of Colima, where he is currently a Research-Professor in telecommunications networks. His current research interests include wireless and sensor networks.

Omar Alvarez-Cardenas received the M.Sc degree in Telematics at the University of Colima, Mexico, in 1999. He is finishing the Ph.D in Telecommunications at the Jose Antonio Echeverria Institute from La Havana, Cuba. Since 1998 he has been teaching networking at the School of Telematics where he is a Research-Professor in telecommunications. His research focuses on Network Simulations, QoS in Metropolitan Ethernet Networks, mobile computing and bandwidth efficiency.

Arthur Edwards is a Senior Professor/Researcher at the College of Telematics of the University of Colima, Mexico, where his primary interest is Computer Assisted Language Learning, multimedia applications, collaborative learning environments, educational information systems, virtual reality applications and wireless and mobile learning systems.

Margarita Mayoral-Baldivia received the M.Sc degree in Telematics at the University of Colima, Mexico, in 1999. Since 1996, she has been working in the networking area and actually she is a Research-Professor in telecommunications networks and CCNA Cisco Networking Academy instructor. Her main research focuses in security, wire and wireless networks and bandwidth efficiency.

Nadia N Qadri is currently a PhD student at School of Computer Science and Electronics Engineering, University of Essex, UK. She received her Masters of Engineering (Communication Systems and Networks) and Bachelors of Engineering (Computer Systems), from Mehran University of Engineering and Technology, Jamshoro, Pakistan in 2004 and 2002 respectively. She has more than 4 years of teaching and research experience at renowned universities of Pakistan viz Mehran University of Engineering & Technology, Fatima Jinnah Women University and COMSATS Institute of Information Technology. Her research interests include Mobile Ad hoc Networks, Vehicular ad hoc Networks, Peer-to-Peer Systems, P2P file sharing, P2P Streaming, Video Error Resilience Techniques, and Video Streaming. She is a student member of IEEE and IET.

Noha Ibrahim holds a 'Diplôme d'Ingénieur' from the Ecole Nationale Supérieure d'Informatique et de Mathématique Appliquée de Grenoble (ENSIMAG), and a PHD from National Institute for Applied Sciences (INSA Lyon), about service integration in pervasive environments. Her dissertation focused on providing a spontaneous service integration middleware adapted for pervasive environnements. Noha Ibrahim is currently a post doctoral at the Laboratoire d'nformatique de Grenoble (LIG) laboratory where she works on spontaneous mesh networks. Her main interests are pervasive and ambient computing for the middleware but also the network layers.

Frédéric Le Mouël holds a 'Diplôme d'Ingénieur en Informatique et Communication', French equivalence of a master's degree in Languages andOperating Systems, and a Ph.D. degree from the University of Rennes 1, France.His dissertation focused on an adaptive environment for distributed execution of applications in a mobile computing context. Frédéric Le Mouël is currentlyassociate professor in the National Institute for Applied Sciences of Lyon(INSA Lyon, France), Telecommunications Department, Center for Innovation inTelecommunication and Integration of Services (CITI Lab.). He is also member ofthe AMbient Architectures: Service-Oriented, Networked, Efficient, Secureresearch group at INRIA (Amazones Team). His main interests are component andservice-oriented middleware and more specifically on the fields of dynamicadaptation, composition, coordination and trust of services. He is speciallystudying these topics in the domain of ambient intelligence.

Stéphane Frénot holds a 'Diplôme d'Ingénieur' from INSA Lyon, and a PhD from University Lyon I, France about distributed information systems in hospitals. Stéphane Frénot is associate professor at the Center for Innovation in Telecommunication and Integration of Services (CITI Lab.), Telecommunications Department of the National Institute for Applied Sciences of Lyon (INSA Lyon, France). He is also member of the AMbient Architectures: Service-Oriented, Networked, Efficient, Secure research group at INRIA (Amazones Team). His main interests are middlewares, component instrumentation, component security, deployment and management, constrained devices and specially home gateways.

Pramita Mitra is a graduate student in the Computer Science and Engineering Department at the University of Notre Dame. She received her Bachelor of Engineering degree in Computer Science and Engineering from the Jadavpur University, India in 2006. Her research interest includes reliable routing mechanisms in wireless ad-hoc networks such as link stability, link asymmetry, and multi-receiver communications in mobile environments.

Christian Poellabauer is an Assistant Professor in the Computer Science and Engineering Department at the University of Notre Dame. He received his Ph.D. degree in Computer Science from the Georgia Institute of Technology in 2004. He has published over 50 refereed conference and journal papers in areas such as real-time systems, energy-efficient computing, wireless networks, and distributed computing. Dr. Poellabauer is recipient of a 2006 NSF CAREER award and he is a member of IEEE Computer Society, ACM, and USENIX.

Spyros Veronikis holds a Diploma and an MSE in Electronic and Computer Engineering from Democritus University of Thrace, Greece. Currently, he is a PhD candidate at the Department of Archive and Library Sciences, Ionian University, Corfu, Greece. His research interests include wireless communications, mobile computing and integration of diverse information sources.

Kyriaki Zoutsou received her B.Sc. and M.Sc. degrees from the Department of Archive and Library Sciences, Ionian University, Corfu, Greece. She currently works as a librarian in the Library and Information Center at University of Patras, Greece. Her research interests include digital libraries evaluation and digital preservation.

Giannis Tsakonas is a PhD candidate student in the Department of Archive and Library Sciences, Ionian University, Corfu, Greece. He is also a librarian in the Library and Information Center, University of Patras, Greece. His research interests include user centred digital library evaluation, information behaviour and visual communication.

Christos Papatheodorou holds a BSc. and a Ph.D in Computer Science from the Department of Informatics, Athens University of Economics and Business, Athens, Greece. Currently he is an Assistant Professor at the Department of Archive and Library Sciences, Ionian University, Corfu, Greece, where he teaches Information Systems, Information Retrieval and Metadata. His research interests include, Digital Libraries Evaluation, Metadata Interoperability, User Modelling, Personalization and Web Mining. He has been involved in several national and international R&D projects, as well as he has participated to the programme committee of various international conferences and workshops.

Stan Kurkovsky is an associate professor at the Department of Computer Science at Central Connecticut State University. Stan earned his PhD from the Center for Advanced Computer Studies of the University of Louisiana in 1999. The results of his doctoral research have been applied to network planning and industrial simulation. Stan's research interests are in mobile commerce and pervasive computing, distributed systems and software engineering. He published over 40 papers in refereed proceedings of national and international conferences, scientific journals and books. Stan serves as a reviewer and a member of program committees on a number of national and international conferences. During his academic career, Stan received over a million dollars in funding from private and federal sources.

Torsten Braun received diploma and PhD degrees in Computer Science from the University of Karlsruhe, Germany, in 1990 and 1993, respectively. From 1994 to 1995, he was a guest scientist with INRIA Sophia-Antipolis. From 1995 to 1997, he worked as a project leader and senior consultant at the IBM European Networking Center, Heidelberg, Germany. Since 1998, he has been a full professor of Computer Science at University of Bern, Switzerland, heading the Computer Networks and Distributed Systems Research group.

Markus Anwander got his diploma degree from the University of Bern, Switzerland, in 2006. Since October 2006, he is working as a research assistant and Ph.D. student in the Computer Networks and Distributed Systems research group at the University of Bern. His research interests include mobile ad-hoc networks and wireless sensor networks.

Philipp Hurni received his B.S. and M.S. degree in 2006 and 2008 from University of Bern, Switzerland. He received the "KUVS GI/ITG" award for his master thesis "Unsynchronized Energy-Efficient Medium Access Control and Routing in Wireless Sensor Networks" in 2008. He has been working as a Visiting Researcher at Purdue University during Winter/Spring 2008. Since April 2008, he is working towards PhD in the field of energy-efficient MAC and routing protocols in wireless ad hoc and sensor networks at University of Bern.

Markus Wälchli received his Master Degree in Computer Science from University of Bern in January 2005. Since February 2005 he has been employed as a PhD student at University of Bern. He participates in the National Competence Center in Research on Mobile Information and Communication Systems (NCCR-MICS), a center supported by the Swiss National Science Foundation under grant number 5005-67322. His main research interests are event detection, tracking, and classification in wireless sensor networks. Furthermore, he is doing research in medium access control and topology control in sensor networks.

Quan Yuan is a PhD Candidate with the Computer Science and Engineering Department at Florida Atlantic University. He obtained both of his M.S. degree and B.S. degree from Department of Computer Science and Technology, Huazhong University of Science and Technology, Wuhan, China, in Computer Science. His research interests include Wireless Networks and Mobile Computing, Web Service, and Disruption Tolerant Networks.

Mihai Fonoage is a PhD Candidate with the Computer Science and Engineering Department at Florida Atlantic University. He has a B.S. degree in Computer Science from the Technical University of Cluj-Napoca, with a degree thesis in mobile e-commerce. His current areas of interest involve Mobile Platforms and Technologies, Modeling Languages, Semantic Web, and Wireless Sensor Networks.

Ionut Cardei got his Ph.D. in Computer Science in 2003 from the University of Minnesota in Minneapolis. Since 2004 he has been an Assistant Professor at the Department of Computer Science and Engineering at Florida Atlantic University. His main research interests are in the area of automation for model-driven design and in protocols for wireless ad-hoc and sensor networks.

Zhaowen Xing received his B.S. degree in Mathematics from Suzhou Railway Teachers College, China, in 1995, M.S. degree in Mathematics and M.S. degree in Computer Science from the University of Oklahoma, in 2001 and 2002, respectively. From July 1995 to May 1999, he was a MATH teacher at Beijing Nankou Railway Middle School, China. From January 2003 to May 2006, he worked for Oklahoma State Department of Health, USA, as a web developer and an application specialist. As a Ph.D. candidate, His research interests include mobile ad hoc networks, mobile peer-to-peer networks, distributed databases and mobile databases.

Le Gruenwald is the Director and Presidential and Dr. David W. Franke Professor in the School of Computer Science at The University of Oklahoma (OU). She received her Ph.D. in Computer Science from Southern Methodist University in 1990. She was a Program Director at National Science Foundation, a Member of Technical Staff in the Database Management Group at the Advanced Switching Laboratory of NEC, America, a Software Engineer at White River Technologies, and a Lecturer in the Computer Science and Engineering Department at Southern Methodist University. Her major research interests include Mobile Databases, Sensor Databases, Information Privacy and Security, Web-enabled Databases, Real-Time Main Memory Databases, Bioinformatics, Data Warehouse and Data Mining. She has published more than 150 technical articles in journals, books and conference proceedings.

Keat Keong Phang is an Associate Professor at the Faculty of Computer Science & Information Technology, University of Malaya, Malaysia. He obtained his PhD from the University of Malaya in 2004. He teaches object-oriented programming, computer networks and distributed systems. His current research interests include high speed computer networks, network quality of service, grid computing, mobile ad hoc networks, distributed systems, distributed databases and fuzzy logic.

Szabolcs Nováczki was born in Nagykanizsa, Hungary, in 9th February 1982. He received the M.Sc. degree in Electrical Engineering in 2006 from the Budapest University of Technology and Economics (BME), Hungary in the field of mobile communications and computer architectures. He is currently a Ph.D. student at BME as the member of Mobile Communications and Computing Laboratory (MC2L) and Mobile Innovation Center Hungary (MIK). He is also a student member of IEEE. His research interests include advanced network layer mobility solutions, new networking architectures, network simulation programming and network performance analysis.

László Bokor graduated with M.Sc. degree in Computer Engineering from the Budapest University of Technology and Economics (BME) at the Department of Telecommunications, in 2004. He also holds an M.Sc.+ Specialist of Bank Informatics degree from BME's Department of Information and Knowledge Management. He is a Ph.D. student at the same university, student member of the IEEE, member of Mobile Communications and Computing Laboratory (MC2L), and employed as research fellow in the Mobile Innovation Center Hungary (MIK) where he participates in researches of wireless protocols and works on mobility management related projects (e.g. FP6-IST ANEMONE, FP7-ICT OPTIMIX, CELTIC-BOSS). His research interests include IPv6 mobility, mobile computing, next generation networks, mobile broadband networking architectures, network performance analyzing, and heterogeneous networks.

Gábor Jeney was born in Miskolc, Hungary, in 13th January 1975. He received the M.Sc. degree in Electrical Engineering in 1998 from the Technical University of Budapest, Hungary in the field of mobile and optical communications. He received the Ph.D. degree in Electrical Engineering in 2005 from the same university, renamed meanwhile as Budapest University of Technology and Economics, Hungary in the field of multi-user detection. He also holds an M.Sc.+ Engineer-economist degree from the Budapest University of Economic Sciences and Public Administration, Hungary. He is currently a research fellow at the Mobile Innovation Centre, Budapest University of Technology and Economics, Budapest, Hungary. He is involved in several national and European Commission funded project. He was the author/coauthor of several dozens of articles published in conferences and journals. His research interest includes neutral networks, mobile and wide-band communication systems, networking and transport protocols. Dr. Jeney is a member of HTE (Scientific Association for Info-communications), Hungary since 1999, and he is a member of IEEE since 2000. Recently, he was involved in the Technical Program Committee (TPC) of ICC 2007.

Sándor Imre is member of IEEE and HTE. He received the M.Sc. degree in Telecommunication Engineering from the Budapest University of Technology and Economics (BME) in 1993. In 1999 he obtained the Ph.D. degree in Electrical Engineering and D.Sc. in 2007. He is an Associate Professor at the Department of Telecommunications of BME and the R&D Director of the Mobile Innovation Center, Hungary. His research areas include IP mobility, routing, reliability; Wireless access; Software Defined Radio; Quantum Computing.

Haidar Safa received a B.Sc. in Computer Science in 1991 from the Lebanese university, Lebanon, M.Sc. in Computer Science in 1996 from University of Quebec at Montreal (UQAM), and a Ph.D. in Electrical and Computer Engineering in 2001 from Ecole Polytechnique de Montreal. He joined ADC Telecommunications in 2000 then SS8 Networks in 2001 where he worked on designing networking and system software. In 2003, he joined the American University of Beirut where he is currently an assistant professor. Dr. Safa's research interests include computer networks, mobile and pervasive computing, mobile and wireless networks, Quality of Service, and network security.

Zeina Koteiche received a B.Sc. in Computer Science in 2005 from the American University of Beirut where she is currently pursuing her graduate studies. She joined the Consolidated Contractors Company (CCC Intl.) as a software engineer working on designing and developing a just-in time visualization framework for planning mass construction projects. Her research Interests include pervasive computing, networking, and mobile services.

Charalampos Konstantopoulos received the "Diploma" (five year first degree) in computer engineering from the Department of Computer Engineering and Informatics at University of Patras, Greece (1993). He also received his PhD degree in Computer Science from the same department in 2000. Currently, he is a Lecturer in the Department of Informatics, University of Piraeus, Greece. His research interests include parallel and distributed algorithms/architectures and mobile computing.

Grammati Pantziou received the Diploma in Mathematics and the Ph.D. Degree in Computer Science from the University of Patras, Greece, in 1980 and 1991, respectively. She was a Post-Doctoral Research and Teaching Fellow at the University of Patras (1991-1992), a Research Assistant Professor at Dartmouth College, Hanover, NH, USA (1992-1994), an Assistant Professor at the University of Central Florida, Orlando, FL, USA (1994-1995) and a Senior Researcher at the Computer Technology Institute, Patras (1995-1998). Since September 1998, she is a Professor at the Department of Informatics of the Technological Educational Institution of Athens, Greece. Her current research interests are in the areas of parallel computing, design and analysis of algorithms, distributed and mobile computing and multimedia systems.

Damianos Gavalas received his BSc degree in Informatics (Computer Science) from University of Athens, Greece, in 1995 and his MSc and PhD degree in electronic engineering from the University of Essex, U.K., in 1997 and 2001, respectively. Currently, he is an Assistant Professor in the Department of Cultural Technology and Communication, University of the Aegean, Greece. His research interests include distributed computing, mobile code, network and systems management, e-commerce, mobile ad-hoc & sensor networks.

B. Mamalis obtained his diploma (in Computer Engineering and Informatics) as well as his Ph.D. from the University of Patras (Greece), in 1993 and 1997 respectively. He has worked for almost ten years (in several projects and research programs) in Computer Technology Institute (C.T.I.) of Patras as well as an academic teacher in the Hellenic Open University. During the last two years he teaches as an Assistant Professor at the Technological Educational Institution (T.E.I.) of Athens, Dept. of Informatics. His research interests include information retrieval, document & knowledge management, parallel algorithms & architectures, distributed systems & databases, operating systems and mobile computing.

Mohammad Shirali-Shahreza is an undergraduate student in computer science at Sharif University of Technology in IRAN. He got his diploma from Allameh Helli high school, Tehran, IRAN, that is a school for exceptional talents students. He is selected as "Iranian Distinguished University Student" by the Ministry of Science, Research and Technology of Iran in 2008. His project on Steganography won the best prize of 5th Iranian Khwarizmi young festival. He has 90 accepted papers in international conferences, 16 published papers in journals, two book chapters, one book and one Iranian patent. He won the "Best Undergraduate Researcher Award" from Iranian Society of Cryptology (2006), the "Young Researcher Award" of IEEE ICTTA 2006 conference and the "Young Researcher Award" of the 11th International CSI Computer Conference (CSICC'2006). He is a student member of IEEE and Iranian Computer Society. His research interest includes Steganography, CAPTCHA, mobile programming, and e-learning.

M. Hassan Shirali-Shahreza is an assistant professor in the Virtual Education Graduate College of Amirkabir University of Technology, Tehran, IRAN. He was educated at Isfahan University of Technology, Isfahan, IRAN (B.Sc., Computer Hardware Engineering, 1986), Sharif University of Technology, Tehran, IRAN (M.Sc., Computer Hardware Engineering, 1988), and Amirkabir University of Technology (Tehran Polytechnic), Tehran, IRAN (Ph.D. Computer Hardware Engineering, 1996). In 1994 he was a visiting research scholar in the Electrical Engineering Department, Southern Methodist University (SMU), Dallas, Texas, USA. His interesting fields are Genetic Algorithm, Neural Networks, OCR (Optical Character Recognition), and Pattern Recognition. He is a member of Iranian Computer Society, Iranian Society of Cryptology and a senior member of IEEE.

Moez Jerbi received his M.S degree in computer engineering (honors) from INSA Lyon, and Ph.D. degree from the University of Evry, France in 2005 and 2008 respectively. From September 2005 to October 2008, he was working as research Engineer in Orange Labs - France Telecom R&D. He acted and still acts as a technical reviewer for several IEEE conferences and journals (ICC, VTC, VTM, PIMRC, etc). His research interests include wireless Communications, mobile ad hoc networks and inter-vehicular communications. Moez is a member of IEEE.

Sidi-Mohammed Senouci received the M.Sc. degree from the University of Paris 13, France, and the Ph.D. degree from the University of Paris 6, France in 1999 and 2003 respectively. From 2002 to 2004, he was an associate lecturer in the University of Cergy-Pontoise, France. From December 2004, he joined France Telecom R&D Lannion where he is project leader of different European and French projects dealing with infrastructure-less networks. He acted as general chair of Ubiroads2007 workshop (collocated with IEEE GIIS2007) and editor of a special issue at UBICC journal. He is founding co-editor of the IEEE ComSoc Ad Hoc and Sensor Network Technical Committee (AHSN TC) Newsletter. He has been TPC member of various conferences. He acted as TPC Co-chair of VehiCom2009 Workshop associated with IEEE IWCMC'2009. He is a Member of IEEE and is a Senior Member of the French engineering society SEE.

Yacine Ghamri-Doudane is currently an assistant professor (or associate professor depending on translations) at the Ecole Nationale Supérieure d'Informatique pour l'Industrie et l'Entreprise (ENSIIE), a major French post-graduate school located in Evry, France. He is also a member of the Networks and Multimedia Systems Research Group (LRSM). He received an engineering degree in computer science from the National Institute of Computer Science (INI), Algiers, Algeria, in 1998, an M.S. degree in signal, image and speech processing from the National Institute of Applied Sciences (INSA), Lyon, France, and a Ph.D. degree in computer networks from the Pierre & Marie Curie University, Paris 6, France, in 1999 and 2003, respectively. He acted or still acts as TPC Co-Chair of the SSMO'08 Workshop, General Co-Chair of the VehiCom'09 Workshop, IEEE ICC'09 and ICC'10 Selected Area in Communications Symposium Co-Chair. He was also involved in the organizing committees of IEEE GIIS 2007, IEEE MUCS 2008, IEEE IM 2009, and IEEE GIIS 2009. He is member of IEEE. Since December 2007, he is the Secretary of the IEEE ComSoc Technical Committee on Information Infrastructure (TCII). He is founding co-editor of the IEEE ComSoc Ad Hoc and Sensor Network Technical Committee (AHSN TC) Newsletter.

Mohamed oussama Cherif is currently a PhD student at the Université de Technologie de Compiègne (UTC). He received his engineering degree in Telecommunications from the Ecole Nationale d'Electronique, Informatique et Radiocommunication de Bordeaux (ENSEIRB), Bordeaux, France, in 2007 and his M.S degree in Networks & Systems from the University Bordeaux I in 2007.In November 2007, he joined France Telecom R&D Lannion where he works within many projects dealing with vehicular networks.

Index

Symbols

3rd Generation Partnership Project (3GPP) 17, 24, 25, 44, 49, 80, 89
3rd party providers 44, 46, 47, 48

A

Active Badge Location System 157
Adaptive Election Algorithm 167
ad hoc environments 217
ad hoc network applications 227
ad-hoc networks 132, 133, 135, 137, 138, 139, 141, 142, 217, 218, 219, 221, 222, 223, 224, 225, 226, 227, 228, 229, 232, 233
Ad-hoc On-Demand Distance Vector (AODV) 113, 114, 116, 120, 133, 134, 140, 143, 144
ad-hoc radio 253
Ad-hoc routing protocols 133
advertisement and discovery (Ad/D) 220
advice of charge (AoC) 47, 48, 49
ambient conditions 159
ambient control space (ACS) 17, 19, 20, 21, 22, 24, 25, 26
ambient network (AN) 17, 18, 19, 20, 21, 22, 23, 24, 25, 26
ambient network interface (ANI) 19, 21, 26
ambient resource interface (ARI) 19, 21, 26
ambient service interface (ASI) 19, 20, 21, 26
Anchor-based Street and Traffic Aware Routing (A-STAR) 257, 258, 261
Artificial Intelligence (AI) 175, 176, 178, 183, 184, 185
Association_REQuest (AREQ) 9
Associativity Based Routing (ABR) 134, 144
asymmetric link 132, 133, 134, 135, 137, 138, 139, 140, 141, 142, 145
Asynchronous Transfer Mode (ATM) 253
audio/video 159
auditory display 94, 95, 96, 99, 100, 101, 102, 103, 104
auditory feedback 95, 96, 98, 99, 101
automated platooning systems 253
automated web service composition 176, 183, 184
automatic repeat-request (ARQ) 36, 39, 41
Average Connection Time (ACT) 189
average revenue per user (ARPU) 59, 60

B

Base Exchange (BE) 204, 205, 212
battery power 189, 190, 191, 193, 194, 196, 200
billing 43, 44, 45, 46, 47, 48, 49, 50
biological data 159
Bluetooth 147, 149
Business Process Execution Language for Web Services (BPEL4WS) 177, 185

C

care-of-address (CoA) 82, 83, 84, 85, 86, 89, 93
carrier-sense multiple access (CSMA) 72
Car-to-Car Communication Consortium (C2C-CC) 255
cellular networks 253
centralized administration 217
channel 28, 29, 30, 31, 32, 33, 34
channel assignment 29, 30, 31, 32, 33
charging 43, 44, 45, 46, 47, 48, 49, 50

charging data record (CDR) 44, 46, 50
clear to send (CTS) 30
cluster-based protocol 167
cluster based routing protocols (CBRP) 8, 220, 224
cluster head (CH) 166, 167, 187, 188, 189, 190, 191, 192, 193, 194, 195, 196, 197, 198, 199, 200, 228, 234, 235, 240
clustering algorithms 187, 188, 189
collaborative learning 151, 154
collision domain 29, 30, 32
Common Object Request Broker Architecture (CORBA) 123, 131
Composition Engine 177, 179
computer evolution 147
computer terminal 150
computing devices 146, 147, 156
computing system 156, 164
connection-oriented networks 68
Connectivity-Aware Routing (CAR) 257, 258
constant bit rate (CBR) 62
content addressable network (CAN) 107, 110, 120
content distribution networks (CDN) 60, 61
context data 158, 159, 161, 162
controlled load (CL) 63
cooperation-based hierarchical routing protocol (CBHRP) 7
co-operative traffic management 254
Cooperative Vehicle Infrastructure Systems (CVIS) 253
COOPERS 253, 254
correspondent nodes (CN) 79, 82, 83, 84, 85, 89, 93
custody transfer protocol (CTP) 259
customer relationship management (CRM) 45, 49

D

deafness problem 29, 30
Dedicated Short Range Communications (DSRC) 253, 254, 260, 261
Defense Advanced Research Projects Agency (DARPA) 106, 119
delay jitter 40, 42
Dempster-Shafer (DS) theory 235, 236
denial-of-service (DoS) 98, 99, 103
departments of transportation (DOT's) 254
depth first search (DFS) 53, 54
destination-sequenced distance-vector routing (DSDV) 113, 116, 120
differentiated services (DiffServ) 63
digital spaces 147
directional propagation protocol (DPP) 259
disaster recovery 217
Distributed Clustering Algorithm (DCA) 189
Distributed Component Object Model (DCOM) 123, 131
Distributed Computing Environment (DCE) 123
domain-independent 178
domain name system (DNS) 23, 24
domain name system security extensions (DNS-SEC) 23
duplicate address detection (DAD) 82, 83, 84, 86, 87, 92
dynamic host configuration protocol (DHCP) 83
Dynamic Source Routing (DSR) 12, 13, 113, 114, 115, 116, 119, 133, 140, 143

E

E911 (enhanced 911) System 243, 249
earcons 96, 99, 102
electronic information 147, 149, 150, 153, 154
electronic publishing 147
emergency services 217
Encapsulating Security Payload (ESP) 204, 212, 214
environment monitoring 217
Eulerian graph 57
Event Manager 179

F

fine granularity scalability (FGS) extension 36
flat topology 167
forward error correction 42
Framework for Open Application Management (FOAM) 253
frequency division medium access (FDMA) 72

G

Gabriel graph 54
game theory 2, 3, 4
generalized multiprotocol label switch (GM-PLS) 63, 64, 67
general packet radio service (GPRS) 44, 46
Geographic Source Routing (GSR) 257
Global Positioning System (GPS) 52, 53, 57, 147, 148, 150, 152, 154, 155, 157, 159, 160, 188, 190, 191, 200, 229, 230, 231, 233, 235, 236, 240, 241, 242, 243, 244, 246, 247, 248, 249, 252, 255, 257
Global System for Mobile communications (GSM) 157, 159, 160, 161
Global Systems for Telematics (GST) 253, 254
GPS sensors 160
graphical models 159
GreedY Traffic-Aware Routing (GyTAR) 257, 258
Grid Services and Execution Engine 179
GSM network 160
'GVGrid 258

H

half-space proximal 54
handoff management 81, 89, 93
haptic data 159
heterogeneous environments 201
heterogeneous technologies 201
Hierarchical Task Network (HTN) 178
high-cluster change (Hi-CC) region 233
hop 28, 29, 30, 32, 33, 34
Host Identifiers (HIs) 204, 205
Host Identity Protocol (HIP) 23, 24, 26, 201, 202, 203, 204, 205, 206, 207, 208, 209, 210, 211, 212, 213, 214, 215, 216
Host Identity Tag (HIT) 23, 204, 205, 207, 208, 209, 210, 212
human-computer interaction (HCI) 95, 96, 101, 102, 104
hybrid adaptive protocol for integrated discovery (HAID) 220
hybrid environment 151
hybrid information space 150
hybrid routing 1, 7, 14, 109
hybrid space 150, 151, 152

I

implicit earliest deadline first (I-EDF) 72, 73, 74
information and communications technologies (ICT) 18, 252
Information Services 179
infra-red (IR) 253
integrated safety systems 252
integrated services digital network (ISDN) 62
integrated services (IntServ) 62, 63, 67
intelligent integrated safety systems 252
intelligent solution 252
Intelligent Transportation Systems (ITS) 252, 260, 261
inter-cluster communication 166
inter-domain mobility 203
interface 28, 29, 30, 31, 32, 33, 34
International Standardisation Organisation (ISO) 61
International Telecommunications Union (ITU) 61, 64, 65, 66, 67
internet control message protocol (ICMP) 36, 39
Internet Engineering Task Force (IETF) 63, 67, 80, 82, 89, 91, 92, 93
internet multimedia subsystem (IMS) 46
internet protocol television (IPTV) 59, 62, 66, 67
Inter-Vehicle Communication (IVC) 252, 253, 255, 257, 259, 260
intra-domain mobility 203

J

Java 2 Enterprise Edition (J2EE) 123
Java Platform Micro Edition (JavaME) 241, 243, 245, 246, 249

K

key/value pairs 159

L

law enforcement 217

Lightweight MAC (LMAC) 165, 167, 168, 170, 171, 172, 173
local algorithms 51, 52, 56
local area network (LAN) 71, 157
local mobility 189
location aware networks 52
location-based broadcast (LBB) 259
location management 81, 93
location update 81, 84
logical sensors 159
logic based context models 159
low-cluster change (Low-CC) region 232, 233
Low Energy Adaptive Clustering Hierarchy (LEACH) 167

M

MAC protocol 165, 166, 167, 171, 173, 174
man-in-the-middle (MITM) 98
markup scheme 159
Medium Access Control (MAC) 30, 32, 33, 34, 36, 37, 38, 39, 40, 41, 69, 70, 71, 72, 73, 74, 75, 76, 77, 165, 166, 167, 168, 169, 170, 171, 172, 173, 174
Message Oriented Middleware (MOM) 123, 131
metric 29, 31, 34
middleware 122, 123, 124, 125, 126, 127, 128, 129, 130, 131
minimum spanning tree (MST) 54, 57, 58
mobile ad-hoc networks (MANET) 105, 106, 108, 109, 110, 111, 112, 113, 114, 115, 116, 117, 119, 120, 132, 133, 142, 152, 187, 188, 190, 193, 194, 196, 198, 199, 200, 217, 218, 219, 221, 222, 223, 224, 226, 227, 228, 229, 230, 232, 233, 234, 235, 237, 240
Mobile Computing (MC) 147, 218
mobile entity 202
Mobile Innovation Center (MIK) 48
mobile networks 188, 203, 209, 210, 211, 212
mobile nodes (MN) 79, 81, 82, 83, 84, 85, 86, 87, 88, 89, 91, 93, 187, 226, 227, 228, 229, 230, 231, 232, 233, 234, 235, 236, 237, 240
mobile phones 241, 242, 243, 244, 245, 246, 247, 248, 249, 250
mobile services 201
Mobile Virtual Network Operator (MVNO) 46, 49
mobility enhancements 202
mobility factor 35
mobility management 201, 202, 203, 204, 206, 210, 213, 216
mobility-only-based 188, 189, 190, 192
mobility predictions 226, 227, 228, 229, 230, 231, 232, 234, 235, 236, 237, 238
mobility prediction schemes 226, 227, 237, 238
modified zone routing protocol (M-ZRP) 220
motion/acceleration 159
motion detectors 159
multi-hop ad-hoc communication 78
multi-interface and multi-channel (MIMC) 28, 30, 31, 32
multimedia data 35, 36, 37, 42
Multimedia Messaging Service (MMS) 242
multiprotocol label switch (MPLS) 63, 64
multi radio resource management (MRRM) 20
musical instrument digital interface (MIDI) 99, 100, 101

N

National Office for Research and Technology (NKTH) 48
near-field-communications (NFC) 151, 155
Neighbor Protocol 167
neighbor solicitation (NS) 83
Network Computing Architecture (NCA) 123
network connection 123
network intrusion detection (NID) 94, 95, 96, 98, 99, 101, 102, 104
network intrusion detection system (NIDS) 95, 96, 97, 98, 99, 100, 101
network level 202, 203
network-level quality of service (NQoS) 63
network mobility 201, 202, 203, 204, 206, 209, 210, 211, 215, 216
Network-on-Wheels (NoW) 253
network partitioning 219
network technology 123
network topology 227, 228, 229, 230, 236

next-generation wireless networks (NGWN) 80, 81, 82, 89
no-cluster change (No-CC) region 232, 233
node mobility 226, 227, 229, 230, 233, 234, 235, 240

O

object oriented models 159
Object Request Broker middleware (ORB) 123, 131
offline charging 44, 45, 48
on-line advertising 226, 227
online charging 44, 45, 46, 47, 48
On-the-Path Service Discovery 205
ontology based context models 159
ontology reasoning methods 159
Ontology Server 179
Ontology Web Language (OWL) 176, 177, 178, 179, 181, 183, 185
optical data 159
optimized routing independent overlay network (ORION) 112, 114
outer-planar graph 58
overlay network 107, 111

P

packet loss rate 42
passive distributed indexing (PDI) 112
peak signal-to-noise ratio (PSNR) 61, 65
peer-to-peer (P2P) 59, 60, 61, 65, 67, 105, 106, 107, 108, 109, 110, 111, 112, 113, 114, 115, 116, 117, 118, 119, 120, 121
personal digital assistants (PDA) 79, 147, 152
Personal Information Management (PIM) 147, 152, 155
personalized wireless services 238
pervasive computing 124, 127, 130, 156, 157, 160, 161, 162, 163
phone call 242, 248
phone lines 242, 248
physical (PHY) 36, 37, 38, 39, 41
physical sensors 159
piggybacking 98, 104
planar face 58
planar graphs 53, 56, 58
planar spanner 52, 57, 58
Police Station 241, 242, 243, 244, 245, 246, 247, 248
post-paid 44, 45, 48, 49, 50
Power Decrease Rate (PDR) 190, 191, 200
Power, Mobility and Workload (PMW) 187, 188, 190, 192, 193, 194, 196, 197, 198, 199
power-only-based 188, 189
pre-paid 44, 45, 46, 48, 49, 50
Principal 1, 5, 7, 8, 9, 10, 11, 12
proactive routing protocols 11, 16, 109, 133
protocol phases 167
proximity sensor 149

Q

qbjective video quality assessment 68
quality of experience (QoE) 59, 60, 64, 65, 66, 67
quality of service (QoS) 36, 37, 38, 39, 40, 41, 42, 60, 63, 64, 66, 67, 68, 80, 81, 90, 91, 93, 226, 227, 236
quick response code (QRC) 150

R

rating 49, 50
rating schemes 47, 48
reactive routing protocols 11, 16, 109, 132, 133, 137
real-time applications 70
real-time streaming protocol (RTSP) 36, 41, 42
real-time transport control protocol (RTCP) 36, 39, 42
real-time transport protocol (RTP) 36, 41, 42
Received Signal Strength Indication (RSSI) 135, 136, 144
received signal strength (RSS) 189, 191, 193
Regional Service Discovery 205
Relative Mobility (RM) 189, 191, 200
Remaining Power (RP) 191, 192
Remote Procedure Call middleware (RPC) 123, 131
remote procedure calls (RPC) 175, 176
Rendezvous Server (RVS) 205, 206, 207, 208, 209, 210, 211, 213, 216
request to send (RTS) 30

Resource Definition Language (RDF) 176, 179, 185
Roadside-to-Vehicle Communication (RVC) 252
route guidance 226, 227
router advertisements (RA) 82, 83, 84, 87, 89
Route_REQuest (RRQ) 7, 8
Route_RESponse (RRES) 7
router solicitation (RS) 82, 85, 89
routing 28, 29, 31, 32, 33, 34
routing protocols 132, 133, 134, 135, 137, 138, 140, 141, 142
Rules Engine 179

S

SAFESPOT 254
safety systems 252
SATOClient (SC) 22
SATOPort (SP) 22, 26, 27
SATOServer (SS) 22
Schedule Exchange Protocol 167
search and rescue 217
Security Association (SA) 204, 212
self-organising ad-hoc radio 253
Self-Organizing MAC for Sensor Networks (SMACS) 167
self-organizing network 217
semantic routing 218, 220, 223
Semantic Web 176, 179, 184, 185
sensor networks 132, 133, 142
sensor node 70, 73, 78
service Ad/D protocol with independent zones (SPIZ) 220, 224
Service Announcement Packets (SAP) 205, 207, 212
service-aware adaptive transport overlays (SATO) 17, 20, 21, 22, 24, 26, 27
service discovery 205, 207, 208, 212, 217, 218, 219, 220, 221, 222, 223, 224, 225
Service Discovery Packets (SDP) 205, 212
service level agreement (SLA) 19
Service Location Protocol (SLP) 217, 224
Service-Oriented Architecture middleware (SOA) 123, 131
service oriented architectures (SOAs) 149
session initiation protocol (SIP) 36, 41, 42
shortest-path 29, 34
Short Message Service Center (SMSC) 242
Short Message Service (SMS) 241, 242, 243, 244, 245, 246, 247, 248, 249, 250
signal to interference plus noise ratio (SINR) 31
signal to noise ratio (SNR) 35, 42
Simple Hierarchical Ordered Planner (SHOP2) 178
simple network management protocol (SNMP) 99
Simple Object Access Protocol (SOAP) 123, 131, 176, 185
sociology 237
soft real-time 70, 71, 74
Sparse Topology and Energy Management (STEM) 168, 169
Stable Cluster Algorithm (SCA) 189
strategy without protocol (S-P) 12, 13
strategy with protocol (S+P) 12, 13
stretch factor of a spanner 58
structured P2P overlays 108, 113
subjective video quality assessment 66, 68

T

telecommunications office 241, 243
tightening techniques 178
time division medium access (TDMA) 72, 73
time to market (TTM) 48
topological location 202, 204
touch sensors 159
Traffic Adaptive Medium Access Protocol (TRAMA) 167
Traffic Aware Energy Efficient MAC (TEEM) 165, 169, 170, 171, 172
traffic engineering 60, 63, 64
traffic engineering (TE) 63, 64, 68
traversal 51, 52, 53, 54, 56, 58
two-phase communication process 132, 133

U

ubiquitous computing 122, 124, 125, 126, 127, 128, 129, 130
Unified Modeling Language (UML) 159, 164
unit disk graph (UDG) 51, 52, 54, 55, 56, 58

Universal Description, Discovery and Integration (UDDI) 176, 181, 185
unstructured P2P overlays 107
Urban Multi-hop Broadcast (UMB) 259

V

variable bit rate (VBR) 62
Vehicle-Assisted Data Delivery (VADD) 257, 258, 261
Vehicle Information and Communication System (VICS) 254
Vehicle-Infrastructure Initiative (VII) 251, 254
Vehicular Communication Networks (VCNs) 251, 252, 255, 256, 257, 258, 259, 261
vehicular safety 251, 253
Vehicular Safety Consortium (VSC) 255
video-on-demand (VoD) 59
VigilNet 71
virtual network operators (VNO) 46, 48
virtual sensors 159

W

wardriving 98, 104
Web Service Definition Language (WSDL) 176, 178, 179, 185
Web Service Description Language (WSDL) 149
weight computation 189
Weighted Clustering Algorithm (WCA) 189, 190, 199
weighted cumulative expected transmission time (WCETT) 31
WiFi access point 157
Windows, Icons, Menus and a Pointing device (WIMP) 149
wireless ad-hoc networks 132, 133, 135, 137, 138, 139, 141, 142
wireless communications 124, 251, 253
wireless device 132, 135
wireless intrusion detection system (WIDS) 104
Wireless Local Area Networks (WLAN) 80, 93, 97, 98, 104, 152, 165, 166, 253, 254, 262
wireless networks 176
wireless sensor network (WSN) 69, 70, 71, 72, 73, 74, 75, 76, 78, 165, 166, 170, 172, 173, 174
wireless signals 149, 151
wireless technologies 201, 202, 203, 253, 260
World Wide Web (WWW) 123, 129

Z

Zone Routing Protocol (ZRP) 134

Breinigsville, PA USA
17 October 2010
247133BV00008B/1/P

9 781605 662503